Computer Algebra
Concepts and Techniques

COMPUTER ALGEBRA

Concepts and Techniques

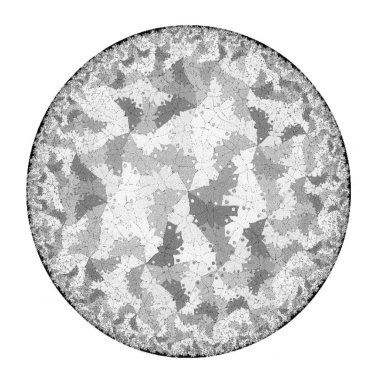

Edmund A. Lamagna

CRC Press
Taylor & Francis Group
Boca Raton London New York

CRC Press is an imprint of the
Taylor & Francis Group, an **informa** business

Cover image: Doug Dunham, *Butterflies (7,3)*. Tessellation inspired by the work of M. C. Escher. Seven butterflies meet at left front wing tips and three meet at right rear wings. There are eight colors; each group of seven butterflies meeting at left front wing tips is surrounded by a ring of butterflies of the eighth color. The entire pattern is composed of interlocking rings of butterflies.

CRC Press
Taylor & Francis Group
6000 Broken Sound Parkway NW, Suite 300
Boca Raton, FL 33487-2742

First issued in paperback 2020

© 2019 by Taylor & Francis Group, LLC
CRC Press is an imprint of Taylor & Francis Group, an Informa business

No claim to original U.S. Government works

ISBN-13: 978-1-138-09314-0 (hbk)
ISBN-13: 978-0-367-51045-9 (pbk)

Library of Congress Cataloging-in-Publication Data

Names: Lamagna, Edmund A., author.
Title: Computer algebra : concepts and techniques / Edmund A. Lamagna.
Description: Boca Raton, Florida : CRC Press, [2019] | Includes
bibliographical references.
Identifiers: LCCN 2018030134| ISBN 9781138093140 (hardback ; alk. paper) |
ISBN 9781315107011 (ebook)
Subjects: LCSH: Algebra--Data processing.
Classification: LCC QA155.7.E4 L36 2019 | DDC 512--dc23
LC record available at https://lccn.loc.gov/201803013

Visit the Taylor & Francis Web site at
http://www.taylorandfrancis.com

and the CRC Press Web site at
http://www.crcpress.com

COMPUTER ALGEBRA

Concepts and Techniques

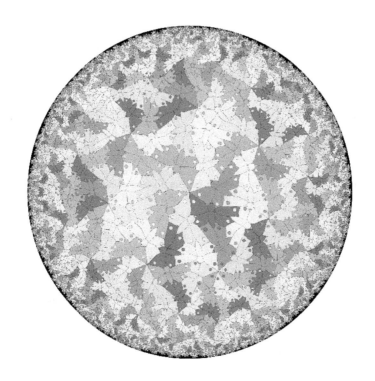

Edmund A. Lamagna

 CRC Press

Taylor & Francis Group

Boca Raton London New York

CRC Press is an imprint of the
Taylor & Francis Group, an **informa** business

Cover image: Doug Dunham, *Butterflies (7,3)*. Tessellation inspired by the work of M. C. Escher. Seven butterflies meet at left front wing tips and three meet at right rear wings. There are eight colors; each group of seven butterflies meeting at left front wing tips is surrounded by a ring of butterflies of the eighth color. The entire pattern is composed of interlocking rings of butterflies.

CRC Press
Taylor & Francis Group
6000 Broken Sound Parkway NW, Suite 300
Boca Raton, FL 33487-2742

First issued in paperback 2020

© 2019 by Taylor & Francis Group, LLC
CRC Press is an imprint of Taylor & Francis Group, an Informa business

No claim to original U.S. Government works

ISBN-13: 978-1-138-09314-0 (hbk)
ISBN-13: 978-0-367-51045-9 (pbk)

Library of Congress Cataloging-in-Publication Data

Names: Lamagna, Edmund A., author.
Title: Computer algebra : concepts and techniques / Edmund A. Lamagna.
Description: Boca Raton, Florida : CRC Press, [2019] | Includes
bibliographical references.
Identifiers: LCCN 2018030134| ISBN 9781138093140 (hardback ; alk. paper) |
ISBN 9781315107011 (ebook)
Subjects: LCSH: Algebra--Data processing.
Classification: LCC QA155.7.E4 L36 2019 | DDC 512--dc23
LC record available at https://lccn.loc.gov/201803013

**Visit the Taylor & Francis Web site at
http://www.taylorandfrancis.com**

**and the CRC Press Web site at
http://www.crcpress.com**

DEDICATION

To my wonderful wife
and
my wondrous son,
without whose love, support and patience
this book would not have become a reality.

Contents

Preface

To raise new questions, new possibilities, to regard old problems from a new angle, requires creative imagination and marks real advance in science.

Logic will get you from A to B. Imagination will take you everywhere.

Albert Einstein (1879–1955)

Our imagination is the only limit to what we can hope to have in the future.

Charles F. Kettering (1876–1958)

Computer algebra systems (CAS) are unique mathematical software programs that deal with formulas as well as numbers and graphs. Maple and Mathematica are perhaps the most familiar commercial systems. These programs include facilities for the manipulation of polynomials and rational functions, for analytic differentiation and integration, and for the simplification of mathematical expressions. For example,

```
> y := x^2 * cos(x);
```

$$y := x^2 \cos(x)$$

```
> diff(y,x);
```

$$2x\cos(x) - x^2 \sin(x)$$

calculates the derivative of an expression in Maple. These systems also have the distinctive ability to perform exact mathematical computations,

$$\frac{1}{2} + \frac{1}{3} \to \frac{5}{6}, \qquad \frac{1}{\left(\sqrt{2}-1\right)^2} \to 3 + 2\sqrt{2},$$

as opposed to simple numerical approximations $0.833333\ldots$ and $5.82843\ldots$.

Computer algebra systems provide impressive capabilities that are revolutionizing the way mathematics is done and taught. They can perform without error gargantuan calculations that would be infeasible to attempt by hand. As a result, CAS have become commonplace tools for computation and modeling in the scientific and engineering workplace. They have also been incorporated into the mathematics curricula at many colleges and universities starting with calculus and moving on to more advanced topics such as linear algebra and differential equations.

Despite their widespread application and importance, few users are familiar with the inner workings of CAS. This is unfortunate. One becomes a more proficient user, better able to comprehend the often astounding and sometimes surprising answers that CAS produce, by understanding the mathematical issues involved in their design and the algorithmic techniques used. These methods are not the same as those we learned in math classes, which are generally heuristic rather than algorithmic and often work only for small problems.

Unfortunately, books and courses on computer algebra are rare, and the source materials are largely inaccessible to general audiences. The few texts that have been written require knowledge of abstract algebra to comprehend much of the material. Very few individuals interested in CAS possess the required background.

The purpose of this book is to demystify CAS for a wide audience including students, faculty, and professionals in scientific fields such as computer science, mathematics, engineering, and physics. The only prerequisites are knowledge of first year calculus and a little programming experience—a background that can be assumed of the intended audience. The book is written (hopefully) in a lean and lively style, with many examples to illustrate the issues and techniques discussed. It is not meant to be encyclopedic but rather a short, rigorous introduction to the topics presented.

The focus is on the central issues involved in the design and implementation of CAS. The inherent and practical limitations of these systems, along with the algorithms and data structures used, are presented. Topics include big number arithmetic, manipulation of polynomials and rational functions, algebraic simplification, factoring, integration, and solution of polynomial equations.

The book has evolved from an extensive set of class notes that the author developed for CSC 550, Computer Algebra, at the University of Rhode Island. The course was first taught in 2000 and is usually offered in alternate years. Through these offerings, the class notes and course materials have undergone continual revision and improvement. Students from mathematics, engineering, and physics have taken the course, in addition to computer science majors. While listed at the graduate level, there is nothing inherent in the material that prevents a motivated junior or senior from succeeding, and those who have taken the course have done well.

Content and Organization

A summary of the concepts and techniques addressed in each chapter is given below.

Chapter 1: Introduction

We begin by surveying the capabilities of CAS that distinguish them from other mathematical software and follow with a few applications to motivate the importance and use of CAS.

We next present a brief historical overview of the major systems. In the early years (1961-1980), the systems were largely experiments used almost exclusively by their developers, although several systems saw limited production use by scientists, engineers, and applied mathematicians. Hardware was expensive and slow; memories were small. As a result, the systems required very significant computer resources by the standards of the day, which greatly limited access. Examples of important systems from this period include FORMAC (Sammet and Tobey, IBM), Altran (Brown, Bell Labs), SAC (Collins, IBM and University of Wisconsin), MATHLAB (Engelman, MIT and MITRE), SIN (Moses, MIT), REDUCE (Hearn, Stanford and Rand Corp.), Scratchpad (Griesmer and Jenks, IBM), Macsyma (Moses and Martin, MIT), and muMath (Stoutemyer and Rich, University of Hawaii).

The 1980s saw the appearance, first, of very powerful minicomputers and, later, of personal computers. The widespread availability of such machines made CAS accessible to a much larger community. The two major systems in use today, Maple (Geddes and Gonnet, University of Waterloo and Maplesoft) and Mathematica (Wolfram, Wolfram Research), were launched during the decade. Derive (Stoutemyer and Rich, SoftWarehouse and Texas Instruments) and Axiom/Scratchpad-II (Jenks, Trager and Watt, IBM) are two other influential systems from this period. More recent systems include the TI-calculator based CAS, MuPAD (Fuchssteiner, University of Paderborn, SciFace and MathWorks), and SageMath (Stein, University of Washington).

Chapter 2: Computer Algebra Systems

The goals of this chapter are 1) to provide an overview of the capabilities and power of CAS and 2) to allow the reader to get started with the two most popular systems: Maple and Mathematica. First, we show how each system can be used as a symbolic calculator to do exact arithmetic and to manipulate polynomials and other common mathematical functions. Commands for equation solving and the basic operations of calculus are also presented. How CAS go about performing many of these computations is the subject of the remainder of the book.

Implementing some of the algorithms presented in the book with a CAS is an important way to learn the concepts and techniques. Therefore, a portion of this chapter is devoted to the programming language embedded in each system. Topics discussed include the semantics of names and variables, evaluation rules, and the elementary data structures and basic operations for their manipulation. Finally, the control structures for selection and iteration are introduced by example.

The chapter is not meant to be replacement for a user's guide to the systems. Rather, it is intended as an introduction so the reader will know how to interact with the software to do basic calculations and to write simple programs. On-line help and other references should be consulted when additional features are required.

Chapter 3: Big Number Arithmetic

Perhaps the first thing that strikes us about CAS is their ability to perform arithmetic on numbers of essentially unlimited size. Who can fail to be impressed by typing 1000! and having all 2568 digits appear instantaneously? We consider how large numbers are represented and manipulated in computers having only a relatively small, finite word size. The obvious procedures to add, subtract, and multiply numbers are readily adapted to CAS. Asymptotically faster methods exist, however, for multiplication and one of these is presented. What about division? The method we were taught in school involves a bit of guesswork and ingenuity. How do we get around this?

The calculation of greatest common divisors (gcds) plays a central role in computer algebra. We examine in some detail what is perhaps the oldest non-trivial algorithm still in use today, Euclid's method. We also present a relatively new, alternative technique for computing greatest common divisors.

Chapter 4: Polynomial Manipulation

Efficient manipulation of polynomials and rational functions (i.e., the ratio of two polynomials) is at the heart of every CAS. The algorithms for addition, subtraction, multiplication, and division of numbers are easily adapted to perform the corresponding operations on polynomials. Importantly, Euclid's algorithm, the technique we use when reducing fractions to lowest terms, can be modified to determine the gcd of two polynomials. Euclid's algorithm finds many surprising applications in CAS, so much attention has been devoted to the development of efficient variations of the procedure.

Chapter 5: Algebraic Simplification

The simplification of algebraic expressions presents a number of design issues and challenges to implementers of CAS. While all CAS transform $5x-2x$ to $3x$, some automatically rewrite $(x^2 - 1)/(x + 1)$ as $x - 1$ while others require the user to request this simplification. Often, our notion of what is "simplest" depends on context. For example, $(x+y)^{100} - y^{100}$ is simpler than its expanded form, but $x^{100} - y^{100}$ is simpler than its factored form. We would all agree that expressions equivalent to zero should always be transformed to zero. Unfortunately, a theoretical result by Richardson (1968) shows that, in general, it is impossible to do this once we get just a little beyond the rational functions.

We also consider the simplification of expressions containing radicals and transcendental functions (e.g., exponentials, logarithms, and trigonometric functions). The problem of "rationalizing" the denominators of expressions containing radicals can be solved elegantly with an extension of Euclid's algorithm. Techniques also exist to "denest" radicals that can produce some of

Ramanujan's beautiful identities. CAS treat each occurrence of a transcendental function as a new variable. The difference is that each transcendental function has a few special "side relations," or identities, that are used to simplify expressions in which it appears. For example, the problem of factoring $e^{6x} - 1$ is essentially equivalent to that of factoring $y^6 - 1$ where $y = e^x$, with the special rule that $e^{nx} = (e^x)^n$.

Chapter 6: Factorization

The related problems of factoring polynomials and finding their rational roots are important capabilities of CAS. We begin by addressing the question of what it means to factor a polynomial. If we ask a CAS to factor $x^2 - 1$, it returns $(x - 1)(x + 1)$. But when asked to factor $x^2 - 2$ and $x^2 + 1$, these polynomials are returned unfactored. Why not answer $(x - \sqrt{2})(x + \sqrt{2})$ and $(x - i)(x + i)$, respectively? When factoring polynomials, we need to do so with respect to some algebraic structure, and CAS choose to factor over the integers unless directed otherwise. Factoring over the rational numbers turns out to be equivalent to factoring over the integers.

The factorization procedure that we were taught in school involves a certain amount of guesswork and works only for polynomials of small degree. Newton (1707) developed a method that works for linear and quadratic factors. Schubert (1793) and Kronecker (c. 1880) independently showed how to extend this method to find factors of any degree. While this technique is easy to understand and to implement, it runs very slowly. How do today's CAS factor? They employ a technique first suggested by Hensel (1908) that is based on factoring the polynomial modulo a suitable small prime number, and then "lifting" this result to the factorization over the integers.

Chapter 7: Symbolic Integration

One of the most impressive capabilities of CAS is their ability to compute difficult integrals (antiderivatives) that would be unwieldy to calculate by hand. Before CAS, complex tables of integrals were consulted to solve such problems. Interestingly, CAS have turned up many errors in such tables.

It is commonly believed that CAS solve all integration problems using a method attributed to Risch (1969) and based on Liouville's theory of integration (c. 1830). In fact, CAS use a multistage approach to compute indefinite integrals. The early stages employ a mix of strategies similar to those taught in introductory calculus. These include table look-up, substitution, integration by parts, and several other heuristics. Perhaps surprisingly, a large number of integrals can be solved by these strategies.

Rational functions that are not integrated in the early stages can always be solved by a refinement of Hermite's method (1872). This technique is an extension and generalization of the familiar partial fractions method but does not require a complete factorization of the denominator. An alternative, perhaps simpler, procedure was developed by Horowitz (1971). The techniques of Hermite and Horowitz produce the rational part of the integral, and an integrand giving the transcendental part. Rothstein (1976) and Trager (1976) independently devised a method to solve this latter integral.

The Risch algorithm is saved for last, when all else fails. The procedure mirrors, in many respects, Hermite's method and the Rothestein-Trager technique for integrating rational functions. We present an introduction to the Risch algorithm, focusing on integration of logarithms and exponentials.

Chapter 8: Gröbner Bases

Many problems in mathematics, science and engineering can be modeled with systems of polynomial equations. Gröbner bases provide a very powerful, systematic approach for solving such problems. Consequently, a Gröbner basis package is a very important tool in every "industrial strength" computer algebra system.

The chapter begins by exploring what Gröbner bases are and how they can be used to solve nonlinear systems of polynomial equations. A comparison is made with the method of Gaussian elimination, which is commonly used to solve systems of linear equations. Several applications of Gröbner bases are given, including integer programming and the problem of finding implicit solutions to sets of parametric equations.

The latter portion of the chapter is concerned with algorithmic methods for computing Gröbner bases. The first problem considered is that of "general polynomial division." How do we divide a multivariate polynomial by a set of polynomial divisors? Armed with such a technique, the construction of Gröbner bases using Buchberger's algorithm is presented and discussed in the final section.

Chapter 9: Mathematical Correctness

Most CAS report that $\int x^k \, dx = x^{k+1}/(k+1)$. This answer is not completely correct mathematically, as any first year calculus student knows. The correct answer includes an arbitrary constant of integration. Moreover, the answer is wrong for the special case when $k = -1$, where the solution is $\ln x$. This example illustrates how implementers of CAS have opted to give what is perhaps the most useful result, rather than one that is strictly correct.

Fractional powers, multivalued functions, and inverse functions pose similar problems for CAS. For example, the transformation $\sqrt{a\,b} = \sqrt{a}\,\sqrt{b}$ that we are taught in school is not valid when a and b are both negative. Most CAS reply that arctan(0) is 0, taking the principal "branch cut" through this multivalued function, although $\arctan(0) = n\,\pi$ for any integer n. The inverse functions sin and arcsin behave in such a way that $\sin(\arcsin(x)) = x$ for all x, but $\arcsin(\sin(x)) = x$ only if $-\pi/2 \le x \le \pi/2$. In recent years, there has been a trend in CAS to allow users to define the domains of variables to obviate such problems. This increases the complexity of the software and places an added burden on the user.

Implementation Issues

Implementers of CAS face a number of difficult design decisions that affect the efficiency of their software. While many computations take small inputs and produce small outputs, the intermediate results during the computation can be huge. It is important to choose algorithms that reduce this phenomenon, termed "intermediate expression swell," as much as possible. Furthermore, techniques like the use of canonical representations for algebraic expressions are often employed to speed processing. Another important issue for implementers is to design their systems in such a way that the code is portable to a variety of different computer platforms. Considerations such as these are discussed throughout the book.

How to Use the Book

The field of computer algebra acts as a bridge between computer science and mathematics. As such, its study is appropriate to either discipline. This book contains enough material for a one-semester course. In his class at the University of Rhode Island, the author spends one week each on Chapters 1 and 2, two weeks each on Chapters 3–5, approximately two and a half weeks each on Chapters 6 and 7, and one week on Chapter 9. Alternatively, an instructor could cover a selection of topics from Chapters 6 and 7, leaving one or two weeks for Chapter 8.

A course in computer algebra is a viable alternative to the numerical methods course that is a standard elective in both the math and computer science curricula. Donald Knuth coined the term "seminumerical" for the kinds of computational techniques presented here that are on the borderline between numerical and symbolic calculation. The book can also be used as a supplementary text in conjunction with a course in computational science or an advanced course on the design and analysis of algorithms.

Alternatively, the book is also intended as a self-study guide for those who are fascinated with computer algebra systems and want to learn more about how they work. This audience includes instructors of mathematics who incorporate computer algebra into courses such as calculus, science and engineering faculty who employ computer algebra in their teaching or research, students in these fields, and interested practitioners who use (or might use) computer algebra in the workplace.

Mathematics and computer science are not "spectator sports." Mastering the material in the book is greatly reinforced by solving problems that use or extend the techniques discussed, or by examining some of the issues presented in greater depth. Therefore, an important feature of the book is the section at the end of each with chapter with exercises and programming projects. Both computational problems and proof-type problems appear in the exercises, and the reader is encouraged to work through a representative sample for each topic studied.

While solutions to the exercises are not provided at the end of the book, the answers to most of the computational problems can be found by feeding them to a CAS. In my computer algebra class, students are to allowed to use their CAS as a "calculator" for many of the exercises to eliminate the need to perform tedious, error-prone calculations by hand.

While writing a complete computer algebra system is an enormous undertaking, computer algebra software may be used to implement many of the central algorithms presented throughout the book. For example, the basic arithmetic algorithms can be programmed in a CAS by manipulating lists of digits. Similarly, the techniques for polynomial arithmetic can be implemented by working with lists of coefficients. This gives the flavor of coding the important parts of a computer algebra system without having to "reinvent the wheel," and also greatly increases proficiency with a specific system.

Edmund A. Lamagna
Department of Computer Science and Statistics
University of Rhode Island
Kingston, RI 02881 U.S.A.

August 2018

Acknowledgments

Many people, other than the authors, contribute to the making of a book, from the first person who had the bright idea of alphabetic writing through the inventor of movable type to the lumberjacks who felled the trees that were pulped for its printing. It is not customary to acknowledge the trees themselves, though their commitment is total.

Richard Forsyth and Roy Rada
Machine Learning, 1986

The idea for my computer algebra course and for this book is the result of several conversations with Benton Leong, one of the original authors of Maple. Teaching the course for the first time involved digging through a vast number of research articles and advanced books, most of which would be inaccessible to the kinds of students enrolled in the class. This convinced me all the more of the need to write the book.

I am grateful to the University of Rhode Island (URI) for providing a sabbatical leave to begin transforming my class notes into a manuscript, and for a second sabbatical to facilitate the book's completion. The Department of Mathematics and Computer Science at Amherst College provided an ideal setting for the first sabbatical. I am indebted to the members of that department, and most especially my host David Cox, for many stimulating discussions. Thanks also to Maling Ebrahimpour, former dean of business at Roger Williams University, for making space available during the summer to work on the book. The Cranston Public Library and the Tiverton Public Library provided quiet working environments for the second sabbatical, which my family jestingly refers to as my "staybatical," when the manuscript was completed.

My thanks are extended to several computer algebraists, most notably David Cox, Michael Monagan and David Stoutemyer, for their thoughtful comments on an earlier draft and for the information they provided. Likewise, suggestions of the anonymous reviewers helped improve the book's content and organization.

Earlier versions of the book were classroom tested in my computer algebra class at URI. The students were asked to annotate the drafts with any errors they found, concepts and language in need of clarification or elaboration, and suggestions for improvement. I thank those students who did this conscientiously as their comments undoubtedly improved the readability and correctness of the book.

I am grateful to Doug Dunham, who has long been interested in the mathematics underlying M. C. Escher's artwork, for creating the computer generated butterfly tessellation appearing on the cover. Emily Greene of the interlibrary loan department at URI and the Dartmouth College library provided the scan of the page from Delaunay's 1867 book. The photo accompanying my biographical sketch was taken by Pauline and Phil of Leduc Studios.

The book you are holding is the product of my project team at CRC Press, an imprint of the Taylor & Francis Group. The enthusiasm and support of my editor, Sarfraz "Saf" Khan, is greatly appreciated. Other key participants include editorial assistant Callum Fraser, project editor Robin Lloyd-Starkes, proofreader Alice Mulhern, and TEXpert Shashi Kumar.

I also wish to recognize my late parents, Armando and Ruth Lamagna, for the nurturing role they played in my life, for stressing the importance of a good education, and for making sacrifices toward this objective. I am grateful for the many excellent teachers encountered along the way who taught me not only their subject matter but how to teach.

Most importantly, I thank my wonderful wife and my wondrous son, without whose love, support and patience this book would not have been possible. I dedicate the book to them. Finally, I thank myself for having the persistence over many years to bring this work to its successful completion.

E.A.L.

Chapter 1

Introduction

> *The last thing one knows when writing a book is what to put first.*
>
> Blaise Pascal (1623–1662)
> *Pensées,* 1660

> *How thoroughly it is ingrained in mathematical science that every real advance goes hand in hand with the invention of sharper tools and simpler methods which, at the same time, assist in understanding earlier theories and in casting aside some more complicated developments.*
>
> David Hilbert (1862–1943)
> International Congress of Mathematicians, Paris, 1900

Computer algebra systems (CAS) provide impressive mathematical capabilities that have revolutionized the way mathematics is done and taught. They can perform without error gargantuan calculations that would be infeasible to attempt by hand. Accordingly, CAS have become commonplace computational and modeling tools in the scientific and engineering workplace. They have also been incorporated into the mathematics curricula at many colleges and universities starting with calculus and moving to more advanced topics such as linear algebra and differential equations. Yet despite their widespread application and importance, few users are familiar with the inner workings of CAS. The goal of this book is to demystify CAS for a wide audience that includes students, faculty, and professionals in scientific fields such as computer science, mathematics, engineering, and physics.

1.1 What is Computer Algebra?

Computer algebra is a branch of scientific computation that lies at the intersection of computer science and mathematics. It is readily distinguished from numerical analysis, the other principal branch of scientific computation, in that it involves the manipulation of formulas rather than only numbers. An article in the *Encyclopedia of Computer Science* (2000) by Gaston H. Gonnet and Dominik W. Gruntz defines computer algebra as "computation with variables and constants according to the rules of algebra, analysis, and other

1

branches of mathematics." The authors go on to describe computer algebra as "formula manipulation involving symbols, unknowns, and formal operations rather than with conventional computer data of numbers and character strings."

Computer algebra systems can be classified as either *special purpose* or *general purpose*. Special purpose systems are designed to solve problems in one specific area of science or mathematics, such as celestial mechanics or group theory. General purpose systems are, of course, intended for use in most scientific and mathematical fields. General systems are typically used interactively and provide the following facilities:

- *symbolic computation*: expansion and factorization of polynomials, differentiation, indefinite and exact definite integration, exact solution of equations and systems of equations, and linear algebra

- *numerical computation*: arbitrary precision numerical computation including definite integration, numerical solution of equations, and evaluation of elementary and special functions

- *graphics*: plotting in two and three dimensions

- *programming*: an embedded programming language and data structures allowing users to extend the system

The focus of this book is on the symbolic capabilities of computer algebra systems. We present several examples that illustrate the types of calculations these systems can perform and which distinguish them from other mathematical software. First, CAS perform exact arithmetic and can make transformations such as

$$\frac{1}{2} + \frac{1}{3} \to \frac{5}{6}, \qquad \frac{1}{\left(\sqrt{2} - 1\right)^2} \to 3 + 2\sqrt{2}.$$

Most calculators would give only the numerical approximations $0.833333\ldots$ and $5.82843\ldots$ to these quantities. Moreover, CAS perform calculations with essentially unlimited precision and can therefore produce results like

$$100! \to 933262154439441526816992388562667004907159682643816214685929638952175999932299156089414639761565182862536979208272237582511852109168640000000000000000000000000.$$

Another important capability of computer algebra systems is their ability to manipulate polynomials and rational functions (ratios of polynomials). CAS can expand polynomials by multiplying out products,

$$(x^2 - 4x + 4)(x^2 + 4x + 4) \to x^4 - 8x^2 + 16,$$

factor polynomials,

$$(x^2 - 4x + 4)(x^2 + 4x + 4) \to (x - 2)^2(x + 2)^2,$$

and simplify polynomial ratios by removing the greatest common divisor,

$$\frac{x^4-1}{x^8-1} \to \frac{1}{x^4+1}.$$

Computer algebra systems can also perform exact computations with elementary transcendental functions (exponentials, logarithms, trigonometric and inverse trigonometric functions). For example,

$$\sin(\pi/3) \to \frac{\sqrt{3}}{2}, \qquad \arccos(\sqrt{2}/2) \to \frac{\pi}{4},$$

rather than $0.866025\ldots$ and $0.785398\ldots$. CAS can apply multiangle trigonometric identities in either direction,

$$\sin(3\,x) \leftrightarrow 4\,\sin x\,(\cos x)^2 - \sin x,$$

as well as expand and factor polynomials of functions,

$$(e^{3x}-1)\,(e^{2x}-1) \to e^{5x} - e^{3x} - e^{2x} + 1,$$
$$(e^{3x}-1)\,(e^{2x}-1) \to (e^x-1)^2\,(e^x+1)\,(e^{2x}+e^x+1).$$

Another feature of computer algebra systems is their ability to perform the basic operations of calculus (differentiation, indefinite and exact definite integration, calculation of limits, series expansion, and summation) and linear algebra. For instance, CAS can transform

$$\frac{d}{dx}\,x^2\,\cos x \to 2\,x\,\cos x - x^2\,\sin x$$

and

$$\int x\,e^x\,dx \to x\,e^x - e^x,$$

and determine that $\int e^{x^2}\,dx$ cannot be expressed with only elementary functions.

The term *computer algebra* is almost invariably used today, but several other names were more common during the field's infancy. These include *symbolic* or *algebraic computation*, and either *algebraic* or *formula manipulation*. For example, the name for the Special Interest Group on Symbolic and Algebraic Manipulation (SIGSAM) of the Association for Computing Machinery (ACM) dates back to this period. The early ACM conferences in the field went by the acronym SYMSAC (Symposium on Symbolic and Algebraic Computation). In 1988, these meetings merged with a series of European conferences called EUROSAM (again, SAM stands for Symbolic and Algebraic Manipulation) into what is now the principal annual conference on computer algebra, ISSAC (International Symposium on Symbolic and Algebraic Computation).

Computer algebra is generally studied from one of three principal perspectives:

- *algorithms:* the design of efficient mathematical algorithms

- *applications:* the needs of particular groups of users

- *systems:* the challenge of creating a coherent, useful software system

In the remainder of this chapter, we survey the second and third perspectives. Several simple but representative applications of computer algebra are described in the next section. Section 1.3 presents a historical survey of the most important developments and systems. Chapter 2 is an introduction to the most popular systems in use today, Maple and Mathematica. Most of the book, Chapters 3–9, focus on algorithms and the mathematical concepts and issues involved in developing computer algebra systems.

1.2   Applications

Computer algebra makes possible enormous mathematical calculations that would not be feasible to attempt with only pencil and paper. It also serves as a valuable complement to numerical computation. Computer algebra systems may, for example, be used to derive exact expressions that can then be evaluated numerically. This makes it possible to use exact calculation where it is relatively easy or where ill-conditioning makes numerical work error-prone, while more rapid numerical computation can be used when it is adequate. A mixed symbolic-numeric solution often combines the best of both approaches. For instance, symbolic differentiation and numerical integration are both usually well-behaved operations, whereas numerical differentiation and symbolic integration can be troublesome.

Computer algebra systems have proven useful in virtually all fields that use mathematics. These include such disciplines as engineering, the physical and biological sciences, mathematics, computer science, statistics, economics, business and finance, and education. A few representative applications are described below, with emphasis on the advantages that computer algebra provides over strictly numerical computation.

1.2.1   Celestial Mechanics

Theoretical physics encompasses such fields as celestial mechanics, electrodynamics, general relativity, and quantum mechanics. These areas have long traditions of lengthy non-numerical calculations that would also classify them as branches of applied mathematics. It is not surprising, therefore, that several of the first computer algebra programs were written by physicists to assist them in their research. In fact, one of the first and most striking successes of computer algebra was in celestial mechanics.

According to Newton's theory of universal gravitation, the orbit of a point mass around a spherical mass of uniform density is an ellipse. The orbit of the moon about the earth is, however, more complex due to the effect of the sun. First, the plane of the moon's orbit around the earth is inclined at a small angle to the plane of the earth's orbit around the sun due to the perturbing influence of the sun's gravitational field. Second, the sun also causes the point where the moon's orbit is closest to the earth to wobble (precess) slowly with respect to the stars.

Charles E. Delaunay (1816–1872) was a noted French astronomer and mathematician. In 1846, he published a landmark paper in which he considered the movement of the moon as a function of time. He was the first to model this motion as a three-body problem involving the gravitational forces of the sun as well as those of the earth and moon. Delaunay spent much of the remainder of his life working through the details of his theory. It took him ten years to produce by hand a trigonometric series expansion for the motion of the moon as a function of time accurate to seventh order terms, and another ten years to check his results. These monumental calculations are presented in two volumes, *La Théorie du Mouvement de la Lune,* published in 1860 and 1867. The books total over 900 pages, one of which appears in Figure 1.1.

The Apollo lunar exploration program of the 1960s and 70s lent great importance to the problem of accurately calculating the position of the moon over time. This information was required to analyze dynamical data returned from various lunar experiments (radar echoes, orbiters, Apollo missions, and laser retroranging reflectors). The digital computers of the time were much slower than those of today and had very limited memory capacity. While numerical methods could be used to calculate the position of the moon over short periods, errors accumulated over time. Consequently, this approach required continuous monitoring of the moon's position, which was both time-consuming and costly given the available computing power.

Delaunay's approach provided a viable alternative as it required only the evaluation of a set of formulas over time. The principal drawback was that the series used converged slowly and so more terms would be required than Delaunay originally produced. Moreover, several researchers had reported errors in Delaunay's work without ever pinpointing them.

In 1970, three scientists—André Deprit, Jacques Henrard and Arnold Rom—working at the Boeing Scientific Research Laboratories in Seattle set out to remedy the situation. With customized computer algebra software they wrote themselves, the team reproduced Delaunay's calculation to order ten. The computations were so large that they required six continuous runs of approximately three hours each on an IBM 360-44 computer with a core memory of 128K bytes and two disks of one megabyte capacity.

Miraculously, the computation turned up only one error in Delaunay's work. His mistake appears in Figure 1.1. The boxed term should be $\frac{33}{16}\gamma^2 e'^2$

234   THÉORIE DU MOUVEMENT DE LA LUNE.

la fonction R ne contient plus aucun terme périodique; elle se trouve donc réduite à son terme non périodique seul, terme qui, en tenant compte des parties fournies par les opérations 129, 260, 349 et 415, a pour valeur

$$R = \frac{\mu}{2a}$$

$$+ m'\frac{a^2}{a'^2}\left|\frac{1}{4} - \frac{3}{2}\gamma^2 + \frac{3}{8}e^2 + \frac{3}{8}e'^2 + \frac{3}{2}\gamma^4 - \frac{9}{4}\gamma^2 e^2 - \frac{9}{4}\gamma^2 e'^2 + \frac{9}{16}e^2 e'^2 + \frac{15}{32}e^4 - \frac{33}{2}\gamma^4 e^2\right.$$

$$+ \frac{9}{4}\gamma^2 e^4 + \frac{75}{16}\gamma^2 e^4 - \frac{27}{8}\gamma^2 e^2 e'^2 - \frac{45}{16}\gamma^2 e'^4 + \frac{45}{64}e^4 e'^2$$

$$+ \left(\frac{9}{16}\gamma^2 + \frac{225}{64}e^4 - \frac{27}{16}\gamma^4 - \frac{387}{32}\gamma^4 e^2 + \boxed{\frac{23}{16}\gamma^2 e'^2} - \frac{225}{128}e^4 + \frac{825}{64}e^2 e'^2 + \frac{9}{8}\gamma^2\right.$$

$$+ \frac{3897}{64}\gamma^4 e^2 - \frac{99}{16}\gamma^4 e'^2 - \frac{1431}{256}\gamma^2 e^4 - \frac{1419}{32}\gamma^2 e^2 e'^2 - \frac{225}{512}e^4 - \frac{825}{128}e^2 e'^2\left)\frac{n'}{n}\right.$$

$$- \left(\frac{31}{32} - \frac{33}{8}\gamma^2 - \frac{971}{32}e^2 + \frac{465}{64}e'^2 + \frac{273}{64}\gamma^4 + \frac{5709}{64}\gamma^2 e^2 - \frac{117}{4}\gamma^2 e'^2 + \frac{4989}{256}e^4\right.$$

$$\left. - \frac{1905}{8}e^2 e'^2 + \frac{3255}{128}e'^4\right)\frac{n'^2}{n^2}$$

$$- \left(\frac{255}{32} - \frac{31515}{1024}\gamma^2 - \frac{551115}{4096}e^2 + \frac{6885}{64}e'^2 + \frac{20511}{512}\gamma^4 + \frac{927831}{2048}\gamma^2 e^2\right.$$

$$\left. - \frac{218115}{512}\gamma^2 e'^2 + \frac{1622985}{16384}e^4 - \frac{4069635}{2048}e^2 e'^2\right)\frac{n'^3}{n^3}$$

$$- \left(\frac{5515}{192} - \frac{296779}{3072}\gamma^2 - \frac{6380965}{12288}e^2 + \frac{16285}{24}e'^2\right)\frac{n'^4}{n^4}$$

$$- \left(\frac{28841}{288} - \frac{113818307}{294912}\gamma^2 - \frac{168190105}{1179648}e^2 + \frac{1303609}{384}e'^2\right)\frac{n'^5}{n^5}$$

$$- \frac{9814775}{36864}\frac{n'^6}{n^6} - \frac{428268199}{663552}\frac{n'^7}{n^7}$$

$$+ \left[\frac{9}{64} - \frac{45}{16}\gamma^2 + \frac{45}{64}e^2 + \frac{15}{128}e'^2\right.$$

$$\left.+ \left(\frac{225}{512} - \frac{1935}{256}\gamma^2 + \frac{7425}{1024}e^2 + \frac{225}{64}e'^2\right)\frac{n'}{n} + \frac{869}{512}\frac{n'^2}{n^2} - \frac{10391}{8192}\frac{n'^3}{n^3}\right]\frac{a^2}{a'^2}\Big|$$

FIGURE 1.1: A page from Delaunay's *La Théorie du Mouvement de la Lune*.

instead of $\frac{23}{16}\gamma^2 e'^2$, this coefficient arising from the sum $-\frac{45}{16} + \frac{147}{32} + \frac{9}{32}$. The error is not a misprint and propagates consistently into three other formulas.

It was certainly a great triumph of computer algebra that a machine could produce in twenty hours a computation that took Delaunay twenty years to perform! Using Moore's Law, that computing power doubles every eighteen months, this same calculation could be repeated today in just seconds.

1.2.2 Chemical Stoichiometry

Whenever a chemical reaction occurs, it can be described by an equation. The left side shows the chemicals that react (called the *reactants*), and the right side shows the chemicals that are produced (called the *products*). Unlike a mathematical equation, the two sides are separated by an arrow indicating that the reactants form the products and not the other way around.[†]

You have probably balanced such equations in a chemistry course. The key is the Law of Conservation of Mass which states that in an ordinary (non-nuclear) chemical reaction, mass is neither lost nor gained. This rule applies to the individual atoms of any element in a chemical reaction. It also applies to the charges of any ionic materials.

Mathematically, the process of balancing a chemical equation is achieved by solving a system of linear simultaneous equations with integer coefficients. The solution to such a system, when it exists, is always expressed in terms of integers and rational numbers (fractions). In this regard, computer algebra systems have a distinct advantage over numerical software, which would express the solution using inexact floating-point numbers.

Example 1.1. Balance the chemical reaction

$$a\,Cu_2S + b\,H^+ + c\,NO_3^- \rightarrow d\,Cu^{2+} + e\,NO + f\,S_8 + g\,H_2O$$

by determining the values of the coefficients.

Solution. The Law of Conservation of Mass gives the follow equations for each of the elements in the reaction and the charge:

Cu (copper)	$2\,a = d$
S (sulfur)	$a = 8\,f$
H (hydrogen)	$b = 2\,g$
N (nitrogen)	$c = e$
O (oxygen)	$3\,c = e + g$
Q (charge)	$b - c = 2\,d$

Solving these six equations with seven unknowns in terms of a, we have

$$a = a, \quad b = \tfrac{16}{3}a, \quad c = \tfrac{4}{3}a, \quad d = 2a, \quad e = \tfrac{4}{3}a, \quad f = \tfrac{1}{8}a, \quad g = \tfrac{8}{3}a.$$

[†]Some reactions can be run in either direction. These reversible reactions are indicated by a double-headed arrow.

It is customary to write chemical equations with integer coefficients. We can clear the fractions by finding the least common multiple of the denominators, $\text{lcm}(1, 3, 3, 1, 3, 8, 3) = 24$, giving the balanced reaction,

$$24\,Cu_2S + 128\,H^+ + 32\,NO_3^- \rightarrow 48\,Cu^{2+} + 32\,NO + 3\,S_8 + 64\,H_2O. \qquad \square$$

McLaughlin (1985) presents a procedure for the Macsyma computer algebra system that completely automates the process of balancing chemical reactions for students in an introductory college chemistry course. The user specifies each reactant and product molecule by entering the number of atoms of each constituent element. In the case of an ion, the charge is also specified. The system then calculates and reports the balanced chemical equation.

1.2.3   Geometry

The ability of CAS to solve polynomial equations and to manipulate algebraic numbers (radicals) makes them well suited for solving many types of geometric problems. We will examine two. The first is a simple example illustrating how CAS can be used to assist in proving a geometric theorem. Proving geometric theorems has been an active area of computer algebra research. The book by Shang-Ching Chou (1988), for example, surveys some of the automated techniques that have been developed.

Example 1.2 (Heron's Formula). Heron's formula for the area A of a triangle with sides of length a, b, c is

$$A = \sqrt{s\,(s-a)\,(s-b)\,(s-c)}, \tag{1.1}$$

where $s = (a+b+c)/2$ is one-half the perimeter. In proving this formula, we assume the Pythagorean theorem and the formula for the area of a triangle, $A = \frac{1}{2}bh$, where b is the length of a base and h is the height to that base (i.e., the length of a perpendicular line segment to the base from the vertex opposite it). The situation is illustrated in Figure 1.2, where the origin has been chosen to reduce the amount of algebraic manipulation in the derivation.

We first obtain a formula for x in terms of a, b, c. From the Pythagorean theorem, we know $h^2 + x^2 = a^2$ and $h^2 + (b-x)^2 = c^2$. Eliminating h^2, we have $a^2 - x^2 = c^2 - (b-x)^2$ and, solving for x,

$$x = \frac{a^2 + b^2 - c^2}{2\,b}. \tag{1.2}$$

The next step is to obtain an expression for h in terms of a, b, c. Substituting (1.2) into $h^2 = a^2 - x^2$, we have

$$h^2 = a^2 - \left(\frac{a^2 + b^2 - c^2}{2\,b}\right)^2$$

$$= \frac{4\,a^2\,b^2 - (a^2 + b^2 - c^2)^2}{4\,b^2}.$$

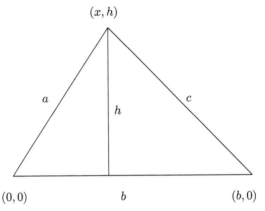

FIGURE 1.2: Heron's formula. $A = \frac{1}{2}bh = \sqrt{s(s-a)(s-b)(s-c)}$, where $s = (a+b+c)/2$.

The numerator factors as

$$(a+b+c)\,(-a+b+c)\,(a-b+c)\,(a+b-c)$$

which, with the algebraic substitution $a+b+c = 2s$, gives

$$h^2 = \frac{4\,s\,(s-a)\,(s-b)\,(s-c)}{b^2}.$$

Taking the square root and simplifying, we have

$$h = \frac{2\,\sqrt{s\,(s-a)\,(s-b)\,(s-c)}}{b}.$$

Finally we obtain Heron's formula (Equation 1.1) by substituting this value for h into the formula for the area, $A = \frac{1}{2}bh$. □

A CAS, like a human problem solver, is capable of performing the symbolic algebraic manipulations described at each step. These include isolating subexpressions, solving a symbolic equation, grouping terms over a common denominator, factoring a polynomial, substituting for an algebraic quantity, and simplifying a square root. Other types of mathematical software are incapable of computations such as these.

We turn our attention to a second example, exploring a computational problem from analytic geometry for which exact results are sought.

Example 1.3. Find the equation of the line through the points $(\sqrt{2}, \sqrt{3})$ and $(\sqrt{3}, \sqrt{2})$. Also determine the distance between these points.

Solution. The equation of the line passing through the points $(x_1, y_1), (x_2, y_2)$ is given by the *point-slope formula*,

$$y - y_1 = \frac{y_2 - y_1}{x_2 - x_1}\,(x - x_1).$$

Substituting the coordinates of the points and rearranging gives

$$y = \frac{\sqrt{2} - \sqrt{3}}{\sqrt{3} - \sqrt{2}}\left(x - \sqrt{2}\right) + \sqrt{3}$$
$$= -x + \sqrt{3} + \sqrt{2},$$

where the last equation for the line is the *slope-intercept formula*. Computer algebra systems can simplify the ratio for the slope to -1 and maintain the y-intercept in the form $\sqrt{3} + \sqrt{2}$, which conveys more information in the context of this problem than a numerical approximation.

The distance between the points $(x_1, y_1), (x_2, y_2)$ is given by

$$\sqrt{(x_2 - x_1)^2 + (y_2 - y_1)^2} = \sqrt{10 - 4\sqrt{6}}$$
$$= \sqrt{2}\sqrt{5 - 2\sqrt{6}}.$$

This expression contains a nested square root that can be eliminated,

$$\sqrt{5 - 2\sqrt{6}} = \sqrt{3} - \sqrt{2},$$

as easily verified by squaring both sides. With this transformation, we arrive at a compact, exact form for the distance between the points,

$$\sqrt{2}\left(\sqrt{3} - \sqrt{2}\right) = \sqrt{6} - 2. \qquad \Box$$

Computer algebra systems are capable of all the transformations performed in the last example. The solution involves several algebraic manipulations we examine later. The problems of "denesting" square roots and other manipulations involving radicals are considered in Chapter 5. The identity $\sqrt{ab} = \sqrt{a}\sqrt{b}$ used in the derivation is appropriate only when a or b (or both) is non-negative. Issues relating to the correctness of various mathematical transformations are discussed in Chapter 9.

1.3   History

The idea that symbolic mathematical computation might be automated dates back to Charles Babbage (1791–1871), the great ancestral figure of computer science. Babbage worked on two types of mechanical computing devices. The first, called the Difference Engine, was designed to construct tables of functions from finite differences. A small prototype was built between 1820 and 1822. After a decade of toil, however, Babbage abandoned plans in 1833 to build a larger model due to personal and financial difficulties.

Babbage's second design was a more general purpose computing device he termed the Analytical Engine. Hoping to secure recognition and monetary support to build the machine, Babbage toured Italy in 1840. Having listened to him lecture in Turin, a young mathematician and military engineer named Luigi F. Menabrea (1809–1896)—who was later to become Prime Minister of Italy—wrote a description of the Analytical Engine. This paper was translated from French into English by Augusta Ada Byron, Countess of Lovelace (1815–1852), and daughter of the famous poet Lord Byron.[†] Ada had an interest in mathematics and was an admirer of Babbage and his ideas. In consultation with him, she produced in 1844 a set of extensive notes to accompany her translation of Menabrea's paper. One of her annotations reads,

> Many persons who are not conversant with mathematical studies imagine that because the business of the engine is to give its results in numerical notation, the nature of its process must consequently be arithmetical rather than algebraic and analytical. This is an error. The engine can arrange and combine its numerical quantities exactly as if they were letters or any other general symbols, and, in fact, it might bring out its results in algebraic notation were provisions made accordingly.

Unfortunately, the Analytical Engine was never built and symbolic mathematical computation did not become a reality until over one hundred years later with the advent of the electronic digital computer.

1.3.1 Early Systems (1961–1980)

Most of the earliest computer algebra systems (or languages) can rightly be regarded as experiments by their designers to investigate the feasibility of mechanizing certain symbolic mathematical computations. Only a handful were eventually placed into "production use" for scientists, engineers, and applied mathematicians other than members of the development team. The computers of the time were extraordinarily slow by today's standards and had extremely limited memory capacities. Moreover, personal computers did not exist and most scientists had only batch, as opposed to interactive, access to a central mainframe computer at their university or industrial site. It is also important to remember that the first high level programming languages—Fortran, Algol, and Cobol—did not come into existence until the late 1950s and early 1960s. We present brief descriptions of several influential early systems.

SAINT (Symbolic Automatic INTegrator)

SAINT was developed by James Slagle for his Ph.D. dissertation at MIT (1961). The goal was to produce a program that could solve integration problems at the level of a freshman calculus student. The program might more

---

[†]Ada was raised by her mother, who separated from Lord Byron just one month after her birth. During the 1980s, the U.S. Department of Defense named a programming language in honor of Ada's early contributions to computing.

properly be regarded as a triumph of artificial intelligence rather than computer algebra since a heuristic, as opposed to an algorithmic, approach was taken in designing the software. Nonetheless, the system had to perform, out of necessity, many of the automatic mathematical simplifications we normally expect of a computer algebra system. SAINT was developed using the programming language Lisp, as were many of the other early systems.

It is worth noting here that the drastically simpler problem of symbolic integration had been solved many years earlier. Two M.A. theses on symbolic differentiation were written independently in 1953—one by H. G. Kahrimanian (Temple University) and the other by J. Nolan (MIT).

FORMAC (FORmula MAnipulation Compiler)

This software was developed from 1962–64 by a team at IBM in Cambridge, Massachusetts, headed by Jean Sammet and Robert Tobey. The package performed many basic manipulations on elementary functions including polynomials, rational functions, exponentials, logarithms, and trigonometric functions. It could expand polynomials but could not factor them. Similarly, it could differentiate elementary functions but could not integrate.

The system was written as an extension to Fortran. Statements beginning with the keyword **LET** were converted by a preprocessor into subroutine calls to perform the various algebraic manipulations. A version of FORMAC based on the PL/I language appeared later. IBM support for the project ended in 1971, although the software continued to be used for some time afterward and was distributed through IBM's SHARE user library.

Altran

An early version of this system called Alpak was first made available in 1964. Created by a team at Bell Laboratories headed by William S. Brown, it consisted of a collection of Fortran-callable subroutines written in assembly language. A language called Altran was developed from 1964–66 as a front-end to Alpak. Altran was implemented in Fortran, making it highly portable.

The Altran system could manipulate polynomials and rational functions, as well as truncated power series. It was characterized by a set of compact tabular data structures, and was the host for important research on the calculation of polynomial greatest common divisors and sparse polynomial arithmetic. The entire package continued to evolve during the late 1960s and 1970s.

PM and SAC

PM, a series of Fortran-callable subroutines for polynomial manipulation, was developed during the period 1961–66 by George Collins at IBM and the University of Wisconsin. A successor called SAC-I, an acronym for Symbolic and Algebraic Computation, incorporated more advanced algorithms for rational functions and was in use from 1966–71. Unfortunately, SAC-I proved to be quite difficult to use as no front-end language was available and users were responsible for their own memory management.

As a consequence of such shortcomings, Collins and Rüdiger Loos developed SAC-II, along with a driver language called Aldes. The SAC-II/Aldes combination saw use throughout most of the 1970s. Aldes was based on a formalization of Knuth's notation for describing algorithms, and allowed for the use of Greek letters as well as symbols decorated with primes, bars, and hats. A machine-readable version of an Aldes program was obtained by applying a set of simple transcription rules to the publication copy. Similar to Altran in terms of mathematical capabilities, PM and SAC also served as important testbeds for research into algorithms for computer algebra. Importantly, the algorithms used by both systems were completely and carefully documented.

MATHLAB

MATHLAB was another successful early system that performed arithmetic on polynomials and rational functions. Written by Carl Engelman in 1965, the software was implemented in Lisp, the language used for most of the artificial intelligence research taking place at MIT's Project MAC at the time. A successor called MATHLAB-68 was produced in 1968 at MITRE Corporation, where Engelman was then working. MATHLAB is significant in two regards. It was the first interactive computer algebra system. It was also the first to produce two-dimensional output in familiar mathematical notation. It did so through a program called Charybdis, written for MATHLAB by Jonathan Millen. MATHLAB performed only exact arithmetic and had an amusing penchant for issuing the warning "No decimals, please!" whenever a user dared to stray into this domain.

SIN (Symbolic INtegrator)

SIN was an integration program written by Joel Moses in 1966–67 for his doctoral dissertation at MIT. Implemented in Lisp, the system was far more algorithmic in its approach than SAINT. Consequently, it was faster and could solve many more problems.

SIN consisted of three stages in which a combination of heuristic and algorithmic techniques were applied. Elementary transformations like breaking the integral of a sum into a sum of integrals and moving constants outside an integral were employed in the first stage. This stage also included a variant of the method of integration by substitution. In the second stage, the integral was examined to determine whether it was one of eleven special forms, each of which is amenable to a certain type of transformation. Many of these transformations (e.g., trigonometric substitution) are familiar to calculus students. The final stage comprised two additional techniques: integration by parts and a method termed the EDGE (EDucated GuEss) heuristic. The latter applied the method of undetermined coefficients based on the expected form of the solution and was a precursor to the Risch integration algorithm.

Many of the early computer algebra systems made extensive use of pattern matching to recognize the form of an expression and to apply various transforms. SIN was no exception and its extensive pattern matcher, called Schatchen, is well-documented in Moses' thesis.

Symbolic Mathematical Laboratory

This project was the basis for a doctoral dissertation by William Martin at MIT in 1967. The idea was to link several computers to do symbolic computation. One of the project's successes was an advanced user interface. The software supported a light pen for input and produced nicely formatted mathematical output on a CRT screen.

REDUCE

This system was developed principally by Anthony Hearn, beginning at Stanford University in 1968 and later working at the University of Utah and Rand Corporation. It was originally written to assist in high energy physics calculations but soon evolved into a general purpose system. REDUCE is interactive and its underlying math kernel is written in Lisp. The system, however, supports an Algol-like language that allows users to write code in a procedural fashion that avoids Lisp's cumbersome parenthesized notation.

REDUCE was designed to be portable to a variety of platforms, and the source code is provided along with the system. It was the most widely distributed computer algebra system during the 1970s and early 1980s. REDUCE has evolved through numerous versions and development continues to this day. The system at one time had a large user base and was most popular in Europe.

CAMAL

CAMAL was developed around 1970 by David Barton, Steve Bourne, and John Fitch at the University of Cambridge, England., The system started as a collection of machine code subroutines to manipulate Poisson series, and was geared toward calculations in celestial mechanics and general relativity. Gradually it grew into a portable, general purpose package and incorporated a user language implemented in BCPL, a forerunner of the C programming language.

One of the design criteria for the system was to make it small and fast by incorporating tightly packed data structures and finely tuned code. Moreover, the authors refused to provide any facilities that could not be supported efficiently. For example, CAMAL had no special primitives to manipulate rational functions. Instead, the designers provided facilities to extend the language by inspecting and dismantling expressions. As a result, CAMAL could be programmed to solve problems that at first might seem beyond its scope.

Scratchpad

The Scratchpad project began at IBM Research in 1970 under the direction of James Griesmer and Richard Jenks. The software they produced was interactive and integrated significant portions of previous Lisp-based systems including REDUCE-2, MATHLAB-68, Symbolic Mathematical Laboratory, and SIN. Scratchpad took two-dimensional mathematical notation seriously and supported it for both input and output. Its algebra engine was driven by the specification of sets of rules, rather than by collections of assignments or procedure definitions. This evaluator implemented a Markov algorithm for transforming expressions based on both user-defined and system rules. While strictly a research project and never released to the public, Scratchpad was a large and powerful system that had great influence on the field of computer algebra.

Macsyma

Macsyma was by far the largest and most ambitious computer algebra system of this period. Development began in 1971 by a team at MIT under the direction of Joel Moses and William Martin, who built upon the work of MATHLAB-68. In addition to the basic algebraic manipulations provided by earlier systems, Macsyma was the first to offer support for the integration of elementary functions, computation of limits, and the solution of many types of equations. The system served as a testbed for the development of many important algorithms of computer algebra including those for symbolic integration, factorization (Hensel lifting), and sparse modular computation.

Macsyma had a large base of users, many of whom accessed the software remotely on the project's mainframe Digital Equipment Corporation PDP-10 computer at MIT. In the late 1970s and early 1980s, the system was ported from Maclisp to a dialect called Franz Lisp that ran under Unix. Macsyma has an enormous library of mathematical routines, contributed by the developers and users alike.

In 1980, a split occurred and two separate versions of the software evolved. The first was distributed by Symbolics, a company that manufactured and sold Lisp machines, and whose founders included several of Macsyma's principal authors. In 1992, a company named Macsyma, Inc., was formed that bought Symbolics' financial rights to the software and ported it to run on personal computers under Microsoft Windows. The company disbanded in 1999, and the commercial version of Macsyma is no longer available.

The second version was called DOE-Macsyma after the Department of Energy, the agency that funded the original development at MIT. The DOE version was distributed free of charge, and maintained and enhanced by William Schelter at the University of Texas from 1982 until his death in 2001. A descendant, called Maxima, is currently available under GNU public license and runs on Linux, Windows, and Mac OS X platforms.

muMATH

Released in 1979, muMATH was the first comprehensive system to run on a small, personal computer. The goal of its developers, David Stoutemyer and Albert Rich at the University of Hawaii, was to determine how much serious computer algebra could be accomplished on a microcomputer. The system was implemented in a small subset of Lisp called muLISP and distributed with its own programming language called muSIMP. It ran on Intel 8080 and Zilog Z80-based processors having 8-bit architectures, and was available for the popular Radio Shack TRS-80.

To make the best possible use of the limited address space, muMATH was structured as a collection of modules. These included packages for high-precision rational arithmetic, manipulation of elementary (trigonometric, exponential, and logarithmic) functions, differentiation and integration, and equation solving. Users could write their own packages fairly easily. While much slower than the mainframe systems of the time and definitely limited in the size of problems it could solve, muMATH was widely distributed and helped to popularize computer algebra.

Special Purpose Systems

Many interesting special purpose systems were developed in the 1970s. A few of the more significant ones are listed here, along with the names of the developers and the field toward which each was geared.

- *Sheep*. This system for tensor component manipulation was developed by Inge Frick and others at the University of Stockholm.

- *Trigman*. This was a system for Poisson series computation, implemented in Fortran by William H. Jeffreys at the University of Texas.

- *Schoonschip*. Written by Martinus Veltman of the Netherlands, this system was originally developed for high energy physics. It was capable of performing huge calculations in which the size of the expressions manipulated exceeded the amount of main memory available.

- *Analitik*. This was a general purpose system implemented in hardware. The development team at the Institute for Cybernetics in Kiev, USSR, was headed by V. M. Glushkov.

1.3.2 More Recent Systems (1981–present)

The computer field saw two important developments during the 1980s. Early in the decade, powerful minicomputers like the Digital Equipment Corporation VAX series began to appear. This started the migration away from a model where a university or corporate site would have a single large mainframe computer, often operating in batch mode, that served all its users. The trend toward a more interactive and distributed model of computer operations had clearly begun. This movement was accelerated in the latter part of the decade with the appearance of reasonably priced personal computers of moderate power. It then became feasible for users to have their own dedicated machines in the office and at home. Memory at the time, however, was very expensive and still fairly limited, dictating the use of space efficient algorithms.

The decade of the 1980s was significant for computer algebra, as well. The two major computer algebra systems in use today, Maple and Mathematica, were launched during this period. We survey some of the most important recent systems here.

Maple

Maple had its inception at the University of Waterloo, Canada, in 1981. The implementation team was led by Keith Geddes and Gaston Gonnet. The initial goal of the project was to provide wider access to computer algebra for a diverse audience ranging from math students to researchers.

Maple has its own interpreted, dynamically typed command and programming language. The software was originally implemented in a preprocessed version of C called Margay and developed in a time-sharing environment on the VAX. The first version of Maple was released in 1982 and, in the system's early years, the software was distributed either free or for a nominal distribution charge.

The system has a modular structure consisting of three primary components. The first is a modest sized kernel implemented in C code. The kernel includes the command interpreter, provides memory management, and performs integer and rational arithmetic, basic simplification and some elementary polynomial manipulations.

The second component is a large main library where almost all of the mathematical knowledge resides. This library is written in Maple itself, and users can supplement it by writing or sharing packages.

The third component is the user interface, called the Iris. The interface can vary from implementation to implementation, and it is even possible for users and other software systems to communicate with Maple through their own interfaces. The author's NEWTON system, a pedagogic interface for college-level students, and several commercial systems—including, at one time, MATLAB® and Scientific Notebook—communicated with Maple in this way. To present a uniform appearance across various computers and operating systems, the interface provided in the standard distribution is currently implemented in Java.

Since 1988, Maple has been developed and sold commercially by Maplesoft, now a subsidiary of Cybernet Systems, Japan. There are active research groups at several Canadian universities including the University of Waterloo, the University of Western Ontario, and Simon Fraser University that work on enhancing the mathematical functionality of the system.

A detailed introduction to Maple is presented in Section 2.1.

Mathematica

The principal designer of Mathematica, Stephen Wolfram, had written an earlier computer algebra system called SMP (Symbolic Manipulation Program) while at Caltech during the period 1979–81. SMP took a rule-based, pattern-directed approach to symbolic computation and was marketed commercially by Interface Corporation of Los Angeles from 1983 to 1988. In 1986 Wolfram began work on Mathematica, and the next year he co-founded Wolfram Research to market and develop the software. The initial version of the system was released in 1988 and ran on the Apple Macintosh and NeXT computers.

Mathematica was the first computer algebra system to incorporate many capabilities that we now expect of a CAS. It was the first to provide a truly integrated mathematical environment for symbolics, numerics, and graphics. It also included a flexible programming language supporting both functional and procedural styles of programming. Mathematica's plotting capabilities (two- and three-dimensional graphs, and animations) vastly exceeded those of any prior computer algebra system. It was also the first CAS to feature a graphical user interface. Another novel feature was its support of "notebooks." These hierarchical documents can contain a mixture of text, attractively formatted mathematics, and Mathematica commands and graphics. Mathematica also marked the beginning of the commercialization of computer algebra. Before its appearance, no one had tried to make serious money by marketing a computer algebra system. CAS were generally distributed by their developers for free or at a nominal fee intended to recoup the cost of media, documentation, and support.

The Mathematica system consists of millions of lines of source code written in C/C++, Java, and Mathematica itself. While the original version consisted of only 150,000 lines of code, programs written for that version will almost always run correctly on the latest release. The code for the kernel is nearly identical for all systems on which Mathematica runs. Roughly 30% of the kernel supports the language and system, 20% algebraic computation, 20% numerical computation, and 30% graphics and kernel output. In contrast, the front end incorporates a substantial amount of specialized code to support each different user interface environment. The computational portion of the system is separate from the user interface, which can access a kernel running on another computer. Users can write software that communicates with Mathematica through MathLink, passing either Mathematica commands or using an industry standard for mathematical markup called MathML.

Section 2.2 contains an introduction to the symbolic capabilities of Mathematica and its embedded programming language.

Derive

Developed by David Stoutemyer and Albert Rich of the Soft Warehouse, Derive is the successor to muMATH. The system was originally designed for machines considered small at the time—16-bit personal computers—with the goal of making computer algebra widely available to the masses. Derive was written in a dialect of Lisp and first released for MS-DOS in 1988. The Windows version, supporting a 32-bit math kernel and a graphical user interface, appeared in 1996. Development of Derive ceased in 2006.

While far smaller than Maple and Mathematica, Derive provided very impressive mathematical capabilities. Under Windows, Derive was an interactive, menu-driven system supporting both two- and three-dimensional graphics. Its programming environment, however, was somewhat limited.

Texas Instruments Calculator-Based CAS

In 1999, the Soft Warehouse sold its interest in Derive to Texas Instruments, whose intent was to develop a line of calculators based on CAS technology. The software powering the TI products is not simply a rewrite of Derive code, but an entirely new system implemented from scratch in the C programming language.

Due to their small physical size, calculators struggle with restrictions of memory and screen space. The original TI-92, released in 1995, came with 68K bytes of RAM (expandable to 188K), a screen with 240×128 pixels, and sported a small QWERTY keyboard. The TI-89 has a keypad layout and a smaller size typical of most scientific calculators. Unlike Derive, the calculators are not menu driven but feature a command line interface similar to those of most computer algebra systems. Commands are also available through pull-down menus or special keys.

Since the TI-89, Texas Instruments has introduced Mathcad-like PC software called TI-InterActive!, the TI-Voyage 200 calculator, and the TI-Nspire handheld CAS. The companion TI-Nspire CAS software offers the same mathematical capabilities for PC and Macintosh computers.

Axiom

Axiom was developed by a group at the IBM Thomas J. Watson Research Center. The project team was headed by Richard Jenks and included Robert Sutor, Barry Trager, Stephen Watt, James Davenport, and Scott Morrison. Work on the system, originally called Scratchpad–II, began in the mid-1970s. It was not until 1991, however, that the software was released as a commercial product, distributed by the Numerical Algorithms Group (NAG) of Oxford, England. NAG discontinued its support for Axiom in 2001 and it is now available as free, open-source software. Axiom is written in Lisp and runs under Unix and on personal computers with Microsoft Windows.

Axiom is a "strongly typed" computer algebra system. Users work in specific algebraic domains such as rings, fields, and polynomials. Whenever an operation is applied to an object, the object's type determines the behavior of the operation. This paradigm is similar to that of object-oriented programming. Multiplication in a field, for example, inherits and extends the properties of multiplication in a ring. Axiom has over 300 different built-in types, some of which are applicable to mathematical objects and others to data structures (e.g., lists, trees, hash tables). The system also includes a separate, associated programming language called A#.

MuPAD

MuPAD was originally developed by a team led by Benno Fuchssteiner at the University of Paderborn, Germany. The project began in 1989 as a collection of M.S. theses written in C++. The earliest versions of the system, now referred to as "MuPAD Light," were available free of charge for research and education. Beginning in 1997, development and distribution of "MuPAD Pro" was continued by SciFace Software, GmbH. In 2008, SciFace was purchased by MathWorks, which included the MuPAD code in its Symbolic Math Toolbox add-on for MATLAB.

MuPAD has a command and programming language syntax that is very similar to Maple. A unique feature is that users can create and manipulate algebraic domains (e.g., the integers modulo 3) in addition to the untyped mathematical objects supported by most CAS. Users can extend the system by writing functions in MuPAD's own programming language. They can also extend the kernel itself by writing dynamically loadable modules in C++.

SageMath

The goal of SageMath, according to the project's website, is to create "a viable free open source alternative to Magma, Maple, Mathematica, and MATLAB." The project was conceived by William Stein, a mathematician at the University of Washington, who continues to lead the development through a non-profit organization, SageMath, Inc. The system was originally called Sage, an acronym for "System for Algebra and Geometry Experimentation," and the first version was released in 2005.

Rather than "reinventing the wheel," SageMath is built on top of nearly 100 open source packages. The philosophy is somewhat akin that of the Scratchpad-I project at IBM Research in the 1970s, which integrated several existing Lisp-based systems. Among the computer algebra packages that Sage-Math builds upon are Maxima (a descendant of Macsyma), GAP, PARI/GP, and SymPy and SciPy (Python libraries for symbolic mathematics, and scientific/technical computing, respectively).

SageMath is written in Python, and users communicate with the software through an interactive command shell based on Python. The system supports procedural, functional, and object-oriented programming paradigms. Another feature is the software's support of browser-based notebooks incorporating text, commands, and graphics. These notebooks are compatible with most popular web-browsers. SageMath can even communicate with third-party applications including Maple and Mathematica.

Special Purpose Systems

To complete our historical tour, we list several of the most widely used special purpose systems, along with their developers and primary application areas.

- *Magma* and *Cayley*. Magma is used in mathematical areas including algebra, number theory, geometry, and combinatorics. The system was developed by the computational algebra group at the University of Sydney, Australia, led by John Cannon. A predecessor of Magma was called Cayley (1982–1993). Both are large systems with a software architecture reminiscent of Macsyma.

- *Form*. This system was developed by Jos Vermaserem for calculations in high energy physics.

- *GAP* (Groups, Algorithms and Programming). This is another system for computational group theory. Its architecture is similar to that of Maple in that it couples a small kernel with a large library written in its own programming language. GAP was developed by a sizable international group of contributors, originally led by Joachim Neubüser. The project was later coordinated by the Mathematical Institute at The University of St. Andrews, Scotland. Today GAP is maintained and distributed by an international consortium that includes St. Andrews.

- *LiE*. This system was developed by A. M. Cohen at the University of Eindhoven, The Netherlands, for Lie algebra computation.

- *Macaulay*. This system was designed for computations in algebraic geometry and commutative algebra. At its heart is a finely tuned set of algorithms for manipulating polynomial equations and performing Gröbner basis calculations. Written by Daniel Grayson and Michael Stillman, Macaulay is implemented in the C programming language for portability and efficiency.

- *PARI/GP*. Developed by a team headed by Henri Cohen, PARI/GP is oriented toward calculations in number theory. Like Macaulay, it is written in C for reasons of portability and efficiency.

- *Theorist/LiveMath*. This system, developed by Allan Bonadio, combines some symbolic capabilities with numerics and graphics. Theorist eschews the command line interface that is typical of a CAS. All user interaction is through a graphical user interface. One novel feature is its drag-and-drop paradigm that provides a natural way to manipulate equations when performing elementary derivations. Theorist, however, lacks some of the functionality normally expected of a general purpose CAS. For example, it is weak in the areas of factoring, integration, and equation solving. Consequently, this system is most attractive in a pre-calculus educational setting.

1.4   Exercises

1. The *center of mass* of the Earth-Moon system is the point where the two bodies would balance if connected by a rigid rod. The mass of the Earth is 5.975×10^{24} kilograms and that of the Moon is 7.348×10^{22} kilograms. The average distance between them is 384,403 kilometers (238,857 miles) center-to-center. The Earth has a diameter of 12,746 kilometers (7,920 miles) and the Moon's is 3,474 kilometers (2,159 miles). Locate the center of mass of the Earth-Moon system. Are you surprised at its location?

2. Astrologers maintain that the position of the planets at the time of birth profoundly affects the course of a person's life. The planet Saturn, in particular, is reputed to have much influence on mood and disposition. If true, this influence must be caused by the gravitational force of the planet because it exerts no other physical forces of even remotely comparable magnitude.

   According to classical Newtonian mechanics, the gravitational force F_g between two objects is given by

   $$F_g = G \frac{m_1 \, m_2}{r^2},$$

   where m_1 and m_2 are the masses of the objects, r is the distance between them, and G is the universal gravitational constant.

   The mass of Saturn is 5.7×10^{26} kilograms and its distance at closest approach (greatest gravitational affect) to the Earth is 1.2×10^{12} meters. Assume the mass of a doctor in a hospital delivery room is 75 kilograms (165 pounds). At what distance does the doctor exert the same gravitational force on a newborn infant as Saturn at closest approach?

3. Which has a greater gravitational effect on the Moon—the Earth or the Sun? By what factor more? The mass of the Sun is 1.988×10^{30} kilograms and its mean distance from the Earth-Moon system is 149,600,000 kilometers (92,950,000 miles). The mass of the Earth and its distance from the Moon are given in Exercise 1, and the formula for Newtonian gravitational force is in Exercise 2.

4. Johannes Kepler (1571–1630) developed three laws describing the motion of the planets. The first is that the planets move in elliptical orbits with the Sun at one focus.

   The polar form for such an orbit is given by

   $$r = \frac{a \, (1 - e^2)}{1 + e \cos \theta},$$

   where r is the distance from the center, e is the eccentricity, and a is the semimajor axis. The *semimajor axis* is the distance from the center through a focus to the edge of the ellipse. The *eccentricity* is the distance from the center to a focus divided by the semimajor axis, and is such that $0 < e < 1$. (An ellipse approaches a circle as the eccentricity goes to zero, and the two foci converge at the center.)

   (a) Find the values of r and θ when the planet is closest to the sun (perihelion) and furthest from the sun (aphelion).

(b) The orbit of Pluto about the Sun has semimajor axis $a = 39.44$ astronomical units. (One AU is the distance from the Earth to the Sun. See Exercise 3.) Its eccentricity $e = 0.2482$ is so great that Pluto is sometimes closer to the Sun than Neptune. (Neptune's orbit is almost circular and 30.06 AU from the Sun.) Find the distance of Pluto from the Sun at perihelion and aphelion.

5. Kepler's third law, a consequence of universal gravitation, states that the square of the period T of any planet is proportional to the cube of the semimajor axis a of its orbital ellipse,

$$T^2 \approx \frac{4\pi^2}{G\, m_{sun}}\, a^3 = k\, a^3.$$

When a is measured in astronomical units and T is given in Earth years, then $k = 1$.

Estimate the time it takes each planet to orbit the sun.

Planet	a (AU)	Planet	a (AU)	Planet	a (AU)
Mercury	0.3871	Mars	1.524	Uranus	19.18
Venus	0.7233	Jupiter	5.203	Neptune	30.06
Earth	1.000	Saturn	9.539	Pluto	39.44

6. The orbit of the planet Pythagoras about its sun is given in parametric form by

$$x(t) = 5\,\cos(3\,t), \qquad y(t) = 4\,\sin(3\,t),$$

where the time t is measured in Earth years and distance is measured in astronomical units.

(a) Show that the orbit can be expressed in non-parametric Cartestian form as

$$\frac{x^2}{25} + \frac{y^2}{16} = 1.$$

Hint. Make use of the Pythagorean (and terrestrial) identity $\sin^2 \theta + \cos^2 \theta = 1$.

(b) What are the possible locations of the planet's sun?

(c) Give the planet's distance from its sun at perihelion (closest approach) and aphelion (furthest approach).

(d) How many Earth years does it take Pythagoras to orbit its sun?

(e) Does the planet orbit its sun in a clockwise or counterclockwise direction?

7. Balance each of these chemical reactions by setting up and solving a set of linear simultaneous equations.

(a) The burning of aluminum in oxygen.

$$a\,Al + b\,O_2 \rightarrow c\,Al_2O_3$$

(b) Aluminum, when heated, reacts with copper oxide to produce copper metal and aluminum oxide.

$$a\,Al + b\,CuO \rightarrow c\,Al_2O_3 + d\,Cu$$

(c) A simple ion exchange involving iron and iodine.

$$a\,Fe^{3+} + b\,I^- \rightarrow c\,Fe^{2+} + d\,I_2$$

(d) Another ion exchange in which the dichromate (VI) ion is reduced to chromium (III) ions. {-} denotes a free electron.

$$a\,Cr_2O_7{}^{2-} + b\,H^+ + c\,\{\text{-}\} \rightarrow d\,Cr^{3+} + e\,H_2O$$

(e) Ethane burns in oxygen yielding carbon dioxide and steam (water vapor).

$$a\,C_2H_6 + b\,O_2 \rightarrow c\,CO_2 + d\,H_2O$$

(f) A similar reaction involving propane.

$$a\,C_3H_8 + b\,O_2 \rightarrow c\,CO_2 + d\,H_2O$$

(g) Potassium ferrocyanide reacts with sulfuric acid and water to form sulfates of potassium, iron and ammonium plus carbon monoxide.

$$a\,K_4Fe(CN)_6 + b\,H_2SO_4 + c\,H_2O$$
$$\rightarrow d\,K_2SO_4 + e\,FeSO_4 + f\,(NH_4)_2SO_4 + g\,CO$$

(h) Iron (III) sulfate reacts with potassium thiocyanate to form an alum, potassium iron (III) thiocyanate.

$$a\,Fe_2(SO_4)_3 + b\,KSCN \rightarrow c\,K_3Fe(SCN)_6 + d\,K_2SO_4$$

When balancing this reaction, treat the sulfate (SO_4) and thiocyanate (SCN) ions as single units.

8. Algebra decides whether a chemical reaction is plausible. If the value of a coefficient is negative, the reactant (or product) belongs on the opposite side of the equation.

Determine whether the following reactions are possible:

(a) $a\,CO + b\,H_2O \rightarrow c\,CO_2 + d\,H_2$

(b) $a\,CO_2 + b\,H_2O \rightarrow c\,CO + d\,H_2$

(c) $a\,P_2I_4 + b\,H_3PO_4 \rightarrow c\,P_4 + d\,H_2O + e\,H_4I$

(d) $a\,(NH_4)_2SO_4 \rightarrow b\,NH_4OH + c\,SO_4$

(e) $a\,K_3MnO_5 + b\,Mn_3O_7 + c\,O_3 \rightarrow d\,KMnO_4$

9. Some chemical reactions can be balanced by more than one set of coefficients. Show that the chemical equation

$$a\,CO + b\,CO_2 + c\,H_2 \rightarrow d\,CH_4 + e\,H_2O$$

has an infinite number of solutions. In particular, each of the reactions

$$CO + CO_2 + 7\,H_2 \rightarrow 2\,CH_4 + 3\,H_2O$$
$$2\,CO + CO_2 + 10\,H_2 \rightarrow 3\,CH_4 + 4\,H_2O$$
$$CO + 2\,CO_2 + 11\,H_2 \rightarrow 3\,CH_4 + 5\,H_2O$$

has been observed to occur under different conditions.

10. Balance the following rather complicated chemical reactions. You may wish to consult a mathematical program to solve the large, resulting systems of equations.

(a) $a\,\left[Cr(N_2H_4CO)_6\right]_4 \left[Cr(CN)_6\right]_3 + b\,KMnO_4 + c\,H_2SO_4$
$\rightarrow d\,K_2Cr_2O_7 + e\,MnSO_4 + f\,CO_2 + g\,KNO_3 + h\,K_2SO_4 + i\,H_2O$

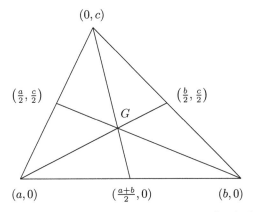

FIGURE 1.3: Medians of a triangle with vertices at $(a, 0)$, $(b, 0)$, and $(0, c)$.

(b) $a\,\mathrm{H_2} + b\,\mathrm{Ca(CN)_2} + c\,\mathrm{NaAlO_4} + d\,\mathrm{FeSO_4} + e\,\mathrm{MgSiO_3} + f\,\mathrm{KI}$
$+\,g\,\mathrm{H_3PO_4} + h\,\mathrm{PbCrO_4} + i\,\mathrm{BrCl} + j\,\mathrm{CF_2Cl_2} + k\,\mathrm{SO_2}$
$\rightarrow l\,\mathrm{PbBr_2} + m\,\mathrm{CrCl_3} + n\,\mathrm{MgCO_3} + o\,\mathrm{KAl(OH)_4} + p\,\mathrm{Fe(SCN)_3}$
$+\,q\,\mathrm{PI_3} + r\,\mathrm{NaSiO_3} + s\,\mathrm{CaF_2} + t\,\mathrm{H_2O}$

11. Find the greatest common divisor (gcd) of the following pairs of integers.

   (a) 24, 60   (b) 22, 50   (c) 360, 640   (d) 654, 2322

   (e) 220, 284   (f) 1029, 1071   (g) 1225, 3267   (h) 12345, 67890

12. Find the least common multiple (lcm) of the integer pairs in Exercise 11.

13. (a) Describe how you found the gcds and lcms in Exercises 11 and 12. Give precise algorithms.

   (b) Find a relationship between $\gcd(m, n)$ and $\operatorname{lcm}(m, n)$ so that one can be computed easily from the other.

14. The *median* of a triangle is a straight line through a vertex and the midpoint of the opposite side.

   (a) Find the equations for the medians of a triangle with vertices at $(a, 0)$, $(b, 0)$, and $(0, c)$. See Figure 1.3.

   (b) Find the (x, y) coordinates of the intersection of each pair of lines found in part (a). You should discover that the three medians intersect at a single point G called the *centroid*. This point is the triangle's "center of gravity" in the sense that if the triangle were made of cardboard, it could be balanced on a pin at this point.

   (c) Show that each median divides the triangle into two equal areas. Consequently, if the triangle were made of cardboard, you could balance it on a median (or, in fact, any other line passing through the centroid).

   (d) Show that the distance between a vertex and the centroid is twice the distance along the median from the centroid to the midpoint of the opposite side.

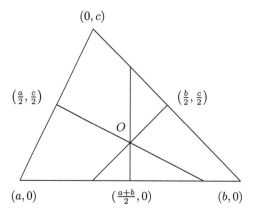

FIGURE 1.4: Perpendicular bisectors of a triangle with vertices at $(a,0)$, $(b,0)$, and $(0,c)$.

15. A *perpendicular bisector* of a triangle is a line passing through the midpoint of a side and perpendicular to it.

    (a) Find the equations of the perpendicular bisectors of a triangle with vertices at $(a,0)$, $(b,0)$, and $(0,c)$. See Figure 1.4.

    (b) Find the (x,y) coordinates of the intersection of each pair of lines found in part (a). You should discover that the perpendicular bisectors intersect at a single point. If all three angles are acute (less than 90°), does this point lie inside, outside, or on the triangle? What if the angle at $(a,0)$ is obtuse (greater than 90°)? What if it is a right angle?

    (c) The point O found in part (b) is called the *circumcenter* and is the center of the circle passing though all three vertices. Verify this by calculating the distance from each vertex to the circumcenter.

16. An *altitude* of a triangle is a line passing through a vertex and perpendicular to the opposite side. The point of intersection is called the *foot* of the altitude.

    (a) Determine the (x,y) coordinates of the feet of a triangle with vertices at $(a,0)$, $(b,0)$, and $(0,c)$. See Figure 1.5.

    (b) Find the equations of the altitudes of the triangle.

    (c) Find the (x,y) coordinates of the intersection of each pair of lines found in part (a). You should discover that the altitudes intersect at a single point H known as the *orthocenter*.

17. The *angle bisectors* of a triangle with vertices at $(a,0)$, $(b,0)$, and $(0,c)$ are shown in Figure 1.6.

    (a) Find the equations of the angle bisectors.

    (b) Find the (x,y) coordinates of the intersection of each pair of lines found in part (a). You should discover that the angle bisectors intersect at a single point I called the *incenter*.

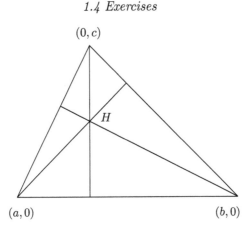

FIGURE 1.5: Altitudes of a triangle with vertices at $(a, 0)$, $(b, 0)$, and $(0, c)$.

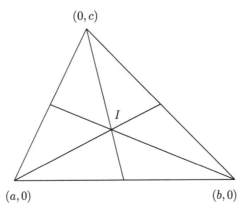

FIGURE 1.6: Angle bisectors of a triangle with vertices at $(a, 0)$, $(b, 0)$, and $(0, c)$.

(c) Find the perpendicular distance from the incenter to each side. You should discover that the incenter is equidistant from the three sides. A circle of this radius drawn through the incenter is tangent to each side of the triangle. This circle is known as the *incircle*.

18. Find the (x, y) coordinates of the centroid, circumcenter, orthocenter, and incenter of the following triangles. These terms are defined in Exercises 14–17. Use exact arithmetic. (No decimal approximations for fractions, square roots, or trigonometric functions.)

   (a) An equilateral triangle with vertices $(-\frac{1}{2}, 0)$, $(\frac{1}{2}, 0)$, and $(0, \frac{\sqrt{3}}{2})$.

   (b) An isosceles triangle with vertices $(-1, 0)$, $(1, 0)$, and $(0, 2)$.

   (c) An isosceles right triangle $(45°\text{-}45°\text{-}90°)$ with vertices $(0, 0)$, $(0, 1)$, and $(1, 0)$.

   (d) A $30°\text{-}60°\text{-}90°$ triangle with vertices $(0, 0)$, $(\sqrt{3}, 0)$, and $(1, 0)$.

   (e) A 3-4-5 right triangle with vertices $(0, 0)$, $(4, 0)$, and $(0, 3)$.

(f) A 5-12-13 right triangle with vertices $(0,0)$, $(12,0)$, and $(0,5)$.

(g) A scalene acute triangle with vertices $(0,0)$, $(3,0)$, and $(2,2)$.

(h) A scalene obtuse triangle with vertices $(0,0)$, $(2,0)$, and $(3,2)$.

19. Simplify the following radicals by denesting.

(a) $\sqrt{5 + 2\sqrt{6}}$ (b) $\sqrt{2 - \sqrt{3}}$ (c) $\sqrt{70 + 20\sqrt{10}}$ (d) $\sqrt{28 - 10\sqrt{3}}$

Hint. Assume the denesting is of the form

$$\sqrt{a \pm b\sqrt{c}} = \sqrt{m} \pm \sqrt{n}.$$

Square both sides, then set up and solve a pair of equations for m, n.

20. When can $\sqrt{a \pm b\sqrt{c}}$ be denested as $\sqrt{m} \pm \sqrt{n}$, where $m > n$?

Projects

1. Read one (or both) of these classical, historical articles on computer algebra.

Richard Pavelle, Michael Rothstein, and John Fitch, "Computer Algebra," *Scientific American* **245**, 6 (1981), 136–152.

A. C. Norman, "Algebraic Manipulation," in A. Ralston and E. D. Reilly (eds.), *Encyclopedia of Computer Science and Engineering* (second edition, 1983), VanNostrand Reinhold, 41-50.

Write a report on an aspect of the article(s) that you find interesting.

2. Find out more about one of the early computer algebra systems by locating and reading one or more articles about the package. Write a report describing the important features of the system. Include an assessment of its strengths and weaknesses. Try to include a small sample program giving the flavor of the system and illustrating its principal control structures.

Chapter 2

Computer Algebra Systems

I wish to God these calculations had been executed by steam.

Charles Babbage (1791–1871)
Quoted in H. Eves, *In Mathematical Circles,* 1969

I cannot do it without comp[u]ters.

William Shakespeare (1564–1616)
Winter's Tale

This chapter provides an overview of the capabilities of computer algebra systems. Emphasis is placed on the symbolic mathematical features discussed in subsequent chapters. As a result, there is only passing mention of the numerical and graphical capabilities, and none of matrix operations.

The chapter is organized into two sections describing the two principal computer algebra systems in use today: Maple and Mathematica. For both systems, we provide an overview of the following features:

- exact arithmetic on numbers

- expansion and factorization of polynomials; manipulation of rational functions (ratios of polynomials)

- manipulation and simplification of trigonometric, logarithmic, and exponential functions

- the definition and utilization of variables and functions

- the basic operations of calculus including differentiation, integration (indefinite and definite), summation, and computation of series and limits

- solution of equations and systems of equations

- manipulation of basic data structures (sets, lists, tables)

- the principal control structures (selection, looping, recursion) used in programming

The sections for the systems can be read independently of each other by a reader interested in using only one of the packages. On the other hand, the organization of both sections is the same and many examples (or similar ones) are repeated so the reader can easily compare features. Thus, a reader familiar with one of the systems can readily see how to access similar functionality in the other.

Our purpose is not to provide a user's manual for each system, but to describe and give examples of the most important commands. Enough information is given for the reader to get started using and programming either system. The reference manual, on-line help, or one of the many in-depth books on each system can then be consulted for further information and assistance.

Descriptions of the user interface (notebooks, editing palettes, etc.) has intentionally been omitted. They can vary from one implementation to another and are not relevant to our subsequent discussions. All of the examples are presented here through a basic command line interface.

2.1 Maple

Maple is an interactive system for symbolic mathematical computation. Development began in 1980 at the University of Waterloo in Ontario, Canada. Since 1988, software development and marketing have been taken over by Maplesoft (formerly Waterloo Maple, Inc.).

The Maple system consists of a kernel and a library of mathematical procedures. The kernel, written in the C programming language, is relatively compact. It supports such facilities as integer, floating-point and polynomial arithmetic, as well as interpreting Maple's own commands and programming language. The mathematical library, written in Maple itself, provides most of the mathematical functionality. Library functions are loaded automatically as required so memory requirements are kept as small as possible, growing at a rate that depends on the user's application.

2.1.1 Numbers

Maple can perform arithmetic on rational numbers. Unlike most calculators, the answers given are exact.

```
> 1/2 + 1/3 + 2/7;
```

$$\frac{47}{42}$$

Calculations are performed using as many digits as necessary.

```
> 1153 * (3^58 + 5^40) / (29! - 7^36);
```

$$\frac{13805075714975527214071994314}{536863065731140371445 3245983}$$

Constants in Maple may be approximated by floating-point numbers. The default precision is 10 digits of accuracy. The `evalf` function causes evaluation to a floating-point number.

```
> evalf(%);
```

$$2.571433313$$

The percent sign (%) refers to the last expression evaluated by the Maple kernel, which is not necessarily the last expression on the worksheet. Similarly, %% refers to the next to last expression, and %%% to the third expression back. You cannot go back further using percent signs.

Terminating a command with a colon (:) instead of a semicolon (;) causes Maple to compute a result without reporting it.

```
> 1/2. + 1/3 + 2/7:
```

```
> %;
```

$$1.119047619$$

Observe that floating-point arithmetic takes place automatically whenever a floating-point or decimal number appears in an expression.

The `iquo` and `irem` functions are used to find the quotient and remainder when two integers are divided.

```
> iquo(75,13);
```

$$5$$

```
> irem(75,13);
```

$$10$$

The basic arithmetic operations obey the usual rules of precedence and associativity, except that nested exponentiation must be parenthesized to eliminate ambiguity. Either ^ or ** may be used to denote exponentiation.

```
> 2 ** 3 ^ 2;
```

```
Error, ambiguous use of '^', please use parentheses
```

```
> 2 ** (3 ^ 2);
```

$$512$$

```
> (2 ** 3) ^ 2;
```

$$64$$

Trigonometric, Logarithmic, and Exponential Functions

Maple knows about most common mathematical functions. Trigonometric functions use radians. The constant π is denoted `Pi`.

```
> sin(Pi/4);
```

$$\frac{\sqrt{2}}{2}$$

```
> evalf(%);
```

$$0.7071067810$$

```
> arcsin(1/2);
```

$$\frac{\pi}{6}$$

The global variable `Digits` allows the user to control the number of digits of precision. Its default value is 10.

```
> Digits := 50;
```

$$Digits := 50$$

```
> evalf(sin(Pi/4));
```

$$0.70710678118654752440084436210484903928483593768845$$

```
> Digits := 10:
```

Alternatively, a second parameter specifying the precision can be given to `evalf`, in which case the value of `Digits` remains unchanged.

```
> evalf(sin(Pi/4), 50);
```

$$0.70710678118654752440084436210484903928483593768845$$

If the argument to a mathematical function is a floating-point number, numerical evaluation occurs automatically.

```
> ln(2);
```

$$\ln(2)$$

```
> ln(2.);
```

$$0.6931471806$$

```
> exp(1.);
```

$$2.718281828$$

```
> exp(1);
```

$$e$$

Note that `exp(1)` is used to represent e, the base for natural logarithms.

Integer Functions

Maple knows about integer functions like factorials and binomials.

```
> 75!;
```

$$2480914081139539809194647711659403366092624388657012283779\backslash$$
$$58945126558426775728674094438154240000000000000000000$$

```
> binomial(100,50);
```

$$100891344545564193334812497256$$

Maple can also factor integers and test for primality.

```
> 2^(2^6) + 1;
```

$$18446744073709551617$$

```
> isprime(%);
```

$$false$$

```
> ifactor(%%);
```

$$(67280421310721)\ (274177)$$

```
> nextprime(2^(2^6) + 1);
```

$$18446744073709551629$$

igcd finds the greatest common divisor of two integers, and ilcm calculates the least common multiple.

```
> igcd(48,60);
```

$$12$$

```
> ilcm(48,60);
```

$$240$$

```
> ifactor(48);
```

$$(2)^4\ (3)$$

```
> ifactor(60);
```

$$(2)^2\ (3)\ (5)$$

2.1.2 Polynomials and Rational Functions

Expanding and Factoring Polynomials

Maple can expand and factor polynomials.

```
> (x + 1)^4 * (x + 2)^2;
```

$$(x+1)^4 (x+2)^2$$

```
> expand(%);
```

$$x^6 + 8x^5 + 26x^4 + 44x^3 + 41x^2 + 20x + 4$$

```
> factor(%);
```

$$(x+1)^4 (x+2)^2$$

Its algebra engine can manipulate polynomials of more than one variable just as easily.

```
> (x + 2*y) * (x - z) * (y - z);
```

$$(x+2y)(x-z)(y-z)$$

```
> expand(%);
```

$$x^2 y - x^2 z + 2xy^2 - 3xyz + xz^2 - 2y^2 z + 2yz^2$$

```
> factor(%);
```

$$(x+2y)(x-z)(y-z)$$

It can also manipulate polynomials of functions.

```
> (cos(x) - 2*sin(x))^2 * (cos(x) - tan(x));
```

$$(\cos(x) - 2\sin(x))^2 (\cos(x) - \tan(x))$$

```
> expand(%);
```

$$\cos(x)^3 - \cos(x)^2 \tan(x) - 4\cos(x)^2 \sin(x) + 4\cos(x)\sin(x)\tan(x)$$
$$+ 4\sin(x)^2 \cos(x) - 4\sin(x)^2 \tan(x)$$

```
> factor(%);
```

$$(\cos(x) - 2\sin(x))^2 (\cos(x) - \tan(x))$$

Rational Functions and Normal Forms

The `normal` command has several important applications. First, it is used to group a sum of terms over a single common denominator.

```
> 1/x + 2/y + 1/z;
```

$$\frac{1}{x} + \frac{2}{y} + \frac{1}{z}$$

```
> normal(%);
```

$$\frac{x\,y + 2\,x\,z + y\,z}{x\,y\,z}$$

The `normal` command also reduces multiple level fractions to just two levels.

```
> (1 + x/(x-1)) / (1 - x/(x+1));
```

$$\frac{1 + \dfrac{x}{x-1}}{1 - \dfrac{x}{x+1}}$$

```
> normal(%);
```

$$\frac{(2\,x - 1)\,(x + 1)}{x - 1}$$

Finally, `normal` removes common factors from the numerator and denominator of an expression. For example, let's define a polynomial *p1* that we will factor. The assignment operator is `:=`, as in some programming languages.

```
> p1 := x^3 + 1;
```

$$p1 := x^3 + 1$$

```
> factor(p1);
```

$$(x + 1)\,(x^2 - x + 1)$$

Now let's define another polynomial, *p2*, and factor it.

```
> p2 := x^2 + 3*x + 2;
```

$$p2 := x^2 + 3\,x + 2$$

```
> factor(p2);
```

$$(x + 2)\,(x + 1)$$

Maple does not ordinarily remove common factors from a rational function.

```
> p1/p2;
```

$$\frac{x^3 + 1}{x^2 + 3x + 2}$$

However, **normal** cancels common factors.

```
> normal(p1/p2);
```

$$\frac{x^2 - x + 1}{x + 2}$$

Greatest Common Divisor and Least Common Multiple

There are built-in functions for greatest common divisor (gcd) and least common multiple (lcm). We just used the **normal** command to cancel the gcd from the ratio *p1/p2*.

```
> gcd(p1,p2);
```

$$x + 1$$

Maple can find gcds of multivariate polynomials, as well.

```
> p := 55*x^2*y - 15*x*y^2 + 44*x*y - 12*y^2;
```

$$p := 55\,x^2\,y - 15\,x\,y^2 + 44\,x\,y - 12\,y^2;$$

```
> factor(p);
```

$$y\,(5\,x + 4)\,(11\,x - 3\,y)$$

```
> q := 77*x^2*y - 21*x*y^2 - 22*x^2 + 6*x*y;
```

$$q := 77\,x^2\,y - 21\,x\,y^2 - 22\,x^2 + 6\,x\,y$$

```
> factor(q);
```

$$x\,(7\,y - 2)\,(11\,x - 3\,y)$$

```
> gcd(p,q);
```

$$11\,x - 3\,y$$

Least common multiples work in much the same way.

```
> p1 := 15*(x - 5)*y;
```

$$p1 := 15\,(x - 5)\,y$$

```
> p2 := 9*(x^2 - 10*x + 25);
```

$$p2 := 9\,x^2 - 90\,x + 225$$

```
> factor(p2);
```

$$9\,(x - 5)^2$$

```
> lcm(p1,p2);
```

$$5\,y\,(9\,x^2 - 90\,x + 225)$$

```
> factor(%);
```

$$45\,y\,(x - 5)^2$$

See if you can produce the same result with gcd and normal.

2.1.3 Variables and Functions

Variables

An identifier in Maple is given a value through an assignment statement.

```
> p := x^2 + 3*x + 7;
```

$$p := x^2 + 3\,x + 7$$

Note that x has no value (it represents itself) and that p has been assigned a value.

```
> x;
```

$$x$$

```
> p;
```

$$x^2 + 3\,x + 7$$

To find the value of p for various values of x, use the subs command.

```
> subs(x=2, p);
```

$$17$$

Equations are formed using =, which is different from the assignment operator :=.

```
> t = p;
```

$$t = x^2 + 3\,x + 7$$

While this looks like an assignment, it's an equation. Nothing happened to t.

```
> t;
```

$$t$$

You can assign equations to variables.

```
> s := t = p;
```

$$s := t = x^2 + 3x + 7$$

Whenever p is used in an expression (for example, on the right side of an assignment), its value is substituted into that expression.

```
> p := p - x^2;
```

$$p := 3x + 7$$

Now p has a new value. Watch what happens when we use p after assigning a value to x.

```
> x := 3;
```

$$x := 3$$

```
> p;
```

$$16$$

Here's how to unassign the value of x. Be sure to use the right kind of quotes!

```
> x := 'x';
```

$$x := x$$

Let's see the effect on p.

```
> p;
```

$$3x + 7$$

The `restart` command clears an entire session, including the values of all variables.

Functions

Mathematical functions of one or more variables can be defined in a very natural way. These functions may be evaluated at either numeric or symbolic points.

```
> f := x -> sin(x)/x;
```

$$f := x \mapsto \frac{\sin(x)}{x}$$

```
> f(2);
```

$$\frac{\sin(2)}{2}$$

```
> f(y);
```

$$\frac{\sin(y)}{y}$$

Here's a function of two variables.

```
> g := (x,y) -> (x^2 - y^2)/(x^2 + y^2);
```

$$g := (x, y) \mapsto \frac{x^2 - y^2}{x^2 + y^2}$$

```
> g(1,2);
```

$$-\frac{3}{5}$$

```
> g(1,x);
```

$$\frac{-x^2 + 1}{x^2 + 1}$$

2.1.4 Calculus

Differentiation

Differentiation in Maple is performed with the `diff` command. The first argument to `diff` is the expression to be differentiated, the second is the variable with respect to which the differentiation is performed.

```
> diff(sin(x)*cos(x), x);
```

$$\cos(x)^2 - \sin(x)^2$$

```
> diff(sin(x) * x^(x^x), x);
```

$$\cos(x) \, x^{x^x} + \sin(x) \, x^{x^x} \left(x^x \left(\ln(x) + 1 \right) \ln(x) + \frac{x^x}{x} \right)$$

Sometimes you want to find the third derivative without finding the first and second.

```
> f := x^9 + 3*x^7 - x^5 + 5*x^3 + 1;
```

$$f := x^9 + 3 \, x^7 - x^5 + 5 \, x^3 + 1$$

```
> diff(f, x, x, x);
```

$$504 \, x^6 + 630 \, x^4 - 60 \, x^2 + 30$$

There's a shorthand to make this easier.

```
diff(f, x$3);
```

$$504 \, x^6 + 630 \, x^4 - 60 \, x^2 + 30$$

The `diff` command can also be used to find partial derivatives.

```
g := x^5 - x^2*y - y^2;
```

$$g := x^5 - x^2\,y - y^2$$

```
> diff(g, x$2);
```

$$20\,x^3 - 2\,y$$

```
> diff(g, x$2, y);
```

$$-2$$

Integration

Analytic integration is an important facility in Maple. It is performed with the `int` command.

```
> f := (x^3 + 9*x^2 + 28*x + 27) / (x + 3)^3 / x;
```

$$f := \frac{x^3 + 9\,x^2 + 28\,x + 27}{(x+3)^3\,x}$$

Observe the effect of the two divisions in the input. Let's use `int` to find the indefinite integral and the definite integral over the interval $[1, 2]$.

```
> int(f, x);
```

$$\ln(x) - \frac{1}{2\,(x+3)^2}$$

```
> int(f, x=1..2);
```

$$\frac{9}{800} + \ln(2)$$

When Maple cannot find a closed form solution for an integral, it returns the input expression.

```
> int(exp(x)/ln(x), x);
```

$$\int \frac{e^x}{\ln(x)}\,dx$$

Summation

Maple can also handle summations with both numeric and symbolic ranges. The syntax is similar to the `int` command.

```
> sum(i, i=0..n);
```

$$\frac{(n+1)^2}{2} - \frac{n}{2} - \frac{1}{2}$$

```
> sum(i^2, i=1..9876543210);
```

$$321139442946604922593984029585$$

Here are a few more interesting sums.

```
> sum(binomial(n,i), i=0..n);
```

$$2^n$$

```
> sum(2^i, i=0..n);
```

$$2^{n+1} - 1$$

```
> sum(1/k^2, k=1..infinity);
```

$$\frac{\pi^2}{6}$$

```
> sum(1/k, k=1..n);
```

$$\Psi(n+1) + \gamma$$

The last result is expressed using the digamma function, Ψ, a Maple built-in function, and Euler's constant, γ. We can use `evalf` to see the value of this constant.

```
> evalf(%);
```

$$\Psi(n+1.) + 0.5772156649$$

Limits

The `limit` function computes the limiting value of an expression as a variable approaches a specified value.

```
> f := (x^2 - 2*x + 1)/(x^4 + 3*x^3 - 7*x^2 + x + 2);
```

$$f := \frac{x^2 - 2x + 1}{x^4 + 3x^3 - 7x^2 + x + 2}$$

If we substitute $x = 1$, the result is undefined because the denominator is zero.

```
> subs(x=1, f);
```

`Error, numeric exception: division by zero`

However, the limit as x approaches 1 is finite and defined.

```
> limit(f, x=1);
```

$$\frac{1}{8}$$

```
> plot(f, x=0..2);
```

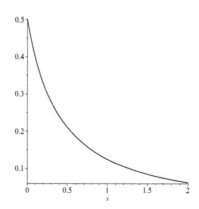

To find directional limits, use **right** or **left** as the third parameter to limit.

```
> limit(1/x, x=0);
```

$$undefined$$

```
> limit(1/x, x=0, right);
```

$$\infty$$

```
> limit(1/x, x=0, left);
```

$$-\infty$$

```
> plot(1/x, x=-1..1, -50..50);
```

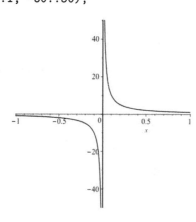

You can also ask Maple to find limits at infinity.

```
> limit(1/x, x=infinity);
```

$$0$$

Series

Maple has facilities for computing Taylor and other series expansions. The expansion point must be given explicitly as the second argument.

```
> taylor(ln(1+x), x=0);
```

$$x - \frac{1}{2}x^2 + \frac{1}{3}x^3 - \frac{1}{4}x^4 + \frac{1}{5}x^5 + O(x^6)$$

```
> taylor(ln(z), z=1);
```

$$(z-1) - \frac{1}{2}(z-1)^2 + \frac{1}{3}(z-1)^3 - \frac{1}{4}(z-1)^4 + \frac{1}{5}(z-1)^5 + O((z-1)^6)$$

```
> taylor(exp(x), x=0);
```

$$1 + x + \frac{1}{2}x^2 + \frac{1}{6}x^3 + \frac{1}{24}x^4 + \frac{1}{120}x^5 + O(x^6)$$

```
> taylor(exp(x), x=1);
```

$$e + e(x-1) + \frac{1}{2}e(x-1)^2 + \frac{1}{6}e(x-1)^3 + \frac{1}{24}e(x-1)^4 + \frac{1}{120}e(x-1)^5 + O((x-1)^6)$$

The O( ) term indicates the truncation of the series. The order of truncation can be controlled by supplying a third argument; the default value is 6.

```
> taylor(sin(x), x=0, 5);
```

$$x - \frac{1}{6}x^3 + O(x^5)$$

```
> taylor(sin(x), x=0, 10);
```

$$x - \frac{1}{6}x^3 + \frac{1}{120}x^5 - \frac{1}{5040}x^7 + \frac{1}{362880}x^9 + O(x^{11})$$

Maple also knows how to compute asymptotic series.

```
> f := n*(n+1)/(2*n-3);
```

$$f := \frac{n(n+1)}{2n-3}$$

```
> asympt(f, n);
```

$$\frac{n}{2} + \frac{5}{4} + \frac{15}{8n} + \frac{45}{16n^2} + \frac{135}{32n^3} + \frac{405}{64n^4} + \frac{1215}{128n^5} + O(\frac{1}{n^6})$$

```
> asympt(f, n, 3);
```

$$\frac{n}{2} + \frac{5}{4} + \frac{15}{8n} + \frac{45}{16n^2} + O(\frac{1}{n^3})$$

The series illustrates the behavior of f as n approaches infinity.

2.1.5   Equation Solving

Maple has a `solve` command that can solve many kinds of equations including single equations involving elementary transcendental functions, and systems of linear and polynomial equations.

Equations in One Variable

Here's how to solve a polynomial with one variable.

```
> poly := 2*x^5 - 3*x^4 + 38*x^3 - 57*x^2 - 300*x + 450;
```

$$poly := 2\,x^5 - 3\,x^4 + 38\,x^3 - 57\,x^2 - 300\,x + 450$$

```
> solve(poly = 0);
```

$$\frac{3}{2},\ \sqrt{6},\ -\sqrt{6},\ 5\,I,\ -5\,I$$

If there is one more than indeterminate, the user must specify which variable to solve for as the second argument.

```
> solve(a*x^2 + b*x + c = 0, x);
```

$$\frac{-b + \sqrt{-4\,a\,c + b^2}}{2\,a},\ -\frac{b + \sqrt{-4\,a\,c + b^2}}{2\,a}$$

For non-polynomials equations, Maple will usually report only one root.

```
> solve(sin(x));
```

$$0$$

Notice that if just an expression is sent to `solve`, Maple sets it to zero and solves.

Numerical Solutions

When `solve` fails to find a closed form, `fsolve` may be able to locate a numerical solution. Note that names generated by Maple begin with an underscore.

```
> solve(sin(x) = exp(-x));
```

$$RootOf(\_Z + \ln(\sin(\_Z)))$$

```
> fsolve(sin(x) = exp(-x));
```

$$0.5885327440$$

```
> plot(sin(x) - exp(-x), x=0..1);
```

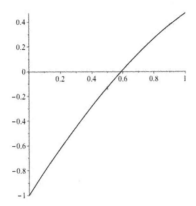

You can often coax the root you want out of `fsolve` by specifying a bounding interval as a third parameter.

```
> fsolve(sin(x)=0, x, 3..4);
```

$$3.141592654$$

Systems of Equations

Here we solve a system of four linear equations in four unknowns.

```
> eq1 := 3*a + 4*b - 2*c + d = -2;
```

$$eq1 := 3\,a + 4\,b - 2\,c + d = -2$$

```
> eq2 := a - b + 2*c + 2*d = 7;
```

$$eq2 := a - b + 2\,c + 2\,d = 7$$

```
> eq3 := 4*a - 3*b + 4*c - 3*d = 2;
```

$$eq3 := 4\,a - 3\,b + 4\,c - 3\,d = 2$$

```
> eq4 := -a + b + 6*c - d = 1;
```

$$eq4 := -a + b + 6\,c - d = 1$$

```
> SolutionSet := solve({eq1, eq2, eq3, eq4});
```

$$SolutionSet := \left\{ a = \frac{1}{2},\ b = -1,\ c = \frac{3}{4},\ d = 2 \right\}$$

We can check the validity of the answer by using the **subs** command to substitute the solution set back into the original equation.

```
> subs(SolutionSet, [eq1, eq2, eq3, eq4]);
```

$$[-2 = -2,\ 7 = 7,\ 2 = 2,\ 1 = 1]$$

Since exact arithmetic is used, Maple can handle non-square systems. In the case where a system is undetermined, Maple will parameterize the solution by treating one or more of the unknowns as free variables.

```
> solve({eq1, eq2, eq3-eq1}, {a, b, c, d});
```

$$\left\{ a = -\frac{31}{10} + \frac{9\,d}{5},\ b = \frac{29}{5} - \frac{17\,d}{5},\ c = \frac{159}{20} - \frac{18\,d}{5},\ d = d \right\}$$

```
> subs(%, [eq1, eq2, eq3-eq1]);
```

$$[-2 = -2,\ 7 = 7,\ 4 = 4]$$

Equations can have coefficients that are parameters.

```
> eqns := {x + y + z = a, x + 2*y - a*z = 0, a*y + b*z = 0};
```

$$eqns := \{x + y + z = a,\ a\,y + b\,z = 0,\ -a\,z + x + 2\,y = 0\}$$

```
> solve(eqns, {x, y, z});
```

$$\left\{ x = \frac{a\,(a^2 + 2\,b)}{a^2 + a + b},\ y = -\frac{b\,a}{a^2 + a + b},\ z = \frac{a^2}{a^2 + a + b} \right\}$$

Maple also includes routines for solving non-linear systems of polynomial equations.

```
> eqns := {x^2 + y^2 = 1, x^2 + x = y^2};
```

$$eqns := \{x^2 + x = y^2,\ x^2 + y^2 = 1\}$$

```
> solve(eqns, {x,y});
```

$$\{x = -1,\ y = 0\},\ \left\{ x = \frac{1}{2},\ y = \frac{RootOf(\_Z^2 - 3)}{2} \right\}$$

2.1.6 Data Structures and Programming

Maple is a programming language as well as a mathematical system. As such, it supports a variety of data structures in addition to mathematical objects like polynomials and series.

Sequences, Sets, and Lists

A sequence is just a collection of expressions separated by commas.

```
> s := sin, cos, tan;
```

$$s := \sin,\ \cos,\ \tan$$

Sequences can be generated using the $ operator.

```
> i^2 $ i=1..10;
```

$$1, 4, 9, 16, 25, 36, 49, 64, 81, 100$$

Sets are input using braces. The usual set operations of **union**, **intersect**, **minus** (for set difference), and **member** (for testing membership) are available.

```
> a := {exp, sin, cos};
```

$$a := \{\cos,\ \exp,\ \sin\}$$

```
> b := {cos, sin, tan};
```

$$b := \{\cos,\ \sin,\ \tan\}$$

```
> a union b;
```

$$\{\cos,\ \exp,\ \sin,\ \tan\}$$

```
> a intersect b;
```

$$\{\cos,\ \sin\}$$

```
> a minus b;
```

$$\{\exp\}$$

```
> member(tan, a);
```

false

Lists, unlike sets, retain the order and multiplicity of the elements. They are input using square brackets.

```
> c := [1, 2, 3, 2, 1];
```

$$c := [1,\ 2,\ 3,\ 2,\ 1]$$

Selection of elements from sets and lists (and more generally, selection of operands from any expression) is accomplished using the **op** and **nops** functions. **op(i,expr)** yields the *i*-th operand in **expr**; **nops(expr)** gives the number of operands in **expr**. Selection can also be accomplished by indexing an item's position in the expression.

```
> nops(c);
```

5

```
> op(4,c);  c[4];
```

2

2

It can be very useful to apply **op** to just one argument, for example to create a new list.

```
> op(c);
```

1, 2, 3, 2, 1

```
> [0, op(c), 0];
```

$$[0,\ 1,\ 2,\ 3,\ 2,\ 1,\ 0]$$

The subsop function may be used to replace the value of an element in a list.

```
> subsop(4=0, c);
```

$$[1, 2, 3, 0, 1]$$

The variable containing the original list remains unchanged (unless it is assigned the new list).

```
> c;
```

$$[1, 2, 3, 2, 1]$$

Replacing an element with a NULL value deletes it from a list.

```
> subsop(1=NULL, 5=NULL, c);
```

$$[2, 3, 2]$$

Functions may be applied to the elements of a set or list using the map command.

```
> map(sqrt, c);
```

$$[1, \sqrt{2}, \sqrt{3}, \sqrt{2}, 1]$$

```
> map(x -> x^2, c);
```

$$[1, 4, 9, 4, 1]$$

You can iterate over the elements of a set or list in a natural way with a program loop. The following procedure returns a list of the cubes of the items in a list.

```
> cubes := proc(s)
    local ans, n;
    ans := [ ];
    for n in s do
      ans := [op(ans), n^3]
    end do;
    return(ans)
  end proc;
```

$cubes := \mathbf{proc}(s)$
  $\mathbf{local}\ ans, n;\ ans := [\ ];\ \mathbf{for}\ n\ \mathbf{in}\ s\ \mathbf{do}\ ans := [\mathrm{op}(ans), n\char`^3]\ \mathbf{end\ do};\ \mathbf{return}\ ans$
  $\mathbf{end\ proc}$

We invoke the procedure to obtain the cubes of the numbers in the list c.

```
> cubes(c);
```

$$[1, 8, 27, 8, 1]$$

Programming in Maple

An important component of the Maple system is its embedded programming language. You should already be familiar with the principal control structures: selection with if-then-else, for- and while-loops, and recursion—so we present just a few examples to give the flavor.

Example 1. The Fibonacci numbers are defined by the recurrence

$$F(n) = F(n-1) + F(n-2)$$

with initial conditions $F(0) = 0$ and $F(1) = 1$. Each number in the sequence is the sum of the previous two. The sequence starts out $0, 1, 1, 2, 3, 5, 8, 13, 21, 34, 55, \ldots$.

Here's an iterative (non-recursive) program to compute the n-th Fibonacci number.

```
> F := proc(n)
    local fib, prev, temp, i;
    if n = 0 then
      fib := 0
    elif n = 1 then
      fib := 1
    else
      prev := 0;
      fib := 1;
      for i from 2 to n do
        temp := fib;
        fib := fib + prev;
        prev := temp
      end do
    end if;
    return(fib)
  end proc:

> F(10);
```

$$55$$

```
> F(35);
```

$$9227465 \qquad \square$$

Example 2. This is a much shorter, recursive procedure to compute Fibonacci numbers.

```
> Fib := proc(n)
    if n=0 or n=1 then n
    else Fib(n-1) + Fib(n-2)
    end if
  end proc:
```

The value returned by a procedure is the last expression evaluated. Alternatively, values are passed back through the **return** label, as in the previous example.

```
> Fib(10);
```

$$55$$

Although the program is simpler than our iterative one, it takes a while to compute $F(35)$. The recursion is killing us!

```
> Fib(35);
```

$$9227465$$

The **remember** option is a way around this problem. If this designation is added to a procedure, all results are stored in a table for quick retrieval.

```
> Fib := proc(n)
    option remember;
    if n = 0 or n = 1 then n
    else Fib(n-1) + Fib(n-2)
    end if
  end proc:

> Fib(10);
```

$$55$$

The **remember** table can be accessed as the fourth operand of Maple's data structure for the procedure.

```
> op(4,eval(Fib));
```

$$table([0 = 0, 1 = 1, 2 = 1, 3 = 2, 4 = 3, 5 = 5, 6 = 8, 7 = 13, 8 = 21,$$
$$9 = 34, 10 = 55])$$

Now we can compute even big Fibonacci numbers very quickly.

```
> Fib(200);
```

$$280571172992510140037611932413038677189525 \qquad \square$$

Example 3. Finally, here's a Maple procedure that produces a list of the prime factors of an integer. The program introduces several new features including type-checking of the argument passed, while-loops, and comments.

```
> factorlist := proc(n::integer)  # n must be an integer
    local flist, m, t;
    if abs(n) < 2 then             # special cases: n = 0, 1, -1
      return([n])
    elif n < 0 then                # n negative, add -1 to flist
      m := -n;
      flist := [-1]
    else                           # work locally with m, not n
      m := n;
      flist := [ ]
    end if;

    while irem(m,2) = 0 do         # divide out factors of 2
      flist := [op(flist), 2];
      m := m/2
    end do;

    for t from 3 to evalf(sqrt(m)) by 2 while m > 1 do
      while irem(m,t) = 0 do       # an odd t divides m
        flist := [op(flist), t];
        m := m/t                   # divide out factors of t
      end do
    end do;

    if m <> 1 then                 # m is prime, add to flist
      flist := [op(flist), m]
    end if;
    return(flist)
  end proc:
```

We compare the output of our procedure with that of Maple's built-in ifactor function.

```
> factorlist(2^4*17), ifactor(2^4*17);
```
$$[2, 2, 2, 2, 17], (2)^4 (17)$$

```
> factorlist(-997), ifactor(-997);
```
$$[-1, 997], -(997)$$

```
> factorlist(10!), ifactor(10!);
```
$$[2, 2, 2, 2, 2, 2, 2, 2, 3, 3, 3, 3, 5, 5, 7], (2)^8 (3)^4 (5)^2 (7)$$

```
> factorlist(4923230976914281), ifactor(4923230976914281);
```
$$[1009, 1201, 1409, 1601, 1801], \ (1009)\,(1201)\,(1409)\,(1601)\,(1801) \qquad \square$$

2.2 Mathematica

Mathematica is an integrated environment for scientific, mathematical, and technical computation. The system provides symbolic and numeric computation, graphics, and a sophisticated notebook environment for producing technical documents with typeset quality mathematics. Mathematica is available for the major computing platforms including Macintosh, Windows, and Linux systems.

Mathematica is a large, complex system and it requires considerable practice to become proficient in its use. This introduction is intended as both a brief overview of its symbolic capabilities, its programming features, and as a quick start guide for new users. Serious users should consult Stephen Wolfram's *An Elementary Introduction to the Wolfram Language* or many of the other excellent books on Mathematica. In addition, the software comes with an extensive on-line help system.

2.2.1 Numbers

Mathematica performs exact arithmetic on rational numbers using as many digits as necessary.

```
In[1]:= 1/2 + 1/3 + 2/7
```
$$\text{Out[1]} = \frac{47}{42}$$

```
In[2]:= 1153 * (3^58 + 5^40) / (29! - 7^36)
```
$$\text{Out[2]} = \frac{13\,805\,075\,714\,975\,527\,214\,071\,994\,314}{5\,368\,630\,657\,311\,403\,714\,453\,245\,983}$$

The N function can be used to obtain a numerical answer. Alternatively, numerical arithmetic takes place automatically whenever a floating-point or decimal number appears in an expression.

```
In[3]:= N[%]
```
```
Out[3]= 2.57143
```

```
In[4]:= 1/2. + 1/3 + 2/7;
```

```
In[5]:= %
```
```
Out[5]= 1.11905
```

% refers to the last expression, %% to the second last, and %%...% (k times) to the k-th last. Terminating a command with a semicolon (;) causes Mathematica to compute a result without reporting it.

Either a space or ∗ may be used to denote multiplication. Arithmetic operations follow the usual rules of precedence and associativity. Exponentiation, unlike the other operations, takes place from right to left.

```
In[6]:= 2 3*2
Out[6]= 12
```

```
In[7]:= 2^3^2
Out[7]= 512
```

```
In[8]:= (2^3)^2
Out[8]= 64
```

Quotient and Mod are used to find the quotient and remainder when two integers are divided.

```
In[9]:= Quotient[75, 13]
Out[9]= 5
```

```
In[10]:= Mod[75, 13]
Out[10]= 10
```

Trigonometric, Logarithmic, and Exponential Functions

Mathematica uses the convention that the names of built-in functions, including the standard mathematical functions, begin with a capital letter. Function arguments are enclosed in square brackets. The constant π is denoted Pi and trigonometric functions work with radians.

```
In[11]:= Sin[Pi/4]
```
$$Out[11]= \frac{1}{\sqrt{2}}$$

```
In[12]:= N[%]
Out[12]= 0.707107
```

```
In[13]:= ArcSin[1/2]
```
$$Out[13]= \frac{\pi}{6}$$

The accuracy of a numerical result can be specified by supplying a second parameter to N.

```
In[14]:= N[Sin[Pi/4], 50]
Out[14]= 0.70710678118654752440084436210484903928483593768847
```

Numerical evaluation occurs automatically if the argument to a numerical function contains a decimal point.

```
In[15]:=  Log[2]
Out[15]=  Log[2]

In[16]:=  Log[2.]
Out[16]=  0.693147

In[17]:=  Exp[1.]
Out[17]=  2.71828

In[18]:=  Exp[1]
Out[18]=  e
```

The base for natural logarithms can be written as E.

```
In[19]:=  N[E]
Out[19]=  2.71828
```

Integer Functions

Mathematica knows about integer functions like factorials and binomials.

```
In[20]:=  75!
Out[20]=  24 809 140 811 395 398 091 946 477 116 594 033 660 926 243 886\
             570 122 837 795 894 512 655 842 677 572 867 409 443 815 424\
             000 000 000 000 000 000

In[21]:=  Binomial[100, 50]
Out[21]=  100 891 344 545 564 193 334 812 497 256
```

Mathematica also has built-in functions to test for primality and to factor integers.

```
In[22]:=  2^2^6 + 1
Out[22]=  18 446 744 073 709 551 617

In[23]:=  PrimeQ[%]
Out[23]=  False

In[24]:=  FactorInteger[%%]
Out[24]=  {{274 177, 1}, {67 280 421 310 721, 1}}
```

GCD and LCM are used to find greatest common divisors and least common multiples.

```
In[25]:=  GCD[48, 60]
Out[25]=  12

In[26]:=  LCM[48, 60]
Out[26]=  240
```

2.2.2 Polynomials and Rational Functions

Expanding and Factoring Polynomials

Mathematica can expand and factor polynomials of one or more variables, and polynomials of functions.

In[1]:= Factor[x^6 - 1]
Out[1]= $(-1 + x)(1 + x)\left(1 - x + x^2\right)\left(1 + x + x^2\right)$

In[2]:= Expand[%]
Out[2]= $-1 + x^6$

In[3]:= Factor[x^6 - y^6]
Out[3]= $(x - y)(x + y)\left(x^2 - x\,y + y^2\right)\left(x^2 + x\,y + y^2\right)$

In[4]:= Expand[%]
Out[4]= $x^6 - y^6$

In[5]:= Factor[Exp[2x]^3 - Sin[x]^6]
Out[5]= $(e^x - Sin[x])\,(e^x + Sin[x])\left(e^{2x} - e^x\,Sin[x] + Sin[x]^2\right)$
$\left(e^{2x} + e^x\,Sin[x] + Sin[x]^2\right)$

In[6]:= Expand[%]
Out[6]= $e^{6x} - Sin[x]^6$

Manipulating Rational Functions

The Together command can be used to group a sum of terms over a single common denominator.

In[7]:= 1/x + 2/y + 1/z
Out[7]= $\dfrac{1}{x} + \dfrac{2}{y} + \dfrac{1}{z}$

In[8]:= Together[%]
Out[8]= $\dfrac{x\,y + 2\,x\,z + y\,z}{x\,y\,z}$

Together can also be used to reduce multiple level fractions to just two levels.

In[9]:= (1 + x/(x - 1))/(1 - x/(x + 1))
Out[9]= $\dfrac{1 + \dfrac{x}{-1 + x}}{1 - \dfrac{x}{1 + x}}$

In[10]:= Together[%]
Out[10]= $\dfrac{(1 + x)(-1 + 2\,x)}{-1 + x}$

Apart plays somewhat the reverse role, decomposing a rational function into partial fractions. This involves long division when the degree of the numerator is greater than or equal to that of the denominator.

In[11]:= Apart[%]

Out[11]= $3 + \dfrac{2}{-1+x} + 2x$

In[12]:= Apart[1/%%]

Out[12]= $\dfrac{2}{3(1+x)} - \dfrac{1}{3(-1+2x)}$

The Cancel command removes common factors from the numerator and denominator of a rational function. For example, we define two polynomials, $p1$ and $p2$, and factor them. The assignment operator in Mathematica is =.

In[13]:= p1 = x^6 - 1

Out[13]= $-1 + x^6$

In[14]:= Factor[p1]

Out[14]= $(-1+x)(1+x)(1-x+x^2)(1+x+x^2)$

In[15]:= p2 = x^9 - 1

Out[15]= $1 - x^9$

In[16]:= Factor[p2]

Out[16]= $(-1+x)(1+x+x^2)(1+x^3+x^6)$

Mathematica does not ordinarily remove common factors from a rational function.

In[17]:= p1/p2

Out[17]= $\dfrac{-1+x^6}{-1+x^9}$

Cancel, however, removes such factors.

In[18]:= Cancel[p1/p2]

Out[18]= $\dfrac{1+x^3}{1+x^3+x^6}$

Greatest Common Divisor and Least Common Multiple

There are built-in functions for greatest common divisor (gcd) and least common multiple (lcm). We just used the Cancel command to remove the gcd from the ratio $p1/p2$.

In[19]:= Factor[PolynomialGCD[p1, p2]]

Out[19]= $(-1+x)(1+x+x^2)$

The lcm contains each of the factors in either polynomial.

In[20]:= Factor[PolynomialLCM[p1, p2]]

Out[20]= $(-1+x)(1+x)(1-x+x^2)(1+x+x^2)(1+x^3+x^6)$

Mathematica computes gcd's and lcm's of multivariate polynomials, as well.

```
In[21]:= p = 15 (x - 5) y
Out[21]= 15 (-5 + x) y
```

```
In[22]:= q = Factor[9 (x^2 - 10 x + 25)]
Out[22]= 9 (-5 + x)²
```

```
In[23]:= PolynomialGCD[p, q]
Out[23]= 3 (-5 + x)
```

```
In[24]:= PolynomialLCM[p, q]]
Out[24]= 45 (-5 + x)² y
```

2.2.3 Variables and Functions

Variables

An identifier in Mathematica is given a value through an assignment statement.

```
In[1]:= p = x^2 + 3x + 7
Out[1]= 7 + 3 x + x²
```

x has no value (it represents itself), while p has been assigned a value.

```
In[2]:= x
Out[2]= x
```

```
In[3]:= p
Out[3]= 7 + 3 x + x²
```

To find the value of p for various values of x, we use a *transformation rule* of the form `/. x -> value`.

```
In[4]:= p /. x -> 3
Out[4]= 25
```

Whenever p is used in an expression (e.g., on the right side of an assignment), its value is substituted.

```
In[5]:= p = p - x^2
Out[5]= 7 + 3 x
```

Now p has a new value. Watch what happens when we use p after assigning a value to x.

```
In[6]:= x = 3
Out[6]= 3
```

```
In[7]:= p
Out[7]= 16
```

Here's how to unassign a value to x.

```
In[8]:= x = .
```

Alternatively, the command `Clear[x]` could be used. Now let's see the effect on p.

```
In[9]:= p
Out[9]= 7 + 3 x
```

The `==` operator is used to construct an equation.

```
In[10]:= t == p
Out[10]= t == 7 + 3 x
```

This looks like an assignment, but it's an equation! Nothing happened to t.

```
In[11]:= t
Out[11]= t
```

You can assign equations to variables.

```
In[12]:= s = t == p
Out[12]= t == 7 + 3x
```

Functions

Simple math functions of one or more variables can be defined as expressions. Notice the underscore character _ (known as *blank* in Mathematica) on the left side and that `:=` is used to define a function.

```
In[13]:= f[x_] := Sin[x]/x
```

Functions may be evaluated at either numeric or symbolic points.

```
In[14]:= f[2]
```
$$Out[14]= \frac{Sin[2]}{2}$$

```
In[15]:= f[y]
```
$$Out[15]= \frac{Sin[y]}{y}$$

Here's a function of two variables.

```
In[16]:= g[x_, y_] := (x^2 - y^2)/(x^2 + y^2)
```

```
In[17]:= g[1, 2]
```
$$Out[17]= -\frac{3}{5}$$

```
In[18]:= g[1, x]
```
$$Out[18]= \frac{1 - x^2}{1 + x^2}$$

2.2.4   Calculus

Differentiation

Differentiation in Mathematica is peformed with the D command. The first argument to D is the expression to be differentiated, the second is the variable with respect to which the differentiation is performed.

In[1]:= D[Sin[x] Cos[x], x]
Out[1]= $\mathrm{Cos}[x]^2 - \mathrm{Sin}[x]^2$

In[2]:= D[Sin[x] x^(x^x), x]
Out[2]= $x^{x^x} \mathrm{Cos}[x] + x^{x^x} \left(x^{-1+x} + x^x \mathrm{Log}[x] (1 + \mathrm{Log}[x])\right) \mathrm{Sin}[x]$

Sometimes you want to find the third derivative without finding the first and second.

In[3]:= f = x^9 + 3 x^7 - x^5 + 5 x^3 + 1
Out[3]= $1 + 5 x^3 - x^5 + 3 x^7 + x^9$

In[4]:= D[f, x, x, x]
Out[4]= $30 - 60 x^2 + 630 x^4 + 504 x^6$

Of course, there's a shorthand to make this easier.

In[5]:= D[f, {x, 3}]
Out[5]= $30 - 60 x^2 + 630 x^4 + 504 x^6$

The D command can also be used to find partial derivatives.

In[6]:= g = x^5 - x^2 y - y^2
Out[6]= $x^5 - x^2 y - y^2$

In[7]:= D[g, {x, 2}]
Out[7]= $20 x^3 - 2 y$

In[8]:= D[g, {x, 2}, y]
Out[8]= -2

Integration

An important facility in Mathematica is analytic integration, which is performed with the **Integrate** command.

In[9]:= f = (x^3 + 9 x^2 + 28 x + 27)/(x + 3)^3/x
Out[9]= $\dfrac{27 + 28 x + 9 x^2 + x^3}{x (3+x)^3}$

Observe the effect of the two divisions. Let's use **Integrate** to find the indefinite integral and the definite integral on the interval $[1, 2]$.

```
In[10]:= Integrate[f, x]
```

$$\text{Out}[10] = -\frac{1}{2\,(3+x)^2} + \text{Log}[x]$$

```
In[11]:= Integrate[f, {x, 1, 2}]
```

$$\text{Out}[11] = \frac{9}{800} + \text{Log}[2]$$

When Mathematica cannot find a closed form solution for an integral (definite or indefinite), it returns the input expression.

```
In[12]:= Integrate[Exp[x]/Log[x], x]
```

$$\text{Out}[12] = \int \frac{e^x}{\text{Log}[x]}\, dx$$

Summation

Mathematica can also handle summations with both numeric and symbolic limits. The syntax is the same as that of the **Integrate** command.

```
In[13]:= Sum[i, {i, 0, n}]
```

$$\text{Out}[13] = \frac{1}{2}\, n\,(1+n)$$

```
In[14]:= Sum[i^2, {i, 1, 9876543210}]
Out[14]= 321 139 442 946 604 922 593 984 029 585
```

Here are a few more interesting sums.

```
In[15]:= Sum[Binomial[n, i], {i, 0, n}]
Out[15]= 2^n
```

```
In[16]:= Sum[2^i, {i, 0, n}]
Out[16]= -1 + 2^{1+n}
```

```
In[17]:= Sum[1/k^2, {k, 1, Infinity}]
```

$$\text{Out}[17] = \frac{\pi^2}{6}$$

```
In[18]:= Sum[1/k, {k, 1, n}]
Out[18]= HarmonicNumber[n]
```

Limits

The `Limit` function computes the limiting value of an expression as a variable approaches a specified value.

In[19]:= f = (x^2 - 2 x + 1)/(x^4 + 3 x^3 - 7 x^2 + x + 2)

$$\text{Out[19]} = \frac{1 - 2x + x^2}{2 + x - 7x^2 + 3x^3 + x^4}$$

If we substitute $x = 1$, the result is undefined because the denominator is zero.

In[20]:= f /. x -> 1

    Power: Infinite expression 1/0 encountered.

    Infinity: Indeterminate expression 0 ComplexInfinity encountered.

Out[20]= Indeterminate

However, the limit as x approaches 1 is finite and defined.

In[21]:= Limit[f, x -> 1]

$$\text{Out[21]} = \frac{1}{8}$$

In[22]:= Plot[f, {x, 0, 2}]

Out[22]=

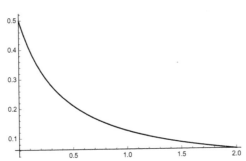

Now let's ask Mathematica to find the limit of $1/x$ as x approaches 0.

In[23]:= Limit[1/x, x -> 0]
Out[23]= Indeterminate

The result is indeterminate since the limits from the left and from the right are not identical. For directional limits, we include a third parameter. Use `Direction -> -1` to find the limit from the right and `Direction -> +1` to find the limit from the left.

In[24]:= Limit[1/x, x -> 0, Direction -> -1]
Out[24]= ∞

In[25]:= Limit[1/x, x -> 0, Direction -> +1]
Out[25]= −∞

`In[26]:= Plot[1/x, {x, -1, 1}, PlotRange -> {-50, 50}]`

`Out[26]=`

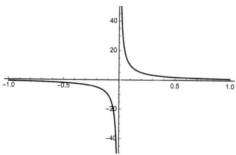

Mathematica can also determine limits at infinity.

`In[27]:= Limit[1/x, x -> Infinity]`

`Out[27]= 0`

Series

Mathematica has facilities for computing Taylor series and other series expansions of expressions. The second parameter gives both the expansion point and the order of the series.

`In[28]:= Series[Log[1 + x], {x, 0, 5}]`

$$\text{Out[28]}= \; x - \frac{x^2}{2} + \frac{x^3}{3} - \frac{x^4}{4} + \frac{x^5}{5} + O[x]^6$$

`In[29]:= Series[Log[x], {x, 1, 3}]`

$$\text{Out[29]}= \; (x-1) - \frac{1}{2}(x-1)^2 + \frac{1}{3}(x-1)^3 + O[x-1]^4$$

`In[30]:= Series[Sin[x], {x, 0, 5}]`

$$\text{Out[30]}= \; x - \frac{x^3}{6} + \frac{x^5}{120} + O[x]^6$$

`In[31]:= Series[Sin[x], {x, 0, 10}]`

$$\text{Out[31]}= \; x - \frac{x^3}{6} + \frac{x^5}{120} - \frac{x^7}{5040} + \frac{x^9}{362880} + O[x]^{11}$$

Mathematica knows how to find asymptotic series as well. These series illustrate the behavior of a function at infinity.

`In[32]:= f = n(n + 1)/(2 n - 3)`

$$\text{Out[32]}= \; \frac{n(1+n)}{-3+2n}$$

`In[33]:= Series[f, {n, Infinity, 3}]`

$$\text{Out[33]}= \; \frac{n}{2} + \frac{5}{4} + \frac{15}{8n} + \frac{45}{16n^2} + \frac{135}{32n^3} + O\left[\frac{1}{n}\right]^4$$

2.2.5 Equation Solving

The Mathematica `Solve` and `Reduce` commands can solve many kinds of equations including polynomials, elementary transcendental functions, and systems of linear and polynomial equations.

Equations in One Variable

The `Solve` command requires an equation that includes Mathematica's equal sign, `==`.

In[1]:= `poly = 2 x^5 - 3 x^4 + 38 x^3 - 57 x^2 - 300 x + 450`
Out[1]= $450 - 300\,x - 57\,x^2 + 38\,x^3 - 3\,x^4 + 2\,x^5$

In[2]:= `Solve[poly == 0]`
Out[2]= $\left\{ \{x \to -5\,i\},\ \{x \to 5\,i\},\ \left\{x \to \dfrac{3}{2}\right\},\ \{x \to -\sqrt{6}\},\ \{x \to \sqrt{6}\} \right\}$

The result returned is a list of lists. The replacement operator can be used to get a simple list of the solutions for x.

In[3]:= `x /. %`
Out[3]= $\{-5\,i,\ 5\,i,\ \dfrac{3}{2},\ -\sqrt{6},\ \sqrt{6}\}$

If there is more than one indeterminate in an equation, the user must specify which variable to solve for as the second argument.

In[4]:= `Solve[a x^2 + b x + c == 0, x]`
Out[4]= $\left\{ \left\{x \to \dfrac{-b - \sqrt{b^2 - 4\,a\,c}}{2\,a}\right\},\ \left\{x \to \dfrac{\sqrt{b^2 - 4\,a\,c} - b}{2\,a}\right\} \right\}$

For non-polynomial equations, Mathematica sometimes reports only one root.

In[5]:= `Solve[Exp[x] Sin[x] == 0, x]`

    Solve: Inverse functions are being used by Solve, so some solutions
      may not be found; use Reduce for complete solution information.

Out[5]= $\{\{x \to 0\}\}$

Reduce tries to determine all possible solutions and often yields more complete information.

In[6]:= `Reduce[Exp[x] Sin[x] == 0, x]`
Out[6]= $C[1] \in \mathbb{Z}\ \&\&\ (x = 2\,\pi\,C[1]\ ||\ x = \pi + 2\,\pi\,C[1])$

In this case, Mathematica reports that for any integer C_1, $2\pi C_1$ or $\pi + 2\pi C_1$ is a solution. This is the same as saying that the solutions are of the form $n\pi$ for any integer n.

Numerical Solutions

When the Solve command fails to find a closed form, FindRoot may be able to locate a numerical solution. Its second parameter includes a starting value.

```
In[7]:= Solve[Sin[x] == Exp[-x], x]
```
    Solve: This system cannot be solved with the methods available
        to Solve.
$$\text{Out[7]= } \text{Solve}\left[\text{Sin}[x] == e^{-x}, x\right]$$

```
In[8]:= FindRoot[Sin[x] == Exp[-x], {x, 0}]
Out[8]= {x → 0.588533}
```

```
In[9]:= Plot[Sin[x] - Exp[-x], {x, 0, 1}]
Out[9]=
```

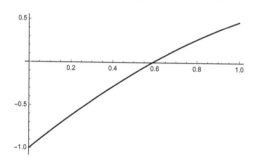

You can often coax the root you want out of FindRoot by specifying a starting point near the desired root.

```
In[10]:= FindRoot[Sin[x] == 0, {x, 3}]
Out[10]= {x → 3.14159}
```

Systems of Equations

Here we solve a linear system of four equations in four unknowns.

```
In[11]:= eq1 = 3 a + 4 b - 2 c + d == -2;
In[12]:= eq2 = a - b + 2 c + 2 d == 7;
In[13]:= eq3 = 4 a - 3 b + 4 c - 3 d == 2;
In[14]:= eq4 = -a + b + 6 c - d == 1;
```

```
In[15]:= SolutionSet = Solve[{eq1,eq2,eq3,eq4}, {a,b,c,d}]
```
$$\text{Out[15]= } \left\{\left\{a \to \frac{1}{2}, \ b \to -1, \ c \to \frac{3}{4}, \ d \to 2\right\}\right\}$$

We check the validity of the answer by substituting the solution set back into the original equations.

```
In[16]:= {eq1, eq2, eq3, eq4} /. SolutionSet
Out[16]= {{True, True, True, True}}
```

Since exact arithmetic is used, Mathematica can properly handle non-square systems. In the case where a system is underdetermined, the user can obtain a parameterized solution by treating one or more of the unknowns as a free variable.

```
In[17]:= eq5 = Part[eq3, 1] - Part[eq1, 1] ==
                 Part[eq3, 2] - Part[eq1, 2]
```
$$Out[17]= \ a - 7\,b + 6\,c - 4\,d == 4$$

eq5 is the difference, *eq3 − eq1*. The **Part** command is described in the next section.

```
In[18]:= Solve[{eq1, eq2, eq5}, {a, c, d}]
```
$$Out[18]= \ \left\{\left\{a \to \frac{1}{34}\,(-1 - 18\,b),\ c \to \frac{3}{68}\,(41 + 24\,b),\ d \to \frac{1}{17}\,(29 - 5\,b)\right\}\right\}$$

Here *b* serves as the free variable.

Equations can have coefficients that are parameters.

```
In[19]:= eqns = {x + y + z==a, x + 2 y - a z==0, a y + b z==0}
```
$$Out[19]= \ \{x + y + z == a,\ x + 2\,y - a\,z == 0,\ a\,y + b\,z == 0\}$$

```
In[20]:= Solve[eqns, {x, y, z}]
```
$$Out[20]= \ \left\{\left\{x \to \frac{a\,(a^2 + 2\,b)}{a + a^2 + b},\ y \to -\frac{a\,b}{a + a^2 + b},\ z \to \frac{a^2}{a + a^2 + b}\right\}\right\}$$

And Mathematica includes algorithms for solving non-linear systems of polynomial equations.

```
In[21]:= eqns = {x^2 + y^2 == 1, x^2 + x == y^2}
```
$$Out[21]= \ \{x^2 + y^2 == 1,\ x + x^2 == y^2\}$$

```
In[22]:= Solve[eqns, {x, y}]
```
$$Out[22]= \ \left\{\{x \to -1,\ y \to 0\},\ \left\{x \to \frac{1}{2},\ y \to -\frac{\sqrt{3}}{2}\right\},\ \left\{x \to \frac{1}{2},\ y \to \frac{\sqrt{3}}{2}\right\}\right\}$$

```
In[23]:= Reduce[eqns, {x, y}]
```
$$Out[23]= \ \left(x == -1 \,\|\, x == \frac{1}{2}\right)\ \&\&\ \left(y == -\frac{\sqrt{1+x}}{\sqrt{2}} \,\|\, y == \frac{\sqrt{1+x}}{\sqrt{2}}\right)$$

2.2.6   Lists

Lists are the most important data structures in Mathematica. Many Mathematica functions expect arguments in the form of lists and return lists as results. A list is a collection of expressions separated by commas.

```
In[1]:= a = {Sin[x], Cos[x], Tan[x]}
```
$$Out[1]= \ \{Sin[x],\ Cos[x],\ Tan[x]\}$$

```
In[2]:= b = {Exp[x], Sin[x], Cos[x]}
```
$$Out[2]= \ \{e^x,\ Sin[x],\ Cos[x]\}$$

Most of Mathematica's algebraic operations can be applied to lists.

```
In[3]:= D[a, x]
Out[3]= {Cos[x], -Sin[x], Sec[x]^2}

In[4]:= a + b
Out[4]= {e^x + Sin[x], Cos[x] + Sin[x], Cos[x] + Tan[x]}
```

Tables and Iterators

Lists can be used as tables of values. The `Table` command creates a list by evaluating an expression for a sequence of different values.

```
In[5]:= sq = Table[i^2, {i, 1, 10}]
Out[5]= {1, 4, 9, 16, 25, 36, 49, 64, 81, 100}
```

The second parameter to `Table` is a special kind of list called an *iterator*. It acts like a for-loop in conventional programming languages and can take one of several different forms. The first parameter is a variable used as a counter, the second is the initial value of the counter, and the third is its final value. An optional fourth parameter is used to specify an increment other than one.

```
In[6]:= Table[i^2, {i, 10, -10, -2}]
Out[6]= {100, 64, 36, 16, 4, 0, 4 ,16, 36, 64, 100}
```

If only two parameters are given, the initial value is assumed to be one.

```
In[7]:= Table[x^i, {i, 10}] /. x -> 2
Out[7]= {2, 4, 8, 16, 32, 64, 128, 256, 512, 1024}
```

`Table` can generate two and higher dimensional lists by using more than one iterator.

```
In[8]:= t = Table[j - i, {i, 1, 3}, {j, 1, 3}]
Out[8]= {{0, 1, 2}, {-1, 0, 1}, {-2, -1, 0}}

In[9]:= TableForm[%]
Out[9]//TableForm=
          0    1    2
         -1    0    1
         -2   -1    0
```

Accessing List Elements

The `Part` command is used to extract portions of a list. An alternative syntax is to access the desired item using a pair of square brackets, in a notation that is similar to the way array elements are accessed in most programming languages.

```
In[10]:= Part[sq, 5]
Out[10]= 25

In[11]:= sq[[5]]
Out[11]= 25
```

The next examples show how to extract portions of a two-dimensional array.

```
In[12]:= Part[t, 3]
Out[12]= {-2, -1, 0}
```

```
In[13]:= Part[Part[t, 3], 2]
Out[13]= -1
```

```
In[14]:= t[[3]]
Out[14]= {-2, -1, 0}
```

```
In[15]:= t[[3, 2]]
Out[15]= -1
```

The `Length` command gives the number of items in a list.

```
In[16]:= Length[sq]
Out[16]= 10
```

```
In[17]:= Length[t]
Out[17]= 3
```

Note that Mathematica regards t as a list of three items, each of which is a list of three numbers.

`Flatten` makes a one-dimensional list from the elements of a multidimensional list.

```
In[18]:= Flatten[t]
Out[18]= {0, 1, 2, -1, 0, 1, -2, -1, 0}
```

```
In[19]:= Length[%]
Out[19]= 9
```

When working with lists, it is important to pay careful attention to the nesting of brackets. Two lists that are otherwise identical, except that one is enclosed in a single set of brackets and the other in a double set, mean two entirely different things to Mathematica. Beginners often don't pay close attention to brackets and get frustrated when Mathematica does not do what they think it should be doing.

Modifying Lists

The `Prepend` and `Append` operations add an item to either the beginning or the end of a list.

```
In[20]:= Append[sq, 11^2]
Out[20]= {1, 4, 9, 16, 25, 36, 49, 64, 81, 100, 121}
```

```
In[21]:= Prepend[sq, 0^2]
Out[21]= {0, 1, 4, 9, 16, 25, 36, 49, 64, 81, 100}
```

Join concatenates lists.

```
In[22]:= c = Join[a, b]
Out[22]= {Sin[x], Cos[x], Tan[x], eˣ, Sin[x], Cos[x]}
```

The assignment statement is used to give a new value to a list element.

```
In[23]:= c[[3]] = Log[x]
Out[23]= Log[x]
```

```
In[24]:= ?c
         Global'c
         c = {Sin[x], Cos[x], Log[x], eˣ, Sin[x], Cos[x]}
```

Set Operations

Mathematica supports set operations on lists.

```
In[25]:= a = {exp, sin, cos}
Out[25]= {exp, sin, cos}
```

```
In[26]:= b = {sin, cos, tan}
Out[26]= {sin, cos, tan}
```

```
In[27]:= Union[a, b]
Out[27]= {cos, exp, sin, tan}
```

```
In[28]:= Intersection[a, b]
Out[28]= {cos, sin}
```

```
In[29]:= Complement[a, b]
Out[29]= {exp}
```

Complement actually gives the set difference, the items in the first set which are not in the second.

MemberQ tests whether a list contains a particular element.

```
In[30]:= MemberQ[a, tan]
Out[30]= False
```

2.2.7 Programming

Mathematica offers a rich and sophisticated programming environment for users to create functions that perform complex computations. The environment supports both procedural and functional programming styles. Our goal in this brief overview is to show how Mathematica implements the principal control structures with which you should be familiar.

Mathematica uses the convention that the names of user-defined functions begin with lower case letters. This distinguishes them from built-in functions which begin with upper case letters.

The principal selection construct is the If statement. Its syntax is

```
If[condition, then-statements, else-statements].
```

If more than one statement is to appear in either the *then* or *else* clause, the statements can be grouped with parentheses and separated by semicolons. The condition can include relational operators (==, !=, <, <=, >, >=) and logical operators (&& and, || or, ! not).

An example of an If statement is

```
In[1]:= If[Random[ ] < 0.5, Heads, Tails]
Out[1]= Heads
```

The parameterless function Rand generates a random number in the range 0 to 1, so Heads and Tails are equally likely to be returned..

Mathematica's principal looping construct is the Do statement, whose syntax is given by

```
Do[body, iterator].
```

As with If, statements in the *body* may be grouped. The *iterator* takes one of the forms described in Section 2.2.6. The next statements compute the n-th harmonic number, $H_n = 1 + 1/2 + 1/3 + \ldots + 1/n$, for $n = 100$.

```
In[2]:= n = 100; hs = 0; Do[hs = hs + 1/i, {i, 1, n}]; hs
```

$$Out[2]= \frac{14\,466\,636\,279\,520\,351\,160\,221\,518\,043\,104\,131\,447\,711}{2\,788\,815\,009\,188\,499\,086\,581\,352\,357\,412\,492\,142\,272}$$

Procedures are usually written using the Module construct,

```
fname[var1_, var2_, ...] := Module[{loc1, loc2, ...}, statements].
```

Don't forget to use :=, not =, when defining a procedure. It's a good idea to use local variables within a procedure so that they do not interfere with any variables used outside the procedure.

Example 1. The Fibonacci numbers are defined by the recurrence relation $F_n = F_{n-1} + F_{n-2}$ with initial conditions $F_0 = 0, F_1 = 1$. The sequence begins $0, 1, 1, 2, 3, 5, 8, 13, 21, 34, 55, \ldots$. The following iterative procedure computes the n-th Fibonacci number.

```
In[3]:= f[n_] := Module[
            {fib, prev, temp, i}, (*local variables*)
            (If[n == 0 || n == 1, (*begin statement list*)
            (*then*) fib = n,
            (*else*) (prev = 0;
                       fib = 1;
                       Do[(temp = fib; (*loop body*)
                           fib = fib + prev;
                           prev = temp),
                          {i, 2, n}] (*iterator*) )]; (*end if*)
            Return[fib])] (*end module*)
```

```
In[4]:=  f[10]
Out[4]=  55
```

```
In[5]:=  f[35]
Out[5]=  9 227 465
```

Of course, we could have written a simpler, recursive procedure to do the computation. We clear the iterative procedure and define the function recursively along with the two initial conditions using Mathematica's syntax.

```
In[6]:=  Clear[f]
```

```
In[7]:=  f[n_]  := f[n - 1] + f[n - 2]
In[8]:=  f[0] = 0;  f[1] = 1;
```

Now let's compute F_{10}.

```
In[9]:=  f[10]
Out[9]=  55
```

It takes a while to find F_{35} due to the tremendous amount of the recursion taking place.

```
In[10]:=  f[35]
Out[10]=  9 227 465
```

Mathematica has stored only the definition of the function and the two initial conditions.

```
In[11]:=  ?f
    Global'f
    f[0] = 0
    f[1] = 1
    f[n_] := f[n - 1] + f[n - 2]
```

To make the code more efficient, we can have Mathematica remember the values of the function it computes. Here's how.

```
In[12]:=  f[n_]  := f[n] = f[n - 1] + f[n - 2]
```

Now let's see what Mathematica has stored after computing F_{10}.

```
In[13]:=  f[10]
Out[13]=  55
```

```
In[14]:=  ?f
    Global'f
    f[0] = 0
    f[1] = 1
    f[2] = 1
    f[3] = 2
    f[4] = 3
    f[5] = 5
```

```
f[6] = 8
f[9] = 13
f[8] = 21
f[9] = 34
f[10] = 55
f[n_] := f[n − 1] + f[n − 2]
```

With this definition, the value of each Fibonacci number is saved when it is first computed. Now even large Fibonacci numbers can be computed instantly.

```
In[15]:= f[200]
Out[15]= 280 571 172 992 510 140 037 611 932 413 038 677 189 525
```
☐

Before presenting our final example, we introduce three additional control structures. Which acts as a case-statement,

$$Which[test1, value1, test2, value2, \ldots].$$

The *tests* are evaluated starting with the first until one is true. The corresponding *value* is returned (or list of statements is executed).

While is a looping construct,

$$While[condition, statements].$$

that evaluates the *statements* in the body repetitively as long as the *condition* holds true.

For is a more complex looping construct that combines the features of Do and While,

$$For[start, test, increment, body].$$

The *start* clause is evaluated to initialize a control variable. Then the loop *body* and *increment* are evaluated repeatedly until the *test* fails.

Example 2. The following procedure returns a list of the prime factors of an integer.

```
In[16]:= primeFactors[n_] := Module[
             {flist, m, t},    (*local variables*)
             (Which[Abs[n] < 2, Return[{n}],
                    n < 0, (m = -n; flist = {-1}),
                    True, (m = n; flist = {})];
              While[Mod[m, 2] == 0,
                 (flist = Append[flist, 2]; m = m/2)];
              For[t = 3, m > 1 && t <= N[Sqrt[m]], t = t + 2,
                 While[Mod[m, t] == 0,
                    (flist = Append[flist, t]; m = m/t)]];
              If[m != 1, flist = Append[flist, m]];
              Return[flist])]    (*end module*)
```

Let's compare the output of our procedure with Mathematica's built-in function FactorInteger.

In[17]:= primeFactors[10!]
Out[17]= {2, 2, 2, 2, 2, 2, 2, 2, 3, 3, 3, 3, 5, 5, 7}

In[18]:= FactorInteger[10!]
Out[18]= {{2, 8}, {3, 4}, {5, 2}, {7, 1}}

In[19]:= primeFactors[-1234567890]
Out[19]= {-1, 2, 3, 3, 5, 3607, 3803}

In[20]:= FactorInteger[-1234567890]
Out[20]= {{-1, 1}, {2, 1}, {3, 2}, {5, 1}, {3607, 1}, {3803, 1}}   □

2.3   Exercises

Use a computer algebra system to solve these exercises. You may also wish to solve some of the problems in Chapter 1 with the aid of a computer algebra system.

1. Compare the "big numbers" that result from the following calculations. Order them from largest to smallest.

   (a) Physicists estimate the age of the universe to be 13.7 billion years old. Calculate the time since the "big bang" in seconds.

   (b) The number of moves required to solve the famous Tower of Hanoi problem for n disks is given by $2^n - 1$. According to legend. Brahmin monks at the time of creation were set to the task of solving the problem for $n = 64$. When they finish the world will come to an end. Assuming the monks move one disk per second, how long will it take them to complete their work?

   (c) Calculate the distance in centimeters that light travels in one year. The speed of light in a vacuum is 299,792,458 meters per second.

   (d) The Andromeda Galaxy, our nearest large neighbor galaxy, is 2.54 million light years away. Find this distance in kilometers. (See the previous item for the speed of light.)

   (e) A Rubik's cube has 8 corner pieces each of which can be oriented 3 ways, and 12 edge pieces each of which can be oriented 2 ways. The orientation of the last corner (or edge) is determined once the other corners (edges) are placed. Since each twist of the cube moves an even number of pieces, only "even" permutations are possible. Consequently, the number of arrangements of the cube is given by

   $$\frac{8! \cdot 3^8 \cdot 12! \cdot 2^{12}}{3 \cdot 2 \cdot 2}.$$

   How many arrangements are there?

(f) According to legend, the creator of the game of chess showed his invention to the king. The king was so pleased that he gave the inventor the right to name his prize. The man was very wise and he asked the king to place one grain of rice on the first square of the 8×8 board, two grains on the second, four on the third, eight on the fourth, etc., doubling the number on each square. How many grains of rice did the man receive?

(g) Find the number of ways to line up 20 people in row for a photograph.

(h) Determine the volume of the sun in cubic meters. The sun's diameter is 1.38 million kilometers, and the volume of a sphere of radius r is given by $V = \frac{4}{3}\pi r^3$.

2. The noted physicist Sir Arthur Eddington (1882-1994) once kiddingly stated what he believed to be the number of protons (and electrons) in the universe. This number appears in one of the quotations at the beginning of Chapter 3. Where did the number come from? Use your computer algebra system to find its prime factorization.

3. (a) Is there a largest integer that can be represented in your CAS, or is the size of an integer unbounded? Check the system documentation and experiment by calculating large factorials.

   (b) Is there a largest (smallest) exponent for floating-point numbers? Consult the documentation and experiment.

4. Comment on the correctness of the plots obtained for the following functions over the range indicated. Can you help your CAS do better by adjusting the window and setting plot parameters?

   (a) $1/x$, $-1 \le x \le 1$

   (b) $\tan x$, $-\pi \le x \le \pi$

   (c) $\dfrac{x\sqrt{x - 2 + \frac{1}{x}}}{x - 1}$, $0 \le x \le 2$ (Note the behavior at $x = 0$ and 1.)

   (d) $\sin\left(\frac{1}{x}\right)$, $-1 \le x \le 1$

   (e) $\sin(e^x)$, $-2 \le x \le 8$

   (f) $\sin(310\,x)$, $0 \le x \le 1$ and $-4 \le x \le 4$

   (g) $\displaystyle\prod_{k=-6}^{6} (x - k)$, $-7 \le x \le 7$

   (h) $1 + x^2 + \frac{1}{8}\ln|(1 - 3(x - 1)|$, $-2 \le x \le 2$ (This curve has a "notch" at $x = \frac{4}{3}$.)

5. Construct fourth-degree polynomials with the roots indicated. Start with linear factored form and then expand the factors. Plot the polynomials to validate your answers.

   (a) -1 (double root), 1 (double root)

   (b) 0 (single root), 1 (triple root)

   (c) 1 (quadruple root)

   (d) $-1, 1, -i, i$

(e) $-i$ (double root), i (double root)

(f) $-\frac{2}{3}$ (double root), $\frac{1}{2}$ (double root)

(g) $1-\sqrt{2},\, 1+\sqrt{2},\, -1-i,\, -1+i$

(h) $-2,\, 0,\, 1\sqrt{2}$. Can you find a polynomial with integer coefficients having these roots? Explain.

6. Use your computer algebra system to explore these polynomials. How many real roots does each have? How many complex roots? What are they? Also find the (x,y)-coordinates of any local maxima, minima, and inflection points. Plot the polynomials to validate your answers.

(a) $6\,x^4 - 35\,x^3 + 22\,x^2 + 17\,x - 10$ (b) $2\,x^4 - 5\,x^3 + 10\,x - 12$

(c) $x^4 + 3\,x^3 - 3\,x^2 - 11\,x - 6$ (d) $8\,x^2 + 2\,x^3 - x^4$

(e) $x^4 - 6\,x^3 + 15\,x^2 - 2\,x - 78$ (f) $x^4 - x^3 + x^2 + x - 1$

7. Explore these rational functions using your computer algebra system. Determine the real roots and the location of the singularities. Evaluate the limit(s) of the function at each singularity and its limits at $-\infty$ and $+\infty$. Plot the rational functions to verify your solutions.

(a) $\dfrac{x+1}{x-1}$ (b) $\dfrac{x^2+2x+1}{x^2-2x+1}$ (c) $\dfrac{x^2+x-2}{x^2-x-2}$

(d) $\dfrac{x^3+3x^2-4}{x^3-3x-2}$ (e) $\dfrac{x^3-2x+2}{x^2-1}$ (f) $\dfrac{3x^3+3x^2-6x+1}{x^2+x-2}$

(g) $\dfrac{x^4+x^3-2x+1}{x^2+x-2}$ (h) $\dfrac{x^2}{x^2+2x+1}$ (i) $\dfrac{x^3}{x^3+3x^2+3x+1}$

(j) $\dfrac{6x^2-5x-2}{4x^2-6x-4}$ (k) $\dfrac{12x^4-7x^3+x^2-2}{3x^3-x^2+2}$ (l) $\dfrac{4x^8+5x^2}{x^3+10}$

8. A rational function can be decomposed into a polynomial (possibly zero) plus a "proper" rational function[†] through polynomial long division. Find the command(s) that your computer algebra system uses to perform this decomposition, then apply it (or them) to the rational functions in Exercise 7. Use this information to determine whether each of these functions has a horizontal, slant (linear), or curvilinear asymptote that describes its limiting behavior as $x \to \pm\infty$. Plot the rational function and an expression for the asymptote over a suitable domain to verify your answer.

9. Construct rational functions $r(x)$ with roots at $x = -1$ (double root) and $x = +1$ (single root), and the vertical asymptotes given below. For each rational function, state the horizontal, slant (linear), or curvilinear asymptote that describes its limiting behavior as $x \to \pm\infty$. Verify your solutions by plotting.

(a) $\lim_{x\to 0} r(x) = -\infty$

(b) $\lim_{x\to 0} r(x) = +\infty$

(c) $\lim_{x\to 0^-} r(x) = +\infty,\ \lim_{x\to 0^+} r(x) = -\infty$

(d) $\lim_{x\to 0^-} r(x) = -\infty,\ \lim_{x\to 0^+} r(x) = +\infty$

[†] A *proper rational function* is one in which the degree of the polynomial in the numerator is strictly less than the degree of the polynomial in the denominator.

10. The derivative of the function $f(x)$ is given by

$$f'(x) = \lim_{h \to 0} \frac{f(x+h) - f(x)}{h}.$$

Calculate the derivatives of the following expressions by setting up and evaluating the appropriate limit. Check your results with the differentiation command of your CAS.

(a) x^3

(b) x^n

(c) $\ln x$

(d) x^x

(e) e^x

(f) e^{x^2}

(g) $\sin x$

(h) $\arctan x$

(i) $\sqrt{(x+1)^3}$

(j) $\dfrac{x^2 - 1}{x^2 + 1}$

(k) $\dfrac{x^2 + x^2 - x - 1}{(x-1)^2 x}$

(l) $\sin(\sin x)$

11. Determine the derivative of the n-th function in the sequences shown by examining the derivatives your CAS produces for small values of n.

(a) $e^x, e^{e^x}, e^{e^{e^x}}, \ldots$

(b) $\ln x, \ln(\ln x), \ln(\ln(\ln x)), \ldots$

(c) $\sin x, \sin(\sin x), \sin(\sin(\sin x)), \ldots$

(d) $x^x, (x^x)^x, ((x^x)^x)^x, \ldots$

(e) $x^x, x^{(x^x)}, x^{\left(x^{(x^x)}\right)}, \ldots$

12. Which of these indefinite integrals can be expressed in terms of *elementary functions* (exponentials, logarithms, trigonometric and inverse trigonometric functions)? When an elementary solution exists, try to derive it by hand.

(a) $\displaystyle\int \frac{\ln x}{x}\, dx$

(b) $\displaystyle\int \ln x \, dx$

(c) $\displaystyle\int \frac{x}{\ln x}\, dx$

(d) $\displaystyle\int \frac{e^x}{x}\, dx$

(e) $\displaystyle\int \frac{x}{e^x}\, dx$

(f) $\displaystyle\int e^x \, \ln x \, dx$

(g) $\displaystyle\int \frac{\sin x}{x}\, dx$

(h) $\displaystyle\int \frac{x}{\sin x}\, dx$

(i) $\displaystyle\int \frac{x}{\sin(x^2)}\, dx$

(j) $\displaystyle\int x \arctan x \, dx$

(k) $\displaystyle\int \arctan x \, dx$

(l) $\displaystyle\int \frac{\arctan x}{x}\, dx$

13. Which of these integrals are elementary? As in Exercise 12, try to derive the solutions having an elementary closed form.

(a) $\displaystyle\int \ln(x^2) \, dx$

(b) $\displaystyle\int \sin(x^2) \, dx$

(c) $\displaystyle\int \arcsin(x^2) \, dx$

(d) $\displaystyle\int \arctan(x^2) \, dx$

(e) $\displaystyle\int e^{x^2} \, dx$

(f) $\displaystyle\int x \, e^{x^2} \, dx$

(g) $\displaystyle\int \cos(x^2) \sin(x^2) \, dx$

(h) $\displaystyle\int x \cos(x^2) \sin(x^2) \, dx$

(i) $\displaystyle\int e^{x^2} \cos(x^2) \sin(x^2) \, dx$

(j) $\displaystyle\int x \, e^{x^2} \cos(x^2) \sin(x^2) \, dx$

14. Set up and evaluate a definite integral to determine the area of the regions described below. Plot each of the regions.

   (a) The region between the sine wave, $\sin x$, and its mirror image, $-\sin x$, for one full cycle, $0 \le x < 2\pi$.

   (b) The region lying above the x-axis and under the parabola $y = 4x - x^2$.

   (c) The region between the curve $y = x^3 - 6x^2 + 8x$ and the x-axis.

   (d) The region bounded by the parabola $y^2 = 4x$ and the line $y = 2x - 4$.

   (e) The region bounded by the parabolas $y = 6x - x^2$ and $y = x^2 - 2x$.

   (f) The area between the curve $\dfrac{1}{1+x^2}$ and the x-axis from -1 to 1; also the area from $-\infty$ to ∞.

   (g) The area between the curve $\dfrac{1}{\sqrt{\pi}}e^{-x^2}$ and the x-axis from -1 to 1; the area from -2 to 2; and the area from $-\infty$ to ∞.

   (h) The area common to the circles $x^2 + y^2 = 4$ and $x^2 + y^2 = 4x$.

15. Evaluate the following sequences of summations. Can you detect a pattern?

   (a) $\displaystyle\sum_{i=0}^{n-1} i^k$ for $k = 0, 1, 2, \ldots, 9$.

   (b) $\displaystyle\sum_{i=0}^{n-1} i^{\underline{k}}$ for $k = 0, 1, 2, \ldots, 9$. $i^{\underline{k}}$ denotes the k-th *falling power* of i, $i^{\underline{k}} = i(i-1)\ldots(i-k+1)$. For example, $i^{\underline{4}} = i(i-1)(i-2)(i-3)$. *Hint.* Factor the result.

   (c) $\displaystyle\sum_{i=0}^{n-1} i^{\underline{-k}}$ for $k = 0, 1, 2, \ldots, 9$. The negative k-th falling power of i is $i^{\underline{-k}} = \dfrac{1}{(i+1)(i+2)\ldots(i+k)}$.‡ For example, $i^{\underline{-4}} = \dfrac{1}{(i+1)(i+2)(i+3)(i+4)}$.

16. Use your computer algebra system to determine whether each of the following infinite series converges (sums to a finite limit) or diverges (sums to infinity).

   (a) $1 + \dfrac{1}{2} + \dfrac{1}{3} + \dfrac{1}{4} + \ldots$ (b) $1 - \dfrac{1}{2} + \dfrac{1}{3} - \dfrac{1}{4} + \ldots$

   (c) $1 + \dfrac{1}{2^2} + \dfrac{1}{3^2} + \dfrac{1}{4^2} + \ldots$ (d) $1 - \dfrac{1}{2^2} + \dfrac{1}{3^2} - \dfrac{1}{4^2} + \ldots$

   (e) $1 + \dfrac{1}{3} + \dfrac{1}{5} + \dfrac{1}{7} + \ldots$ (f) $1 - \dfrac{1}{3} + \dfrac{1}{5} - \dfrac{1}{7} + \ldots$

   (g) $\dfrac{1}{0!} + \dfrac{1}{1!} + \dfrac{1}{2!} + \dfrac{1}{3!} + \ldots$ (h) $\dfrac{1}{0!} - \dfrac{1}{1!} + \dfrac{1}{2!} - \dfrac{1}{3!} + \ldots$

   (i) $\dfrac{1}{2^0} + \dfrac{1}{2^1} + \dfrac{1}{2^2} + \dfrac{1}{2^3} + \ldots$ (j) $\dfrac{1}{2^0} - \dfrac{1}{2^1} + \dfrac{1}{2^2} - \dfrac{1}{2^3} + \ldots$

   (k) $1 + \dfrac{2}{2^2} + \dfrac{3}{2^3} + \dfrac{4}{2^4} + \ldots$ (l) $1 + \dfrac{2^2}{2^2} + \dfrac{3^3}{2^3} + \dfrac{4^2}{2^4} + \ldots$

   (m) $\dfrac{1}{1\cdot 2} + \dfrac{1}{2\cdot 3} + \dfrac{1}{3\cdot 4} + \dfrac{1}{4\cdot 5} + \ldots$ (n) $\dfrac{1}{1\cdot 2} - \dfrac{1}{2\cdot 3} + \dfrac{1}{3\cdot 4} - \dfrac{1}{4\cdot 5} + \ldots$

---

‡Observe that $i^{\underline{3}} = i(i-1)(i-2)$, $i^{\underline{2}} = i(i-1)$, $i^{\underline{1}} = i$, $i^{\underline{0}} = 1$. So it is consistent to define $i^{\underline{-1}} = \frac{1}{i+1}$, $i^{\underline{-2}} = \frac{1}{(i+1)(i+2)}$, $i^{\underline{-3}} = \frac{1}{(i+1)(i+2)(i+3)}$. In this way, $i^{\underline{k-1}} = i^{\underline{k}}/(i-k+1)$ for all k.

17. Use your computer algebra system to obtain a closed form expression for the sum of the first n terms of the series in Exercise 16. Then take the limit of this expression as n goes to infinity. Do the results agree with the answers you obtained for Exercise 16?

18. (a) For which non-negative integer values of n can $\int x^n e^{x^2}\, dx$ be expressed in terms of elementary functions? (See Exercise 13.) What is the general form of the integral when it exists? Prove your answer.

   (b) Repeat for $\int x^n e^{x^3}\, dx$.

   (c) Repeat for $\int x^n \cos(x^2)\, dx$.

19. (a) How many terms are required to distinguish the Taylor series expansions about $x = 0$ of the following pairs of functions?

   i) $\sin x$, $\tan x$

   ii) $\sin \tan x$, $\tan \sin x$

   iii) $\tan \sin \tan x$, $\tan \tan \sin x$

   (b) Find a pair of sines and tangents nested to four levels whose expansions are identical through $O(x^8)$ terms.

20. Find the leading term in the asymptotic expansion as n goes to infinity of the following functions:

   (a) $\dfrac{n}{1 - n - n^2}$ (b) $n!$ (c) $\dbinom{2n}{n}$

21. Find the following limits, then verify the results by plotting.

   (a) $\lim\limits_{x \to 0} \dfrac{\sin x}{x}$ (b) $\lim\limits_{x \to 0} \sin(1/x)$ (c) $\lim\limits_{x \to \infty} \sin x$

   (d) $\lim\limits_{x \to 0} \dfrac{\ln(x^2)}{\ln x}$ (e) $\lim\limits_{x \to 0} \dfrac{e^x - 1}{x}$ (f) $\lim\limits_{x \to 1} \dfrac{x^{100} - 1}{x - 1}$

   (g) $\lim\limits_{x \to 0} (1 + x)^{1/x}$ (h) $\lim\limits_{x \to \infty} (1 + 1/x)^x$ (i) $\lim\limits_{x \to \infty} 1/x$

   (j) $\lim\limits_{x \to 0+} 0^x$ (k) $\lim\limits_{x \to 0+} x^0$ (l) $\lim\limits_{x \to 0+} x^x$

22. (a) Ask your computer algebra system to solve the general quadratic equation, $a x^2 + b x + c = 0$.

   (b) Now ask it to solve the general cubic, $x^3 + a x^2 + b x + c = 0$.[§] Compare the complexity of the result with the solution of the quadratic.

   (c) Use the solution of the general cubic to solve $x^3 - x = (x - 1)\, x\, (x + 1)$ by substituting appropriate values for a, b, c.

   (d) Repeat part (c) for $x^3 - 6 x^2 + 11 x - 6 = (x - 1)\, (x - 2)\, (x - 3)$.

   (e) Now repeat part (c) for $x^3 + 3 x^2 + 4 x + 2 = (x + 1)\, (x^2 + 2 x + 2)$.

   (f) Finally repeat for $x^3 - x^2 - x - 1$. The solutions to this very simple appearing cubic are messy!

[§]This form of a cubic can be obtained by dividing both sides of the more familiar general form, $\alpha x^3 + \beta x^2 + \gamma x + \delta = 0$, by α.

23. Use your computer algebra system to determine the solution to the following trigonometric equations over the interval $0 \leq x < 2\pi$. From this information, can you deduce the general solution to the equations? Obtain a plot to verify your answers.

(a) $\sin x + \cos x = 1$

(b) $2\cos(3x) = 1$

(c) $\sin^2 x - \sin x = 2$

(d) $\cos^2 x + \cos x = \sin^2 x$

(e) $\sin x = \sin(2x)$

(f) $3\tan^3 x - 3\tan^2 x - \tan x + 1 = 0$

(g) $\tan(x/2) = 1$

(h) $2\sin x - 2\sqrt{3}\cos x - \sqrt{3}\tan(x) + 3 = 0$

24. Explore the features of your computer algebra system by finding a command or sequence of commands to transform each of the following expressions to zero.

(a) $\dfrac{x^4 + x^3 - 4x^2 - 4x}{x^4 + x^3 - x^2 - x} - \dfrac{x^2 - 4}{x^2 - 1}$

(b) $2\dfrac{x^3 - x^2 y - xy + y^2}{x^3 - x^2 y - x + y} + \dfrac{y-1}{x-1} - \dfrac{y-1}{x+1} - 2$

(c) $\dfrac{\sec x}{\sin x} - \dfrac{\sin x}{\cos x} - \cot x$

(d) $\dfrac{\sin x + \sin(3x) + \sin(5x) + \sin(7x)}{\cos x + \cos(3x) + \cos(5x) + \cos(7x)} - \tan(4x)$

(e) $\sqrt{30} - \sqrt{2}\sqrt{3}\sqrt{5}$

(f) $5 + 2\sqrt{6} - \dfrac{\sqrt{3}+\sqrt{2}}{\sqrt{3}-\sqrt{2}}$

(g) $3 - \sqrt{-2+\sqrt{-5}}\sqrt{-2-\sqrt{-5}}$

(h) $\left(\sqrt[3]{2} + \sqrt[3]{4}\right)^3 - 6\left(\sqrt[3]{2} + \sqrt[3]{4}\right) - 6$

Programming Projects

1. **(Chebyshev Polynomials)** The Chebyshev polynomials are defined by

$$T_k(x) = 2x\,T_{k-1}(x) - T_{k-2}(x)$$

with $T_0(x) = 1$ and $T_1(x) = x$. The Chebyshev polynomials of degree 4 or less are

k	$T_k(x)$
0	1
1	x
2	$2x^2 - 1$
3	$4x^3 - 3x$
4	$8x^4 - 8x^2 + 1$

(a) Write an efficient recursive procedure, T(k,x), that produces the k-th polynomial in the sequence.

(b) Write an iterative procedure to produce $T_k(x)$.

2. **(Digits of a Number)**   Write a procedure

DigitList(<number>)

which, when passed a non-negative integer, returns a list of the decimal digits in the number. DigitList(12345678909876543210), for example, should produce

$$[1, 2, 3, 4, 5, 6, 7, 8, 9, 0, 9, 8, 7, 6, 5, 4, 3, 2, 1, 0].$$

3. **(Sieve of Eratosthenes)**   The algorithm described below separates prime numbers from non-primes (also called *composite* numbers). Implement a procedure

Eratosthenes(<number>)

that returns a list of primes less than or equal to the argument passed.

Suppose we want to find all the primes up to 50. We start with a list of the numbers from 2 (the smallest prime) to 50. The first number in the list is 2. Crossing out every second number after 2 removes the other even numbers.

2	3	~~4~~	5	~~6~~	7	~~8~~	9	~~10~~
11	~~12~~	13	~~14~~	15	~~16~~	17	~~18~~	19	~~20~~
21	~~22~~	23	~~24~~	25	~~26~~	27	~~28~~	29	~~30~~
31	~~32~~	33	~~34~~	35	~~36~~	37	~~38~~	39	~~40~~
41	~~42~~	43	~~44~~	45	~~46~~	47	~~48~~	49	~~50~~

The number after 2 is 3, the next prime. We eliminate the multiples of 3 by crossing out every third number after 3. Some numbers, like 6 and 12, are multiples of 2 as well, but no harm is done by crossing them out again. Seven more composites are found: 9, 15, 21, 27, 33, 39, 45.

2	**3**	4	5	~~6~~	7	8	~~9~~	10
11	~~12~~	13	14	~~15~~	16	17	~~18~~	19	20
~~21~~	22	23	~~24~~	25	26	~~27~~	28	29	~~30~~
31	32	~~33~~	34	35	~~36~~	37	38	~~39~~	40
41	~~42~~	43	44	~~45~~	46	47	~~48~~	49	50

The next number, 4, is composite, so we do not need to remove its multiples. They have already been struck out since they are also multiples of 2. The next prime is 5, so we cross out every fifth number after 5. This removes 25 and 35 from the list.

2	3	4	**5**	6	7	8	9	~~10~~
11	12	13	14	~~15~~	16	17	18	19	~~20~~
21	22	23	24	~~25~~	26	27	28	29	~~30~~
31	32	33	34	~~35~~	36	37	38	39	~~40~~
41	42	43	44	~~45~~	46	47	48	49	~~50~~

Skipping 6, which is composite, we come to the next prime, 7. We eliminate its multiples by striking out every seventh number after 7. This removes 49 from the list.

2	3	4	5	6	**7**	8	9	10
11	12	13	~~14~~	15	16	17	18	19	20
~~21~~	22	23	24	25	26	27	~~28~~	29	30
31	32	33	34	~~35~~	36	37	38	39	40
41	~~42~~	43	44	45	46	47	48	~~49~~	50

We continue crossing out multiples until we reach the square root of the largest number in the list. (Why can we stop at this point?) Since $\sqrt{50} \approx 7.071$, we are done. The numbers that have not been crossed out are the primes less than 50.

	2	3	4	5	6	7	8	9	10
11	12	13	14	15	16	17	18	19	20
21	22	23	24	25	26	27	28	29	30
31	32	33	34	35	36	37	38	39	40
41	42	43	44	45	46	47	48	49	50

4. **(Maxima, Minima, and Inflection Points)**

$f(x)$ has a *local maximum* at $x = a$ if its first derivative is zero and its second derivative is negative at the point a.

$f(x)$ has a *local minimum* at $x = a$ if its first derivative is zero and its second derivative is positive at the point a.

$f(x)$ has an *inflection point* at $x = a$ if its second derivative is zero at the point a.

Write three procedures,

$$\text{Maxima(<expression>)},$$
$$\text{Mimima(<expression>)},$$
$$\text{Inflect(<expression>)},$$

which, when passed an *expression* containing a single variable, return a list of the (x, y)-coordinates of the point(s) where the function has, respectively, a local maximum, minimum, or inflection point. Give exact results, rather than numerical approximations.

Test your procedures by finding the local maxima, minima, and inflection points of the following:

(a) $x^4 - 2x^3 + 1$,

(b) $x^3 + x$,

(c) $\dfrac{1}{x^2(x-1)} + 3$,

(d) $e^{x^2} - ex^2 + 1$.

Produce a Maple or Mathematica worksheet that includes the procedures together with a narrative exploration of the four functions using the symbolic, numeric, and graphic capabilities of the system.

Chapter 3

Big Number Arithmetic

I believe there are 15,747,724,136,275,002,577,605,653,961,181,555,
468,044,717,914,527,116,709,366,231,425,076,185,631,031,296 protons
in the universe and the same number of electrons.

Sir Arthur Eddington (1882–1944)
The Philosophy of Physical Science, 1939

I do hate sums. There is no greater mistake than to call arithmetic
an exact science. There are permutations and aberrations discernible to
minds entirely noble like mine; subtle variations which ordinary accoun-
tants fail to discover; hidden laws of number which it requires a mind
like mine to perceive. For instance, if you add a sum from the bottom
up, and then from the top down, the result is always different.

Mrs. M. P. La Touche
Quoted in *Mathematical Gazette, 12* (1924)

One feature of computer algebra systems that always amazes new users is their ability to perform arithmetic on numbers of essentially unlimited size. Who can fail to be impressed by typing 1000! and then seeing all 2,568 digits of the result appear instantaneously?

In this chapter, we consider how large numbers are represented and manipulated in computers whose hardware performs arithmetic only on relatively small, finite precision integers. As we shall see, the traditional methods to add, subtract, and multiply numbers are easily adapted to this context. Asymptotically faster methods exist for multiplication, however, and one is presented. What about division? The method we are taught in school involves a certain amount of guesswork and ingenuity. How do we get around this?

After considering the four basic arithmetic operations, we examine how to compute the greatest common divisor of two integers. Euclid's technique for solving this problem is perhaps the oldest non-trivial algorithm still in common use. In addition to Euclid's method, we present a fairly new algorithm that is very efficient for binary representations of numbers because it avoids the use of division. The computation of gcds plays a pivotal role in the implementation of computer algebra systems. Not only is it useful in reducing fractions but, as we shall see later, it is key to performing the polynomial manipulations out of which more complex operations like factorization and integration are built.

Once we know how to compute integer gcds, we are in a position to study arithmetic on rational numbers (fractions) since computer algebra systems always reduce fractions to lowest terms. The final arithmetic problem we consider is that of raising a number to a positive integer power efficiently.

3.1   Representation of Numbers

The hardware on which a computer algebra system is implemented provides the facilities to store and manipulate integers of finite precision. We refer to this unit of information as a "word" or "single precision integer." Typical computer word sizes are 8, 16, 32, 36, 48, or 64 bits. How then do computer algebra systems represent and compute with very large numbers?

To represent big numbers, implementers choose a base (or radix) b in such a way that the numbers $0, 1, 2, \ldots, b-1$ can be directly manipulated by the machine. Moreover, b is generally either a power of 2 or (less frequently) 10. A common choice is the largest positive single precision integer that can be manipulated directly by the hardware (for example, $b = 2^{31}$ on a machine with a 32-bit word size).

There are several issues involved in the selection of the radix b. It is clearly desirable to make b large to save both time and space. At the same time, b should be selected so that the result of a multiplication or division fits in a word.[†] Machine independence is another important consideration because most computer algebra systems run on a variety of hardware platforms. The value of b should be selected so that the system operates efficiently on all target computers, and should not be so large that it precludes porting the system to other machines in the future.

Three data structures are useful for storing and manipulating large integers: linked lists, dynamic arrays, and fixed length arrays. Each of these representations, or a minor variant thereof, has been employed in actual computer algebra systems. Suppose we want to store an l-digit base-b number, $N = \pm (d_{l-1} \ldots d_1 d_0)_b$. The value of N is given by $\pm \sum_{i=0}^{l-1} d_i b^i$. We describe how such a number is stored in each of the three representations. As a concrete example, Figure 3.1 illustrates these schemes for the number $N = 1234567890$ assuming a base $b = 1000$.

The *linked list* representation consists of a series of linearly linked nodes. Each node has two fields—one containing a single base-b digit (number between 0 and $b-1$) and the other a link (pointer) to the next node in the list. The digits are stored with the least significant, d_0, in the first node and the most significant, d_{l-1}, in the last. This order facilitates the implementation of the standard algorithms for addition and multiplication, which access the

---

[†]The product of two n-digit numbers can have $2n$ digits. Division of a $2n$-digit number by one with n digits produces a quotient and remainder, both of which can have n digits.

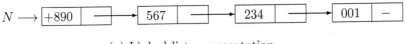

(a) Linked list representation

$$N \longrightarrow \boxed{+\ \ 4\ |\ 890\ |\ 567\ |\ 234\ |\ 001}$$

(b) Dynamic array allocation

$$N \longrightarrow \boxed{+\ 890\ |\ 567\ |\ 234\ |\ 001\ |\ 000\ |\ 000\ |\ 000\ |\ 000\ |\ 000\ |\ 000}$$

(c) Fixed length array representation with $l = 10$ words

FIGURE 3.1: Representations of $N = 1234567890$ with base $b = 1000$.

digits of a number from right to left. The sign is also stored in the first node since the standard procedures for addition and subtraction require knowing the sign at the outset. The link field in the last node contains a null pointer.

In the *dynamic array* allocation, the digits of a number are stored contiguously in memory. The first word contains both the sign of the number and its length l in digits. (Sometimes $l + 1$ is used instead since this gives the total length of the data structure.) The digits themselves are stored, from least to most significant, in the remaining l words.

Fixed length arrays are similar to dynamic arrays in that all the information associated with a number is stored contiguously in memory. The first word contains the sign and the low-order base-b digit. As with the other representations, the digits are stored from least to most significant. The major difference between fixed length and dynamic arrays is that the number of words allocated to any number is set in advance. This implies, of course, there is a largest number that can be represented. In the example of Figure 3.1c, $l = 10$ and since each word stores a digit between 0 and 999, the maximum size number that can be handled is 30 decimal digits.

Some discussion of the relative merits of the three representations is in order. Many of the earlier Lisp-based systems, like Macsyma and REDUCE, use linked lists to store big numbers. Lists have two principal advantages over the other schemes. First, storage management is facilitated since every node in each list has the same size. Second, it is possible to represent arbitrarily large numbers, which is not the case for the array-based representations.

Linked lists, however, have three significant disadvantages. Considerable memory (roughly the amount required for the digits alone) is used for pointers. Moreover, the time to access digits is far greater due to "pointer chasing." Finally, the digits of a number must *always* be accessed sequentially. This restricts the nature of the algorithms that can be implemented efficiently. For example, a divide-and-conquer algorithm like Karatsuba's multiplication scheme requires that numbers be split in half at the outset. This is prohibitively difficult with a linked representation.

Implementers of recent computer algebra systems have generally chosen dynamic arrays to store big numbers. Both Maple and Mathematica employ a variant of this representation. Dynamic arrays use far less memory than linked lists, and achieve a faster processing time because the digits are stored contiguously in memory. While the numbers stored in dynamic arrays can be enormous, their size is in principle limited because a field containing the number of digits must fit in the first word. More sophisticated storage management algorithms are also required due to the varying sized blocks.

Fixed length arrays were used in several older systems, for example, Altran. This representation facilitates memory management because a block of a single fixed size is allocated for each number. While linked lists also allocate memory in fixed size blocks, no "pointer chasing" is required to access the digits of a fixed length array. The most obvious disadvantage is the limited precision of the numbers that can be represented. Moreover, since every integer is allocated an identical size block, memory is wasted for the relatively small numbers that are more typically encountered. Furthermore, processor time is required to access the higher order digits whenever an arithmetic operation is performed since it not known in advance whether they are zeros.

3.2   Basic Arithmetic

We now turn our attention to the four basic arithmetic operations: addition, subtraction, multiplication, and division. To simplify the discussion, we restrict our attention to operations performed on non-negative integers. Moreover, for subtraction, we consider only the case where the smaller number is subtracted from the larger. These restrictions present no loss in the functionality of the arithmetic methods. (Why?)

In developing procedures to perform arithmetic on big numbers, we need to make some assumptions about what "primitive" operations are allowed. We assume that the underlying hardware performs base-b arithmetic for us by supplying the following operations:

1. addition (subtraction) of two single digit numbers, producing a single digit result and a carry (borrow);

2. multiplication of two single digit numbers, giving a two digit result;

3. division of a two digit number by a non-zero single digit number, producing a one digit quotient and a one digit remainder.

procedure $Add(u, v)$
{Input: $u = (u_{n-1} \ldots u_1 u_0)_b$, $v = (v_{n-1} \ldots v_1 v_0)_b$ }
{Output: $w = u + v = (w_n \ldots w_1 w_0)_b$ }
    $k := 0$ {initialize carry k}
    for $i := 0$ **to** $n - 1$ **do**
        $w_i := (u_i + v_i + k) \bmod b$ {compute next digit of sum}
        if $u_i + v_i + k \geq b$ **then** $k := 1$ {determine carry}
                      else $k := 0$ **end if**
    end do
    $w_n := k$ {high-order digit of result}
    return(w)
end procedure

ALGORITHM 3.1: Addition of non-negative numbers

3.2.1 Addition and Subtraction

We begin by examining the classical algorithms for adding and subtracting two numbers in an arbitrary base b. If the numbers are of unequal length, the smaller can be left padded with zeros to make them the same length, say n, before performing the arithmetic. When we are done, it may be necessary to "clean up" the result by removing any extraneous leading zeros.

Suppose we want to add two n-digit numbers, u and v. Note that the sum, w, may be an $n + 1$ digit number. (Why?)

$$
\begin{array}{rl}
u & = & (u_{n-1} & \cdots & u_1 & u_0)_b \\
+ v & = & (v_{n-1} & \cdots & v_1 & v_0)_b \\
\hline
w & = & (w_n & w_{n-1} & \cdots & w_1 & w_0)_b
\end{array}
$$

The standard procedure for addition operates in $O(n)$ time and processes the digits from right to left. At each stage, we add the corresponding digits, u_i and w_i, plus the carry (either 0 or 1) from the previous step. The corresponding digit of the result, w_i, is equal to this sum, reduced modulo b if necessary. If the sum is greater than b, then there is a carry into the next position, otherwise there is none. The high-order digit of the result, w_n, is just the final carry. The procedure is presented formally as Algorithm 3.1.

This procedure can be modified slightly to produce an algorithm for subtracting two n-digit numbers, u and v where $u \geq v$. The difference w has at most n digits. (Why?)

$$
\begin{array}{rl}
u & = & (u_{n-1} & \cdots & u_1 & u_0)_b \\
- v & = & (v_{n-1} & \cdots & v_1 & v_0)_b \\
\hline
w & = & (w_{n-1} & \cdots & w_1 & w_0)_b
\end{array}
$$

procedure *Subtract*(u, v)
{Input: $u = (u_{n-1} \ldots u_1 u_0)_b$, $v = (v_{n-1} \ldots v_1 v_0)_b$ where $u \geq v$}
{Output: $w = u - v = (w_{n-1} \ldots w_1 w_0)_b$ }
   $k := 0$ {initialize borrow k}
   for $i := 0$ **to** $n - 1$ **do**
      $w_i := (u_i - v_i + k) \bmod b$ {compute next digit of difference}
      if $u_i - v_i + k < 0$ **then** $k := -1$ {determine borrow}
                    else $k := 0$ **end if**
   end do
   return(w)
end procedure

ALGORITHM 3.2: Subtraction of non-negative numbers

Again we proceed from right to left, calculating a single digit of the difference plus a borrow (either 0 or −1) at each step. At each iteration, we take the difference $u_i - v_i$ of the corresponding digits, to which we add the borrow from the previous digit. If the result is non-negative, it is assigned to w_i and the borrow into the next digit is zero. Otherwise, the result modulo b is assigned to w_i and there is a borrow at the next step. The entire procedure operates in $O(n)$ time and is given in Algorithm 3.2.

3.2.2 Multiplication

Multiplication is a bit more involved. Suppose we wish to multiply an m-digit multiplicand, u, with an n-digit multiplier, v. The product, w, can have $m + n$ digits.

$$
\begin{array}{rcllll}
u & = & (u_{m-1} & \cdots & u_1 & u_0\,)_b \\
\times\ v & = & (v_{n-1} & \cdots & v_1 & v_0\,)_b \\
\hline
w & = & (w_{m+n-1} & \cdots & w_1 & w_0\,)_b
\end{array}
$$

The paper-and-pencil method we learned in school takes each digit of the multiplier v_j in turn, from right to left, and finds its product with the multiplicand, $(u_{m-1} \ldots u_1 u_0)_b \times v_j$. These products are then added together with appropriate shift (scaling) factors. Since mn digitwise products are formed and we sum n numbers having $m + 1$ digits, the running time is $O(mn)$.

It is more space efficient to perform the additions concurrently with the multiplications, as described in Algorithm 3.3. For each digit of the multiplier v_j, we initialize a carry k to zero. We take the digits u_i of the multiplicand in turn, starting from the right, and form the product of digits, $u_i \cdot v_j$. To this, we add the carry k from the previous digit. This sum contributes to the digit w_{i+j} of the result, so we add its current value to give $t = u_i v_j + k + w_{i+j}$. We

```
procedure Multiply(u, v)
{Input: u = (u_{m-1} ... u_1 u_0)_b, v = (v_{n-1} ... v_1 v_0)_b}
{Output: w = u × v = (w_{m+n-1} ... w_1 w_0)_b }
    w := 0   {initialize w = (0 ... 0)_b}
    for j := 0 to n - 1 do
        if v_j = 0 then next j end if
        k := 0
        for i := 0 to m - 1 do
            t := u_i · v_j + k + w_{i+j}
            w_{i+j} = t mod b
            k := ⌊t/b⌋       {k := floor(t/b)}
        end do
        w_{m+j} := k
    end do
    return(w)
end procedure
```

ALGORITHM 3.3: Multiplication of non-negative numbers

place this back in w_{i+j} after reducing it modulo b to obtain a single digit. The carry k into the next position is just the high order digit of t, if it exists. Note that at each iteration, $t < b^2$ and $0 \le k < b$. (Why?) Any carry that remains after processing all the digits of u is placed in the appropriate position of w.

3.2.3   Karatsuba's Algorithm

The classical algorithm does $O(n^2)$ work to compute the product of two n-digit numbers. Although for centuries no one had developed a better method, the advent of digital computers spurred the development of several asymptotically faster methods. We describe a technique due to Anatolii Karatsuba and Yu. Ofman (1962) that is based on the fact that the product of two 2-digit numbers can be found with three digitwise multiplications instead of the usual four.

Suppose we want to find the product, w, of $u = (u_1 u_0)_b = u_1 b + u_0$ and $v = (v_1 v_0)_b = v_1 b + v_0$,

$$w = (w_2 w_1 w_0)_b = u_1 v_1 b^2 + (u_1 v_0 + u_0 v_1) b + u_0 v_0.$$

In the standard method, we form four products: $u_1 \cdot v_1, u_1 \cdot v_0, u_0 \cdot v_1, u_0 \cdot v_0$. The multiplications by b and b^2 are just shifts and are not computed via digitwise multiplication. With Karatsuba's scheme, we form only three products: $m_1 = (u_1 + u_0) \cdot (v_1 + v_0)$, $m_2 = u_1 \cdot v_1$, $m_3 = u_0 \cdot v_0$. Note that in forming m_1, we add the digits of u and of v together! The product, w, can be expressed as

$$w = m_2 b^2 + (m_1 - m_2 - m_3) b + m_3.$$

This scheme can be applied recursively to larger numbers via an algorithm design technique known as "divide-and-conquer." Suppose we want to multiply two n-digit numbers, U and V. For the sake of simplicity, but without loss of generality, assume that n is some power of two. We begin by dividing the digits of U and V in half, thereby expressing these numbers as $U = U_1 b^{n/2} + U_0$ and $V = V_1 b^{n/2} + V_0$. Note that U_1, U_0, V_1, V_0 are $\frac{n}{2}$-digit numbers. Next, we add the halves of U and V as before, and form the three corresponding products of $\frac{n}{2}$-digit[‡] numbers: $M_1 = (U_1 + U_0) \cdot (V_1 + V_0)$, $M_2 = U_1 \cdot V_1$, $M_3 = U_0 \cdot V_0$. Finally, the product, W, of U and V can be expressed as

$$\begin{aligned} W &= M_2 b^n + (M_1 - M_2 - M_3) b^{n/2} + M_3 \\ &= U_1 V_1 b^n + (U_1 V_0 + U_0 V_1) b^{n/2} + U_0 V_0. \end{aligned} \tag{3.1}$$

So we have multiplied two n-digit numbers by computing the product of three $\frac{n}{2}$-digit numbers along with some additions and shifts. In the classical method, we implicitly calculate four products of $\frac{n}{2}$-digit numbers, as shown in (3.1).

We present an example to illustrate in a concrete way how the method works.

Example 3.1. Calculate the product $5{,}927 \cdot 3{,}141$ with Karatsuba's method.

Solution. Splitting the numbers in half, we have $U_1 = 59$, $U_0 = 27$, $V_1 = 31$, $V_0 = 41$. We add the halves of U and V, and form the three products

$$\begin{aligned} M_1 &= (U_1 + U_0) \cdot (V_1 + V_0) = 86 \cdot 72 = 6{,}192, \\ M_2 &= U_1 \cdot V_1 = 59 \cdot 31 = 1{,}829, \\ M_3 &= U_0 \cdot V_0 = 27 \cdot 41 = 1{,}107. \end{aligned}$$

Note that in forming these products, each of the factors must be further split into two single digit numbers and the method applied recursively. The desired result is formed as

$$\begin{aligned} W &= M_2 \, 10^4 + (M_1 - M_2 - M_3) \, 10^2 + M_3 \\ &= 18{,}290{,}000 + 325{,}600 + 1{,}107 = 18{,}616{,}707. \qquad \square \end{aligned}$$

How much work does Karatsuba's method involve? The recurrence relation

$$T(n) = 3\,T(n/2) + cn$$

captures the overall time, $T(n)$, to multiply two n-digit numbers. The term

---

[‡]In actuality, the sums $U_1 + U_0$ and $V_1 + V_0$ may have $\frac{n}{2} + 1$ digits. This does not materially affect either the algorithm or its analysis.

$3\,T(n/2)$ arises from the three products of $\frac{n}{2}$-digit numbers that are formed when applying the procedure recursively. The term cn results from the addition (subtraction) of numbers whose sizes are linear in n and the shifting. The recursion stops when we reach single digit numbers so an appropriate initial condition is $T(1) = a$, where a is a constant. Letting $n = 2^m$, the solution of this recurrence by back-substitution (telescoping) reveals that

$$
\begin{aligned}
T(n) = T(2^m) &= 3\,T\left(2^{m-1}\right) + c\,2^m \\
&= 3\left[3\,T\left(2^{m-2}\right) + c\,2^{m-1}\right] + c\,2^m = 3^2\,T\left(2^{m-2}\right) + c\,2^m\,(1 + 3/2) \\
&= 3^3\,T\left(2^{m-3}\right) + c\,2^m\left(1 + 3/2 + (3/2)^2\right).
\end{aligned}
$$

Telescoping the recurrence all the way back to the initial condition, we find

$$
\begin{aligned}
T(n) = T(2^m) &= 3^m\,T(1) + c\,2^m\left(1 + 3/2 + \ldots + (3/2)^{m-1}\right) \\
&= 3^m\,a + c\,2^m\,\frac{(3/2)^m - 1}{3/2 - 1} \\
&= 3^{\log_2 n}\,a + 2\,c\,n\left(\frac{3^{\log_2 n}}{2^{\log_2 n}} - 1\right) \\
&= (a + 2\,c)\,3^{\log_2 n} - 2\,c\,n \\
&= (a + 2\,c)\,n^{\log_2 3} - 2\,c\,n.
\end{aligned}
$$

Thus the execution time of Karatsuba's algorithm is $O(n^{\log_2 3}) \approx O(n^{1.59})$, which is asymptotically better than $O(n^2)$ for the classical method. Karatsuba's method, however, results in a larger constant factor and becomes a practical alternative only when the number of base-b digits is around 20. Another consideration is the larger storage requirement for the intermediate calculations.

Several methods that are even faster asymptotically exist. The one whose asymptotic performance is the best to date is due to Arnold Schönhage and Volker Strassen (1971). It is based on the Fast Fourier Transform and uses $O(n \log n \log \log n)$ digitwise operations. Unfortunately, the constant factor associated with this algorithm is so high as to make it impractical for implementation in a computer algebra system.

Finally, we note that the times for division and multiplication are related to one another by a constant factor. As a consequence, a fast algorithm for multiplication can, in principle, be converted into a division algorithm with the same asymptotic running time.

3.2.4 Division

Division is the most involved of the four basic arithmetic operations. It is also the only one that is usually performed from left to right. Moreover, the method that we learned in school involves a certain amount of guesswork and ingenuity. In this section, we develop a framework to place this operation on

Big Number Arithmetic

$$
\begin{array}{r}
q \\
v\ \overline{)\quad u} \\
-q\,v \\
\hline
r
\end{array}
$$

dividend u: $m + n$ digits
divisor v: n digits
quotient q: $m + 1$ digits
remainder r: n digits

FIGURE 3.2: Integer division: $u = q\,v + r$, where $0 \le r < v$.

$$
\begin{array}{r}
66 \\
47\ \overline{)\ 3142} \\
282 \\
\hline
322 \\
282 \\
\hline
40
\end{array}
$$

(a)

$$
\begin{array}{r}
103 \\
47\ \overline{)\ 04852} \\
47 \\
\hline
015 \\
0 \\
\hline
152 \\
141 \\
\hline
11
\end{array}
$$

(b)

FIGURE 3.3: Long division of integers.

a sounder footing and produce an algorithm suitable for implementation with big numbers.

We begin by defining the problem carefully. Suppose we divide an $(m+n)$-digit dividend, u, by an n-digit divisor, v. This operation produces two results: an $(m+1)$-digit quotient, q, and an n-digit remainder, r, where $0 \le r < v$. The Remainder Theorem of Algebra gives the relationship between these quantities, $u = q\,v + r$, where $0 \le r < v$. As with the other arithmetic operations, we simplify our discussion without any loss of generality by assuming that the dividend is non-negative and the divisor is positive. The process of integer division is illustrated in Figure 3.2.

Reflecting on the procedure for long division taught to us as children (Figure 3.3), we can regard each step of the process as dividing an $n+1$ digit piece of the dividend u by the entire n digits of the divisor v. This yields one digit of the quotient plus a remainder r that is less than v. To obtain the piece of the dividend used in the next step, we multiply r by the radix b and add the next digit of u. (Equivalently, we shift r one digit left and append the next digit of u.) Consider what happens when we divide 3142 by 47. (See Figure 3.3a.) We begin by dividing 314 by 47 to get 6 plus a remainder of 32. Then we divide 322 by 47 to obtain 6 along with a remainder of 40.

If the divisor is less than or equal to the first n digits of the dividend, then we begin the process by "left padding" the dividend with a zero. This preserves the paradigm of always using $n + 1$ digits of the dividend at each step. For example, suppose we divide 4852 by 47. (See Figure 3.3b.) Since

$47 \leq 48$, we left pad the dividend with a zero and begin by dividing 47 into 048, producing 1 and a remainder of 1. Next we divide 015 by 47, giving 0 plus a remainder of 15. Finally we divide 152 by 47, yielding 3 and a remainder of 11.

Now let's examine what happens at each step. Let $u = (u_n u_{n-1} \ldots u_1 u_0)_b$ be the $(n+1)$-digit piece of the dividend being processed and $v = (v_{n-1} \ldots v_0)_b$ be the divisor. The digit $q = \lfloor u/v \rfloor^\S$ of the quotient being computed is such that $0 \leq q < b$ and the remainder r is bounded by $0 \leq r < v$.

$$
\begin{array}{r}
q \\
\hline
v_{n-1} \ldots v_1 v_0 \,\big)\, \overline{u_n u_{n-1} \ldots u_1 u_0} \\
-q\,v \\
\hline
r
\end{array}
$$

An obvious approach to the problem of determining q is to make a guess based on the most significant digits of u and v. The standard guess is to divide the leading two digits of u by the leading digit of v. If the result is b or more, we let $b - 1$ be our guess. We call this guess the *trial quotient*, q_t,

$$
q_t = \min\left(\left\lfloor \frac{u_n b + u_{n-1}}{v_{n-1}} \right\rfloor, b - 1 \right).
$$

The two lemmas which follow show that the trial quotient is a remarkably good approximation to the actual quotient as long as the leading digit of the divisor is sufficiently large.

Lemma 3.1. *The trial quotient digit q_t is never too small; i.e., $q \leq q_t$, where q is the actual quotient digit.*

Proof. Since $q\,v \leq u$ and $u_{n-2}\,b^{n-2} + \ldots + u_0 < b^{n-1}$ (why?), we have

$$
q\,v_{n-1}\,b^{n-1} < (u_n b + u_{n-1} + 1)\,b^{n-1}
$$

(again, why?) and, consequently, $q\,v_{n-1} \leq u_n b + u_{n-1}$. By construction q_t is either $b - 1$ or, if $v_{n-1} > u_n$, the largest multiplier of v_{n-1} yielding a product less than or equal to $u_n b + u_{n-1}$. Therefore, $q \leq q_t$. $\qquad\square$

Lemma 3.2. *If the leading digit of the divisor $v_{n-1} \geq \lfloor b/2 \rfloor$, then $q_t - 2 \leq q$.*

Proof. It suffices to show that $(q_t - 2)\,v \leq u$. (Why?) By reasoning similar to that in the proof of the previous lemma, $v_{n-2}\,b^{n-2} + \ldots + v_0 < b^{n-1}$ and

$$
\begin{aligned}
(q_t - 2)\,v &< (q_t - 2)\,(v_{n-1} + 1)\,b^{n-1} = q_t\,v_{n-1}\,b^{n-1} + (q_t - 2 - 2\,v_{n-1})\,b^{n-1} \\
&\leq (u_n b + u_{n-1})\,b^{n-1} + (q_t - 2 - 2\,v_{n-1})\,b^{n-1},
\end{aligned}
$$

$\S \lfloor x \rfloor$ denotes the *floor* of x, or the largest integer less than or equal to x.

where the last inequality follows from the definition of q_t. Since $v_{n-1} \geq b/2$ and $q_t \leq b-1$, we have $q_t - 2 - 2v_{n-1} < 0$. As a consequence,

$$(q_t - 2)\, v \leq (u_n\, b + u_{n-1})\, b^{n-1}$$
$$\leq (u_n\, b + u_{n-1})\, b^{n-1} + u_{n-2}\, b^{n-2} + \ldots + u_1\, b + u_0 = u.$$

□

The following theorem is a direct consequence of combining these lemmas.

Theorem 3.3 (Trial Quotient Theorem). *No matter how large b is, the actual quotient q is either q_t, $q_t - 1$, or $q_t - 2$, provided the leading digit of the divisor $v_{n-1} \geq \lfloor b/2 \rfloor$.*

This result may not seem particularly exciting when the radix $b = 10$ since knowing the trial quotient still leaves 3 out of 10 possibilities for the actual quotient. But when b is 1,000 or 1,000,000, there are still only 3 possibilities and the "guess" given by the trial quotient is incredibly accurate. It is a remarkable fact that, no matter how large the radix, the trial quotient is never off by more than two!

How can we guarantee that $v_{n-1} \geq \lfloor b/2 \rfloor$ in order to take advantage of the Trial Quotient Theorem? If $v_{n-1} < \lfloor b/2 \rfloor$, we multiply both u and v by the single digit number $d = \lfloor b/(v_{n-1} + 1) \rfloor$ to form u' and v'. This process is called *normalization*. The leading digit of v' is guaranteed to be greater than or equal to $\lfloor b/2 \rfloor$ (why?), so Theorem 3.3 applies if we divide u' by v'. The quotient is the same as that obtained by dividing u by v. If we *unnormalize* the remainder by dividing it by d, we obtain the remainder for the original division problem. This final step involves "short division" by a one-place number—a much simpler problem than long divison.

The complete procedure for dividing non-negative integers is presented as Algorithm 3.4. The loop in Step 2 is executed $m + 1$ times and, at each iteration, the number of digitwise operations performed is proportional to the sizes of u and v (which are $n+1$ and n digits, respectively). Consequently, the running time of the algorithm is $O(m\,n)$—the same as the classical method for multiplication. We present an example to illustrate the procedure.

Example 3.2. Divide 27507 by 34 using Algorithm 3.4 with radix $b = 10$ arithmetic.

Solution. The leading digit of the divisor is less than $\lfloor b/2 \rfloor = 5$, so we normalize the inputs by multiplying them by $d = \lfloor 10/(3 + 1) \rfloor = 2$ to obtain $u' = 55014$ and $v' = 68$. Their division is depicted in Figure 3.4.

At the first iteration, we divide $\tilde{u}_2 = 550$ by v'. The trial quotient is obtained by dividing the first two digits of $\tilde{u}_2$ by the first digit of v', $q_{t_2} = \lfloor 55/6 \rfloor = 9$. The trial quotient is too large since $w = 9v' = 612$, which is greater than $\tilde{u}_2$. The actual quotient is $q_2 = q_{t_2} - 1 = 8$, leaving a remainder of $\tilde{u}_2 - 8v' = 6$.

procedure $Divide(u, v)$
{Input: dividend $u = (u_{m+n-1} \ldots u_1 u_0)_b$, $m + n$ digits
    divisor $v = (v_{n-1} \ldots v_1 v_0)_b$, n digits }
{Output: quotient $q = (q_m \ldots q_1 q_0)_b$, $m + 1$ digits
    remainder $r = (r_{n-1} \ldots r_1 r_0)_b$, n digits }
{Step 1. Normalize so that $v_{n-1} \geq \lfloor b/2 \rfloor$. }
    $d := \lfloor b/(v_{n-1} + 1) \rfloor$ { scaling factor }
    $u' := u \cdot d$ { $u' = (u'_{m+n} \ldots u'_1 u'_0)_b$ }
    $v' := v \cdot d$ { $v' = (v'_{n-1} \ldots v'_1 v'_0)_b$ }
{Step 2. Compute a single digit q_j of the quotient by dividing
    $\tilde{u} = (\tilde{u}_n \ldots \tilde{u}_0)_b$ by v'. }
    $\tilde{u} := (u'_{m+n} \ldots u'_m)_b$ { initialize $\tilde{u}$ }
    for $j := m$ **to** 0 **by** -1 **do**
        if $\tilde{u}_n = v'_{n-1}$ **then**   { determine trial quotient }
            $q_j := b - 1$
        else
            $q_j := \lfloor (\tilde{u}_n\, b + \tilde{u}_{n-1})/v'_{n-1} \rfloor$
        end if
        $w := q_j \cdot v'$
        while $w > \tilde{u}$ **do** { if necessary, decrement q_j at most twice }
            $q_j := q_j - 1$; $w := w - v'$
        end do
        $r := \tilde{u} - w$ { remainder at this iteration }
        if $j > 0$ **then**
            $\tilde{u} := r \cdot b + u'_{j-1}$ { left pad $\tilde{u}$ with zeros, as needed }
        end if
    end do
{Step 3. Unnormalize, then return quotient and remainder. }
    $q := (q_m \ldots q_1 q_0)_b$; $r := r/d$
    return(q, r)
end procedure

ALGORITHM 3.4: Division of non-negative numbers

$$\begin{array}{r} 809 \\ 68 \overline{\smash{\big)}\ 55014} \\ 544 \\ \hline 061 \\ 0 \\ \hline 614 \\ 612 \\ \hline 2 \end{array}$$

$q_{t_2} = \lfloor 55/6 \rfloor = 9, \quad q_2 = q_{t_2} - 1 = 8$

$q_{t_1} = \lfloor 06/6 \rfloor = 1, \quad q_1 = q_{t_1} - 1 = 0$

$q_{t_0} = b - 1 = 9, \quad q_0 = q_{t_0} = 9$

FIGURE 3.4: An example of the trial quotient method.

Bringing down a 1 from the dividend, we have $\tilde{u}_1 = 061$ on the second iteration. The trial quotient $q_{t_1} = \lfloor 06/6 \rfloor = 1$, which is again is too big since $w = 1 \cdot v' = 68$. So the actual quotient is $q_1 = q_{t_1} - 1 = 0$, and the remainder is all of $\tilde{u}_1$ or 61.

At the final iteration, we bring down a 4 from the dividend and $\tilde{u}_0 = 614$. Since the first digit of $\tilde{u}_0$ equals the first digit of v', the trial quotient $q_{t_0} = b - 1 = 9$. Continuing as usual, $w = 9v' = 612$ and $\tilde{u}_0 - w = 2$. Since this difference is less than $\tilde{u}_0$, the actual quotient is the same as the trial quotient, $q_0 = q_{t_0} = 9$, and the final remainder is 2.

So dividing $u' = 55014$ by $v' = 68$ gives a quotient of 809 and a remainder of 2. Therefore the quotient when we divide $u = 27507$ by $v = 34$ is also 809, and the remainder obtained by unnormalization is $2/d = 1$. □

3.3   Greatest Common Divisor

The computation of greatest common divisors is at the heart of every computer algebra system. Not only are they required to reduce fractions, but their calculation is a key step in executing many of the polynomial manipulations used to perform more complex operations like factorization and integration. If you look under the hood of any computer algebra system you'll find a computational engine powered by Euclid's algorithm. Find a better way to compute greatest common divisors and you can build a faster computer algebra system!

3.3.1   Euclid's Algorithm

The *greatest common divisor* (gcd) of two non-negative integers is the largest integer that exactly divides (without remainder) both. We can find the gcd of the numbers easily from their prime factorizations by taking the common factors and raising each to the smaller exponent. For example, sup-

---

procedure *Euclid*(*m*, *n*)
{Input: integers *m* and *n* }
{Output: gcd(*m*, *n*), the largest integer dividing both *m* and *n* }
    $u := |m|$; $v := |n|$
   while $v \neq 0$ **do**
       $r := u \bmod v$ {remainder when *u* is divided by *v*}
       $u := v$; $v := r$
   end do
   return(*u*)
end procedure

ALGORITHM 3.5: Euclidean algorithm for integer gcd

---

pose we are interested in determining gcd(7000, 4400). We write each number as a product of primes to integer powers: $7000 = 2^3 \cdot 5^3 \cdot 7$ and $4400 = 2^4 \cdot 5^2 \cdot 11$. The common factors are $\gcd(7000, 4400) = 2^3 \cdot 5^2 = 200$.

Prime factorization is a very inefficient way to compute gcds. A much better way is to use a method attributed to Euclid and described in his *Elements* (Book 7, Propositions 1 and 2), written circa 300 B.C. The procedure is perhaps the oldest nontrivial algorithm that has survived to this day.

Euclid's method for computing the gcd of two non-negative integers can be described recursively as follows:

$$\gcd(u, v) = \begin{cases} \gcd(v, u \bmod v) & \text{if } v \neq 0, \\ u & \text{if } v = 0. \end{cases}$$

The mod function gives the remainder when two integers are divided. By convention, we define $\gcd(0, 0) = 0$. Observe that if $v > u$, the first iteration of the recursive rule merely reverses the arguments, $\gcd(u, v) = \gcd(v, u)$. The relations

$$\gcd(-u, v) = \gcd(u, v) \qquad \text{and} \qquad \gcd(u, 0) = |u|$$

enable us to take gcds involving negative integers by working with absolute values.

An iterative version of Euclid's method is given in Algorithm 3.5. The recursion is replaced by a loop. At each iteration we calculate the remainder, $u \bmod v$, then update *u* and *v*. The iterations conclude when *v* reaches zero.

Example 3.3. Calculate gcd(7000, 4400) with Euclid's algorithm.

Solution. The steps in the recursive version of the algorithm are traced on the left, and the corresponding values of the variables r, u, v in the iterative version (Algorithm 3.5) are given on the right.

	r	u	v
$\gcd(7000, 4400)$		7000	4400
$= \gcd(4400, 2600)$	2600	4400	2600
$= \gcd(2600, 1800)$	1800	2600	1800
$= \gcd(1800, 800)$	800	1800	800
$= \gcd(800, 200)$	200	800	200
$= \gcd(200, 0)$	0	200	0
$= 200$		200	

$\square$

A concept related to the gcd is the least common multiple (lcm). The *least common multiple* of two integers is the smallest positive integer that is a multiple of (or, equivalently, is evenly divisible by) both. As with the gcd, the lcm can be readily determined from the prime factorizations. It is the product of the primes appearing in either number, raised to the larger exponent for primes occurring in both. For example, since $7000 = 2^3 \cdot 5^3 \cdot 7$ and $4400 = 2^4 \cdot 5^2 \cdot 11$, the $\mathrm{lcm}(7000, 4400) = 2^4 \cdot 5^3 \cdot 7 \cdot 11 = 154{,}000$.

We can express the least common multiple of two integers in terms of their gcd as

$$\mathrm{lcm}(u, v) = \frac{u \, v}{\gcd(u, v)}.$$

As with the gcd, we use the convention that $\mathrm{lcm}(0, 0) = 0$.

3.3.2   Analysis of Euclid's Algorithm

What is the maximum number of divisions employed by Euclid's algorithm to compute the gcd of two integers of some given size? The number of divisions is highly sensitive to the integers being processed. For example, the computation

$$\gcd(8, 5) = \gcd(5, 3) = \gcd(3, 2) = \gcd(2, 1) = \gcd(1, 0) = 1$$

uses four divisions to process two single digit numbers, while the calculation

$$\gcd(1234567891, 1234567890) = \gcd(1234567890, 1) = \gcd(1, 0) = 1$$

requires only two divisions to process numbers that are substantially larger.

Gabriel Lamé (1844) showed that the worst case occurs when Euclid's algorithm is applied to two consecutive Fibonacci numbers. These numbers are defined by the recurrence $F_n = F_{n-1} + F_{n-2}$ with the initial conditions $F_0 = 0, F_1 = 1$. The first few numbers in the sequence are

$$0, 1, 1, 2, 3, 5, 8, 13, 21, 34, 55, \ldots .$$

Binet's formula gives a (perhaps surprising) closed form for the n-th Fibonacci number,

$$F_n = \frac{1}{\sqrt{5}} \left[\left(\frac{1 + \sqrt{5}}{2} \right)^n - \left(\frac{1 - \sqrt{5}}{2} \right)^n \right].$$

Letting $\phi = (1 + \sqrt{5})/2 \approx 1.618$ and observing that $(1 - \sqrt{5})/2 \approx -0.618$, we have that $F_n \approx \phi^n/\sqrt{5}$. (Why?)

Applying a single iteration of Euclid's rule reveals that $\gcd(F_n, F_{n-1}) = \gcd(F_{n-1}, F_{n-2})$. Consequently, n divisions are required to determine that $\gcd(F_{n+2}, F_{n+1}) = 1$. Consecutive Fibonacci numbers represent a worst case in the sense that if $u > v > 0$ and Euclid's algorithm performs n divisions to find $\gcd(u, v)$, then $u \geq F_{n+2}$ and $v \geq F_{n+1}$. So if Euclid's algorithm requires n or more divisions and the smaller input $v = N$, then $N \geq F_{n+1} \approx \phi^{n+1}/\sqrt{5}$. Taking logarithms and rearranging, we have

$$n \leq \log_\phi N + \log_\phi \sqrt{5} - 1 \approx 2.078 \ln N + 0.6723$$

$$\approx 4.785 \log_{10} N + 0.6723.$$

In other words, the number of division steps to compute $\gcd(u, v)$ is at most about $2.078 \ln N + 0.6723$, which is less than five times the number of decimal digits in the smaller input. Since the number N has $O(\log N)$ digits in any base, Euclid's algorithm performs $O(d)$ arithmetic operations to process a d-digit number regardless of the base selected. Assuming that multiplication and division take $O(d^2)$ digitwise operations, at most $O(d^3)$ digitwise operations are performed by Euclid's method to find the gcd of two d-digit numbers.

We now turn our attention to a different issue. Suppose we wish to calculate the gcd of n integers, $u_1, u_2, \ldots, u_n$, instead of just two. Assuming all the u_i are non-negative, we can iterate the following generalization of Euclid's rule:

$$\gcd(u_1, u_2, \ldots, u_n) = \gcd(u_1 \bmod u_k, u_2 \bmod u_k, \ldots, u_n \bmod u_k),$$

where $u_k = \min_{u_j \neq 0}(u_1, \ldots, u_n)$—that is, u_k is the minimum of the non-zero u's. The base cases are

$$\gcd(0, \ldots, 0) = 0,$$

$$\gcd(0, \ldots, 0, u_i, 0, \ldots, 0) = u_i.$$

A better way to solve the problem is to work with only two numbers at a time,

$$\gcd(u_1, u_2, u_3, \ldots, u_n) = \gcd(\gcd(u_1, u_2), u_3, \ldots, u_n).$$

The theorem which follows, due to P. G. Lejeune Dirichlet (1849), states that the gcd of two randomly chosen integers is one about 60% of the time. Since $\gcd(1, u_k, \ldots, u_n) = 1$, the computation of the gcd of n random integers can be expected to converge quickly to one with this approach.

Theorem 3.4. *If u and v are integers chosen at random, the probability that $\gcd(u, v) = 1$ is $6/\pi^2 \approx 0.60793$.*

Proof. Let p be the probability that two randomly chosen integers, u and v, have no common factors,

$$p = \mathrm{Prob}\left(\gcd(u, v) = 1\right).$$

For any u, v and positive integer d, $\gcd(u, v) = d$ if and only if:

1. u is a multiple of d,

2. v is a multiple of d, and

3. $\gcd(u/d, v/d) = 1$.

Since every d-th integer is a multiple of d, the probabilities of (1) and (2) are both $1/d$ for random u, v. Letting p denote the probability of (3), we have

$$\text{Prob}\left(\gcd(u, v) = d\right) = \frac{1}{d} \cdot \frac{1}{d} \cdot p = \frac{p}{d^2}.$$

Summing over all d gives

$$1 = \sum_{d \geq 1} \text{Prob}\left(\gcd(u, v) = d\right) = p\left(1 + \frac{1}{2^2} + \frac{1}{3^2} + \frac{1}{4^2} + \cdots\right) \quad (3.2)$$

$$= p\frac{\pi^2}{6},$$

as the sum on the right of (3.2) converges to $\pi^2/6$. Rearranging this equation, we obtain the solution $p = 6/\pi^2 \approx 0.60793$. $\square$

3.3.3   Extended Euclidean Algorithm

In this section, we study a very important modification to Euclid's algorithm. By keeping track of some additional information throughout the course of the computation, we can also find integers s, t such that the $\gcd(u, v)$ can be expressed as a linear combination,

$$\gcd(u, v) = s\, u + t\, v, \quad (3.3)$$

of the original inputs u and v. There are many applications of this technique in computer algebra, as we shall see in the remainder of this book. The relationship (3.3) is referred to as *Bézout's identity*.

Here is the basic idea behind the modified procedure, known as the *extended Euclidean algorithm*. At the first iteration of Euclid's method, the Remainder Theorem gives the following relationship between the original inputs u and v,

$$u = q\, v + r,$$

where q is the quotient and r is the remainder $(0 \leq r < v)$ when u is divided by v. We can rearrange this equation to express r in terms of u and v as

$$r = u - q\, v.$$

r and v are the two numbers processed at the second iteration. The remainder at the next step can likewise be written as a linear combination of r and v. Moreover, since we can express r in terms of u and v, the second remainder can also be written in terms of u and v. We continue the process, expressing the remainder of each division in terms of the original u and v. The final remainder is the $\gcd(u, v)$.

It is perhaps easiest to follow how this is done with an example. Suppose we apply the Euclidean algorithm to determine $\gcd(81, 57)$. At the first iteration, we find $\gcd(81, 57) = \gcd(57, 24)$ and express the remainder in terms of u and v as

$$24 = 1\,(81) - 1\,(57).$$

At the second iteration, we have $\gcd(57, 24) = \gcd(24, 9)$ and write the remainder as

$$9 = 57 - 2\,(24) = -2\,(81) + 3\,(57).$$

The third division gives $\gcd(24, 9) = \gcd(9, 6)$ and

$$6 = 24 - 2\,(9) = 5\,(81) - 7\,(57).$$

The fourth iteration shows that $\gcd(9, 6) = \gcd(6, 3)$, yielding

$$3 = 9 - 6 = -7\,(81) + 10\,(57).$$

The final steps give $\gcd(6, 3) = \gcd(3, 0) = 3$, and so the extended Euclidean algorithm expresses this result as the linear combination

$$\gcd(81, 57) = 3 = -7\,(81) + 10\,(57).$$

It should be noted that the integers s, t found by this procedure are not unique. There are, for example, an unlimited number of ways to express $\gcd(81, 57) = 3$ linearly in terms of 81 and 57. Some possibilities are

$$\begin{aligned} \gcd(81, 57) = 3 &= -7\,(81) + 10\,(57) \\ &= 12\,(81) - 17\,(57) \\ &= -26\,(81) + 37\,(57). \end{aligned}$$

The first combination is the one found by the extended Euclidean algorithm. The other combinations are easily verified.

How do we go about actually implementing the extended Euclidean algorithm? At each step, we represent the two numbers processed and their remainder as a triple of integers. The first component of the triple is simply the value of the corresponding variable in Algorithm 3.5. The other two components express the first as a linear combination of the original inputs. The initial value of the triple for u is $[u, 1, 0]$ since $u = 1\,(u) + 0\,(v)$. Similarly, the initial value of the triple for v is $[v, 0, 1]$, corresponding to $v = 0\,(u) + 1\,(v)$.

At each iteration, we explicitly compute the quotient q using the first components of the current u and v vectors, $q = \lfloor u_1/v_1 \rfloor$. The corresponding remainder is given by the Remainder Theorem, $r_1 = u_1 - q\,v_1$. The other components of the r vector express r_1 as a linear combination of the starting values for u and v, $r_2 = u_2 - q\,v_2$ and $r_3 = u_3 - q\,v_3$. To get ready for the next iteration, the v-vector is assigned to u, and the r-vector is assigned to v. The entire procedure is presented as Algorithm 3.6.

procedure *ExtendedEuclid(m, n)*
{Input: integers m and n }
{Output: $[\gcd(m, n), s, t]$ where $\gcd(m, n) = s\,m + t\,n$ }
 $u := [\,|m|, 1, 0\,]$; $v := [\,|n|, 0, 1\,]$ {initialize u, v vectors}
 while $v_1 \neq 0$ **do**
  $q := \lfloor u_1/v_1 \rfloor$ {quotient when u_1 is divided by v_1}
  $r := [u_1 - q\,v_1, u_2 - q\,v_2, u_3 - q\,v_3]$ {remainder is $u_1 - q\,v_1$}
  $u := v$
  $v := r$
 end do
 return(u)
end procedure

ALGORITHM 3.6: Extended Euclidean algorithm for integers

Example 3.4. Trace the operation of the extended Euclidean algorithm as it determines $\gcd(7000, 4400)$. These are the inputs of Example 3.3.

Solution. We first show how the remainder at each iteration is expressed as a linear combination both of the integers currently processed and of the starting values.

$\gcd(7000, 4400)$
 $= \gcd(4400, 2600)$ $2600 = 7000 - 4400$
 $= \gcd(2600, 1800)$ $1800 = 4400 - 2600 = -7000 + 2\,(4400)$
 $= \gcd(1800, 800)$ $800 = 2600 - 1800 = 2\,(7000) - 3\,(4400)$
 $= \gcd(800, 200)$ $200 = 1800 - 2\,(800) = -5\,(7000) + 8\,(4400)$
 $= \gcd(200, 0)$
 $= 200$ $200 = -5\,(7000) + 8\,(4400)$

Tracing Algorithm 3.6, the quotients and remainders calculated at each step, together with the u and v vectors are:

	q	r_1	u	v
$\gcd(7000, 4400)$			$[7000, 1, 0]$	$[4400, 0, 1]$
$= \gcd(4400, 2600)$	1	2600	$[4400, 0, 1]$	$[2600, 1, -1]$
$= \gcd(2600, 1800)$	1	1800	$[2600, 1, -1]$	$[1800, -1, 2]$
$= \gcd(1800, 800)$	1	800	$[1800, -1, 2]$	$[800, 2, -3]$
$= \gcd(800, 200)$	2	200	$[800, 2, -3]$	$[200, -5, 8]$
$= \gcd(200, 0)$	4	0	$[200, -5, 8]$	$[0, 22, -35]$
$= 200$			$[200, -5, 8]$	

The actual remainder, r_1, is the first component of the r-vector computed at each iteration of the loop. This triple becomes the v-vector that is shown above.   □

3.3.4   Binary GCD Algorithm

Euclid's algorithm has certainly withstood the test of time. Nonetheless, the development of digital computers has spawned interest in a fast, alternative way of computing greatest common divisors. This new technique is known as the *binary method* since an efficient implementation requires a binary representation of the numbers being processed. Its advantage over Euclid's method is that it avoids division entirely by doing bitwise operations. The operations on which it is based are subtraction, testing for odd/even integers, and halving. Most computers have special machine instructions to perform each of these operations efficiently in hardware. Testing for odd/even requires only the examination of the rightmost bit of a number, and halving corresponds to shifting a binary number one bit to the right.

The binary method was first published by Josef Stein in 1967 but, according to Knuth (1998), it may have been known in first century China. It goes about finding the greatest common divisor of two non-negative integers by repeatedly applying these rules:

1. If $u = 0$, then $\gcd(u, v) = v$.

2. If u and v are both even, then $\gcd(u, v) = 2 \gcd(u/2, v/2)$. In this case, 2 is a common divisor.

3. If u is even and v is odd, then $\gcd(u, v) = \gcd(u/2, v)$. In this case, 2 is not a common divisor.

4. If u and v are both odd, then $\gcd(u, v) = \gcd(|u - v|, \min(u, v))$. This follows from the identity $\gcd(u, v) = \gcd(u - v, v)$. Moreover, if u and v are both odd, then $u - v$ is even and $|u - v| < \max(u, v)$.

5. If $v = 1$, then $\gcd(u, v) = 1$.

This last observation is not part of the original algorithm, but the author has discovered that its application often saves several iterations of the other rules.

An iterative version of the binary method is presented as Algorithm 3.7. It begins by taking the absolute value of its arguments so that only non-negative integers are processed in the remaining steps. If either argument is zero, the other is returned immediately as the gcd. Next, a while-loop is entered to calculate and divide out any common factors of two. Once this is done, another while-loop is iterated until either $u = 0$ or $v = 1$. Each time this loop is entered, at most one of u or v is even. If either is even, any factors of two are divided out of that number. Eventually both are odd, and we assign $|u - v|$, an even number, to u and $\min(u, v)$ to v for the next iteration. When either u decreases to 0 or v decreases to 1, we are done. The gcd returned is the value of v times any factors of two found at the outset of the procedure.

procedure $BinaryGCD(m, n)$
{Input: integers m and n }
{Output: gcd(m, n), the largest integer dividing both m and n }
  $u := |m|; \quad v := |n|$
  if $u = 0$ **or** $v = 0$ **then** **return**$(max(u, v))$ **end if**
  $g := 1$ {keeps track of factors of 2}
  while $Even(u)$ **and** $Even(v)$ **do**   {u even, v even}
    $u := u/2; \quad v := v/2; \quad g := 2 \cdot g$
  end do
  while $u > 0$ **and** $v \neq 1$ **do**   {iterate until $u=0$ or $v=1$}
    while $Even(u)$ **do** $u := u/2$ **end do**   {u even, v odd}
    while $Even(v)$ **do** $v := v/2$ **end do**   {u odd, v even}
    $t := |u - v|; \quad v := min(u, v); \quad u := t$ {u odd, v odd}
  end do
  return$(g \cdot v)$
end procedure

ALGORITHM 3.7: Binary algorithm for integer gcd

Example 3.5. Trace the binary method as it finds gcd$(7000, 4400)$.

Solution. The values of the principal variables and the action taken at each step are as follows:

u	v	g	Action		u	v	Action
7000	4400	1	right shift u, v		75	275	$\|u - v\|$
3500	2200	2	right shift u, v		200	75	right shift u
1750	1100	4	right shift u, v		100	75	right shift u
875	550	8	right shift v		50	75	right shift u
875	275		$\|u - v\|$		25	75	$\|u - v\|$
600	275		right shift u		50	25	right shift u
300	275		right shift u		25	25	$\|u - v\|$
150	275		right shift u		0	25	return $g \cdot v = 200$

At each iteration, the magnitude of at least one of the numbers under consideration is halved. As a result, the number of steps required to process two inputs u and v, where $0 \leq u, v < N$, is proportional to the size of the larger number in bits, $\log_2 N$. The subtract and shift operations do not operate in constant time for big numbers, but instead require a time proportional to the size of the numbers (about one operation per word in the representation). As a consequence, $O(d^2)$ digitwise operations are performed to find the gcd of two d-digit numbers.

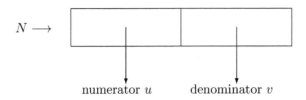

$N \longrightarrow$

numerator u denominator v

FIGURE 3.5: Representation of a rational number.

While the binary method can outperform the traditional Euclidean algorithm on many computers, it is not easy to generalize it to other contexts. For example, it is difficult to produce an extended version that expresses the gcd as a linear combination of the two original inputs. Nor can it be generalized to produce an algorithm to determine the gcd of two polynomials, an important problem that we study in the next chapter.

3.4   Rational Arithmetic

With greatest common divisors under our belts, we are ready to tackle arithmetic on fractions, also called rational numbers. A *rational number* is simply the ratio of two integers, u/v, where $v \neq 0$. Rational arithmetic is pervasive in computer algebra systems, and so it is extremely important to implement the basic arithmetic operations on fractions as efficiently as possible.

Computer algebra systems typically represent a fraction as a node with two links, the first to the numerator and the second to the denominator (see Figure 3.5). Pointers are used rather than the values of the numerator and denominator themselves since the sizes of these numbers are not fixed (i.e., they can be big numbers). The numerator and denominator are always reduced to lowest terms by removing the $\gcd(u, v)$ from both u and v. The sign is stored with the numerator u and the denominator v is always positive.

Multiplication is perhaps the easiest operation to implement. The way we are taught in school to multiply a/b and c/d is to multiply the numerators, multiply the denominators, and then reduce the resulting fraction by dividing the gcd out of these products,

$$\frac{a}{b} \cdot \frac{c}{d} = \frac{a\,c/\gcd(a\,c, b\,d)}{b\,d/\gcd(a\,c, b\,d)}.$$

In a computer algebra system, the fractions a/b and c/d are already reduced, and so

$$\gcd(a\,c, b\,d) = \gcd(a, d) \cdot \gcd(b, c).$$

Division can be performed in a similar fashion since it may be treated as a multiplication problem,

$$\frac{a/b}{c/d} = \frac{a}{b} \cdot \frac{d}{c}.$$

As an example, suppose we want to calculate

$$\frac{4}{9} \cdot \frac{15}{14} = \frac{60}{126} = \frac{10}{21}.$$

The standard way to reduce the result is find $\gcd(60, 126) = 6$ directly. In the alternative method, we compute $\gcd(4, 14) = 2$ and $\gcd(9, 15) = 3$, then take their product.

It is generally preferable to implement the second technique in a computer algebra system. Although it requires two gcd computations, the numbers involved are smaller. If a, b, c and d are all single precision integers, no big number arithmetic is required unless either the numerator or the denominator of the product does not fit in a word. The number of iterations needed to find the gcd grows as the logarithm of the magnitude of the inputs. Since $\log(u\,v) = \log u + \log v$, the total number of iterations for both methods is essentially the same.

Now we turn our attention to addition and subtraction,

$$\frac{a}{b} \pm \frac{c}{d} = \frac{ad \pm bc}{bd}.$$

In school, we are taught to reduce the result by finding the $\gcd(ad \pm bc, bd)$. The trouble with this is that the arguments to the gcd calculation are generally much larger than the original data.

As with multiplication and division, we can do better if we take advantage of the fact that a/b and c/d are already reduced to lowest terms. We begin by finding the greatest common divisor of the denominators of the two fractions, $g_1 = \gcd(b, d)$, and regard the result in terms of integers p and q,

$$\frac{a}{b} \pm \frac{c}{d} = \frac{p}{q} \quad \text{where} \quad p = a\,\frac{d}{\gcd(b, d)} \pm c\,\frac{b}{\gcd(b, d)}, \quad q = \frac{b}{\gcd(b, d)}\,d.$$

(Why do the "fractions" appearing in the representation of p and q evaluate to integers?)

If $\gcd(b, d) = 1$, we are done. Theorem 3.4 shows that this happens about 61% of the time if the denominators of the inputs are randomly distributed, so it is wise to check for this case.

If $\gcd(b, d) > 1$, p and the fraction $b/\gcd(b, d)$, which appears in both p and q are evaluated. To reduce p/q, we calculate $g_2 = \gcd(p, g_1)$ and use this to reduce p/q to lowest terms, $(p/g_2)/(\frac{b}{g_1} \cdot d/g_2)$. Although two gcd computations are required with this method, as was the case with multiplication and division, the inputs are smaller than those in the standard approach.

For example, suppose we want to find

$$\frac{11}{42} + \frac{13}{60} = \frac{11 \cdot 60 + 42 \cdot 13}{42 \cdot 60} = \frac{1206}{2520} = \frac{67}{140}.$$

In the standard approach we proceed as shown, calculating $\gcd(1206, 2520) = 18$ to reduce the result. In the approach outlined above, we begin by determining $\gcd(42, 60) = 6$ and then calculate

$$p = 11 \cdot \frac{60}{6} + 13 \cdot \frac{42}{6} = 11 \cdot 10 + 13 \cdot 7 = 201,$$

$$q = \frac{42}{6} \cdot 60 = 7 \cdot 60.$$

Finally, we compute $\gcd(201, 6) = 3$ to reduce both p and d (and hence q),

$$\frac{201}{3} \bigg/ \left(7 \cdot \frac{60}{3}\right) = \frac{67}{140}.$$

3.5 Exponentiation

The final arithmetic problem we consider is that of raising some quantity to a positive integer power. Merely squaring a number yields a result with approximately twice as many digits, so raising a big number to even a moderate power can produce a gargantuan result. Consequently, it is important to control the amount of intermediate computation performed by devising efficient algorithms for exponentiation.

Cryptography is an important application area where exponentiation is frequently employed. In RSA cryptosystems,[§] enciphering and deciphering involve raising a number to an enormous power. The size of the intermediate results can be controlled in this instance since the computation is performed using modular arithmetic, but the sheer magnitude of the exponent again dictates the need for an efficient strategy to calculate powers.

While it is convenient for us to think of raising a number to a power, the techniques we discuss can be used to raise polynomials and matrices to powers as well. Such computations are far more involved since they require raising many numbers to powers. Once again, we are motivated to devise efficient algorithms for exponentiation.

Suppose we are given a number x and a positive integer n and are asked to find the value of x^n. The obvious way to proceed is to multiply by x a total of $n - 1$ times. For example, to find x^{32} we compute each of the partial results $x^2, x^3, x^4, \ldots, x^{31}, x^{32}$ and arrive at the desired answer after 31

[§]RSA public-key cryptosystems were first proposed by Ronald Rivest, Adi Shamir and Leonard Adelman (1978).

multiplications. A far more efficient way to calculate x^{32} is to square each partial result. At the first step, we square x to obtain x^2. At the second, x^2 is squared to yield x^4, and so on. Using this technique, we arrive at x^{32} with only five multiplications via the sequence of partial results $x^2, x^4, x^8, x^{16}, x^{32}$.

The method of repeated squaring can be used to compute x^n with $\log_2 n$ multiplications when n is a power of two. While it is far more economical than the brute force approach, repeated squaring can only be used directly when n is a power of two.

An algorithm called *binary powering* builds on repeated squaring to work for any n. To apply this technique, we begin with the binary representation of n. For example, if $n = 21$ we write $21 = (10101)_2$. Ignoring the first bit in the binary representation (which is always 1), we replace each remaining 1 by the letters SX and each 0 by the letter S. When $n = 21$ we obtain the sequence of letters $S\ SX\ S\ SX$. This sequence yields a rule for evaluating x^n if we interpret each S to mean "square the result of the previous step" and each X to mean "multiply the result of the previous step by x."

How many multiplications does binary powering use? The number of bits in the binary representation of n is $\lfloor \log_2 n \rfloor + 1$. Since we ignore the first bit of n and each of the other bits produces either one or two multiplications (S or SX), the number of multiplications in the binary method does not exceed $2 \lfloor \log_2 n \rfloor$.

The technique just described processes the bits of n from left to right. Computer programs generally prefer to work in the opposite direction since repeated halving and testing for odd/even numbers can produce the binary representation easily.

Observe that x^n can be constructed by multiplying x raised to powers of two in the same way that n can be constructed by adding together powers of two. This is a consequence of the fact that $x^p \cdot x^q = x^{p+q}$. For example, $x^{21} = x \cdot x^4 \cdot x^{16}$ since $21 = 1+4+16$. So by repeatedly squaring and multiplying together the appropriate powers of two, we can construct x^n by scanning the binary representation of n from right to left.

The *right-to-left binary powering* method is presented as Algorithm 3.8. While it uses the same number of multiplications as the left-to-right scheme, extra storage is necessary since two potentially big numbers (Y and Z in Algorithm 3.8) are required at each step. To illustrate its execution, we give the values of the variables at each iteration of the loop for $n = 21$.

N	Y	Z		N	Y	Z
21	1	x		2	x^5	x^8
10	x	x^2		1	x^5	x^{16}
5	x	x^4		0	x^{21}	

Binary powering does not always use the fewest number of multiplications. The smallest value of n for which it is not optimal is 15. The left-to-right method uses 6 multiplications to evaluate x^{15}, with the sequence $SX\ SX\ SX$ giving rise to the partial results: $x^2, x^3, x^6, x^7, x^{14}, x^{15}$. However, x^{15} can be

---

procedure $RLBinaryPower(x, n)$
{Input: x and a non-negative integer n }
{Output: the value of x^n }
$\quad$ $N := n$; $\ Y := 1$; $\ Z := x$ $\quad$ {initialize N, Y, Z}
$\quad$ **while** $N > 0$ **do**
$\quad\quad$ **if** $Odd(N)$ **then** $Y := Y \cdot Z$ **end if** $\quad$ {multiply in Z}
$\quad\quad$ $N := \lfloor N/2 \rfloor$ $\quad$ {halve N}
$\quad\quad$ **if** $N > 0$ **then** $Z := Z^2$ **end if** $\quad$ {repeatedly square Z}
$\quad$ **end do**
$\quad$ **return**(Y)
end procedure

ALGORITHM 3.8: Right-to-left binary powering

---

calculated in five steps by first finding $y = x^3$ with two multiplications, and then raising y to the fifth power with three more multiplications since $y^5 = (x^3)^5 = x^{15}$.

This method for evaluating x^{15} is based on realizing that 15 can be factored as 3 times 5. In general if the number n can be factored as $n = p \cdot q$, then x^n can be calculated by first evaluating $y = x^p$ and then raising this quantity to the q-th power: $y^q = (x^p)^q = x^{pq} = x^n$.

The following algorithm, called the *factor method*, is based on this principle:

1. If $n = 1$, we have x^n with no calculation.

2. If n is prime, calculate x^n by first finding x^{n-1} using the factor method; then multiply this quantity by x.

3. If n is not prime (i.e., n is composite), express n as $n = p \cdot q$, where p is the smallest prime factor of n and $q > 1$. Calculate x^n by first finding x^p via the factor method; then raise this quantity to the q-th power, again via the factor method.

Example 3.6. Evaluate x^{21} using the factor method.

Solution. First, 21 is factored as $3 \cdot 7$ and x^3 is calculated with two multiplications by repeatedly applying the algorithm. Our problem reduces to calculating y^7, where $y = x^3$. Since 7 is prime, y^7 is computed by first finding y^6 and then multiplying by y. Repeated use of the algorithm reveals that y^6

is found by taking $(y^2)^3$. Letting $z = y^2$, the steps used to evaluate x^{21} are

(1) $x \cdot x = x^2$,

(2) $x^2 \cdot x = x^3 = y$,

(3) $x^3 \cdot x^3 = x^6 = y^2 = z$,

(4) $x^6 \cdot x^6 = x^{12} = z^2$,

(5) $x^{12} \cdot x^6 = x^{18} = z^3 = (y^2)^3$,

(6) $x^{18} \cdot x^3 = x^{21} = y^6 \cdot y$. □

The factor method is practical only if the prime decomposition of the exponent n is easy to compute or known in advance. Unfortunately, it is feasible to factor only small to moderate size integers, or large integers that are highly composite with relatively small factors. In fact, the security of the RSA cryptosystem hinges on the computational intractability of factoring a number that is the product of two large primes.

Although the factor method uses fewer multiplications than the binary method on average, the binary method is superior in some instances. The smallest case is $n = 33$, where the factor method uses seven multiplications and the binary method only six. In fact, infinitely many n exist for which the factor method uses fewer multiplications than the binary method, and vice versa. Moreover, in some cases neither the factor method nor the binary method is always optimal. For example when $n = 23$, both the factor and binary methods use seven multiplications, but x^{23} can be calculated with six multiplications as follows:

(1) $x \cdot x = x^2$,

(2) $x^2 \cdot x = x^3$,

(3) $x^3 \cdot x^2 = x^5$,

(4) $x^5 \cdot x^5 = x^{10}$,

(5) $x^{10} \cdot x^{10} = x^{20}$,

(6) $x^{20} \cdot x^3 = x^{23}$.

Since neither the factor method nor the binary method is always optimal, let's investigate just how good these two methods are. Let $P(n)$ denote the fewest number of multiplications required to calculate x^n regardless of the method used. We next derive a lower bound on $P(n)$.

Theorem 3.5. $P(n) \geq \lceil \log_2 n \rceil$;[§] *i.e., at least $\lceil \log_2 n \rceil$ multiplications are required to evaluate x^n.*

Proof. Observe that the greatest power of x that can be obtained with k multiplications is x^{2^k}, which is achieved by repeated squaring at each step of the computation. Thus to compute a power of x as large as x^n, we need at least k multiplications where $2^k > n$. Since k is an integer, we see by taking logarithms that $k > \lceil \log_2 n \rceil$. Therefore $P(n) \geq \lceil \log_2 n \rceil$. □

The theorem provides a lower bound on the number of multiplications necessary to evaluate x^n but does not indicate whether this number is sufficient. Our earlier analysis of binary powering further reveals that $P(n) \leq 2\lfloor \log_2 n \rfloor$, and hence this method is efficient in the sense that it can never use more

---

[§] $\lceil x \rceil$ denotes the *ceiling* of x, or the smallest integer greater than or equal to x.

than twice the minimal number of multiplications. The operation of the factor method demonstrates that

$$P(n) \leq \begin{cases} 0 & \text{if } n = 1, \\ P(p) + P(q) & \text{if } n = p \cdot q \text{ is composite}, \\ P(n-1) + 1 & \text{if } n \text{ is prime}. \end{cases}$$

Not enough is known about the distribution of primes to obtain a useful solution to this set of inequalities for a general n.

So far in our discussion, we have concentrated on the operation of multiplication. Neither addition nor subtraction are of any help in evaluating powers. But what about division? Both the factor method and the binary method use seven multiplications to evaluate x^{31} and it is impossible to do better with only multiplication. However if division is allowed, x^{31} can be found in six operations by calculating x^{32} with five multiplications using repeated squaring, and then dividing this quantity by x. Unfortunately, the availability of division does not alter our lower bound on $P(n)$.

The problem of finding an optimal computation sequence for x^n has a long and interesting history. Although Arnold Scholz formally raised the question in 1937, before digital computers, algorithms for exponentiation had already been studied for some time. A version of the binary method was expounded by Adrien Legendre in 1798, and it is closely related to a multiplication procedure used by Egyptian mathematicians as early as 1800 B.C. Several authors have published statements that binary powering is optimal but, as we have seen, these claims are false.

3.6   Exercises

1. (a) Addition is traditionally performed by processing the inputs from right to left. Sometimes, however, the digits are more readily accessible from left to right. Design an algorithm that processes the inputs and produces the sum from left to right, going back to change a value whenever a carry makes a previous digit incorrect.

   (b) At first glance, it seems as though the running time of the algorithm in part (a) is quadratic in the length of the numbers. Show, however, that the running time is linear.

2. Show how Karatsuba's algorithm calculates the product 2718 times 4728. Reduce this product of four-digit numbers to three products of two-digit numbers, and then further reduce each of the latter to products of one-digit numbers.

3. Trace the action of our division algorithm when 4100 is divided by 588. This is an instance where three trial quotients are required.

4. Design an algorithm to divide a non-negative integer $u = (u_1, u_2, \ldots, u_n)_b$ by a one-place number v $(0 < v < b)$, producing a quotient $w = (w_1, w_2, \ldots, w_n)_b$ and a remainder r.

5. Design an algorithm to divide an n-place non-negative binary number u by an m-place positive binary number v, producing a binary quotient q and remainder r. Note that trial quotients are not required for binary division.

6. Find the greatest common divisor and least common multiple of each pair of numbers from their prime factorizations.

   (a) 480, 600   (b) 385, 225   (c) 96, 60   (d) 473, 132   (e) 132, 473
   (f) 156, 1740   (g) 169, 273   (h) 999, 2187   (i) 527, 736   (j) 361, 1178

7. Apply the Euclidean algorithm to find the greatest common divisors of the numbers in Exercise 6.

8. Apply the extended Euclidean algorithm to express the greatest common divisors of the numbers in Exercise 6 as a linear combination of the inputs.

9. Use the binary method to compute the greatest common divisors of the numbers in Exercise 6.

10. (a) Calculate gcd(31408, 2718) from the prime factorizations of the numbers.

    (b) Now apply Euclid's algorithm to compute the greatest common divisor.

    (c) Next apply the extended Euclidean algorithm to find integers m and n such that gcd(31408, 2718) = 31408m + 2718n.

    (d) Finally apply the binary method to compute gcd(31408, 2718).

11. (a) How many positive integers divide 130? How many divide 136?

    (b) How many positive integers divide both 130 and 136?

12. How many numbers between 1 and 105 are relatively prime to $105 = 3 \cdot 5 \cdot 7$?

13. Solve the equation $55x + 16y = 3$, where x and y are integers.

14. The extended Euclidean algorithm finds that gcd(30, 18) = 6 = $-1(30) + 2(18)$, but

$$\gcd(30, 18) = 6 = 2(30) - 3(18) \quad \text{and} \quad \gcd(30, 18) = 6 = -4(30) + 7(18)$$

are also valid linear combinations. There are, in fact, an infinite number of linear combinations of the form

$$\gcd(30, 18) = 6 = (3n - 1)(30) + (-5n + 2)(18),$$

where n is any integer. The solution found by the extended Euclidean algorithm is the one where $n = 0$.

Characterize the set of all linear combinations of u and v that are solutions to

$$\gcd(u, v) = d = s\,u + t\,v.$$

15. (a) Find the least common multiple, lcm(39, 102, 75).

    (b) Prove that the least common multiple of three integers u, v, w is given by

$$\mathrm{lcm}(u, v, w) = \frac{u\,v\,w\,\gcd(u, v, w)}{\gcd(u, v)\,\gcd(v, w)\,\gcd(w, u)}.$$

16. Prove by mathematical induction the correctness and termination of Euclid's method for computing the greatest common divisor of two positive integers,

$$\gcd(u, v) = \gcd(u, u \bmod v) \quad \text{with } \gcd(u, 0) = u.$$

17. Show rigorously that $\gcd(u, v) = \gcd(u - v, v)$.

18. (a) What is the probability that $\gcd(u, v) \leq 3$? What is the average value of $\gcd(u, v)$?

    (b) Show that two randomly selected positive odd integers, u and v, are relatively prime with probability $8/\pi^2$.

19. An integer is *squarefree* if its prime factorization contains no factors with an exponent greater than one. Show that every positive integer n can be uniquely represented as the product of a square and a squarefree number, $n = m^2 k$. Characterize the square and squarefree parts.

20. Calculate and reduce to lowest terms the following expressions using the method described in Section 3.4.

    (a) $4/15 + 3/22$ (b) $9/55 - 11/25$ (c) $7/10 + 3/14$

    (d) $45/121 + 64/75$ (e) $19/440 + 17/700$ (f) $17/120 - 27/70$

21. Suggest an efficient computational method for comparing two rational numbers to determine whether $a/b < c/d$.

22. Discuss the merits of using $1/0$ and $-1/0$ as representations for ∞ and $-\infty$.

23. Trace the execution of the right-to-left binary powering method when used to raise some number x to each of the following powers.

    (a) 25 (b) 50 (c) 75 (d) 100 (e) 42 (f) 45

    (g) 48 (h) 49 (i) 63 (j) 65 (k) 99 (l) 101

24. Give the sequence of powers computed when the left-to-left right powering method is used to raise a number x to each of the powers in Exercise 23.

25. Give the sequence of powers computed when the factor method is used to raise a number x to each of the powers in Exercise 23.

26. (a) Evaluate x^{495} using the binary method. How many multiplications are used?

    (b) Now evaluate x^{495} using the factor method. Again, count the number of multiplications.

    (c) Can you do better by observing that $495 = 15 \cdot (2^5 + 1)$ and combining the two methods?

27. (a) Show that the factor method uses fewer multiplications than the binary method (either version) to evaluate x^n for infinitely many values of n.

    (b) Show, likewise, that the binary method uses fewer multiplications than the factor method for infinitely many n.

28. Right-to-left binary powering requires more multiplications on average than the factor method. It also uses the same number of multiplications as left-to-right binary powering but requires more storage. Yet despite these disadvantages, right-to-left powering is the method universally implemented in computer algebra systems. Discuss why.

Programming Projects

1. **(Big Integer Arithmetic)** Write a set of procedures to implement the classical algorithms for non-negative integer arithmetic. All inputs and outputs should be in the `DigitList` format of Programming Project 2 in Chapter 2. Your procedures should operate on lists of digits, one digit at a time.

   The procedures are:

   - `AddInteger(<num1>,<num2>)`, which returns the sum,
   - `SubInteger(<num1>,<num2>)`, which returns the difference if `<num1>` $\geq$ `<num2>` and 0 otherwise,
   - `MultInteger(<num1>,<num2>)`, which returns the product,
   - `DivInteger(<num1>,<num2>)`, which returns a list with the quotient and remainder, `[<quotient>,<remainder>]`, if `<num2>` $\neq 0$,
   - `GCDInteger(<num1>,<num2>)`, an iterative procedure that returns the greatest common divisor.

   You may find it useful to write a utility procedure that compares two integers in `DigitList` format:

   - `CompInteger(<num1>,<num2>)`, returns 0 if `<num1>` = `<num2>`, 1 if `<num1>` > `<num2>`, and -1 if `<num1>` < `<num2>`.

2. **(Rational Arithmetic)** Write a set of procedures that implement arithmetic on signed rational numbers. The procedures are:

   - `AddRational(<num1>,<num2>)`,
   - `SubRational(<num1>,<num2>)`,
   - `MultRational(<num1>,<num2>)`,
   - `DivRational(<num1>,<num2>)`.

   A rational number is represented as a pair of integers, `[<numer>,<denom>]`. While the inputs to the procedures need not be in "standard form," the outputs should be. That is, an input number need not be reduced, but the gcd should be removed from the result. Similarly, signs may appear in either the numerator or denominator (or both) of an input, but a negative result should carry its sign in the numerator. In this representation, an integer has a denominator of 1.

   You may take advantage of your computer algebra system's facilities for integer arithmetic but not those for manipulating rational numbers. Your code should implement the reduction techniques described in Section 3.4 rather than the classical methods you learned in school.

3. **(Karatsuba's Algorithm)** Write a procedure, `Kmult(<num1>,<num2>)`, that implements Karatsuba's algorithm.

   As in Programming Project 1, all inputs should be in the format produced by the `DigitList` procedure, with any leading zeros supressed. The two inputs need not be the same length, nor of a length that is a power of two. However, your internal programming logic may convert numbers to such a form.

   Your program should not rely on the arithmetic capabilities of your computer algebra system but should instead perform arithmetic on the digits of `DigitList` numbers. Your code should keep recursing all the way down to one-digit numbers. Multiplication by a power of 10 should be implemented as a shift as it is not considered a true multiplication.

4. (**Radix Conversion**)   Write a procedure, `ChangeBase(<num>,<b1>,<b2>)`, that converts a non-negative integer in `DigitList`-format from base `b1` to a similarly formatted integer in base `b2`.

   For example, `ChangeBase([4,2,1,3,4],5,13)` takes $(42134)_5 = 2794$ and converts it to base-13, $(1 \ 3 \ 6 \ 12)_{13}$. Note that each digit d in a base-b number is in the range $0 \le d < b$, so "digits" greater than 9 can appear in a list whenever $b > 10$.

Chapter 4

Polynomial Manipulation

In the previous chapter, we studied the arithmetic of numbers. Polynomials, in a certain sense, are the next step up from numbers, and the concepts and techniques developed in Chapter 3 can be extended in a natural way to polynomials.

At the outset, let's explore a bit the analogy between arithmetic on polynomials and on numbers. A single variable polynomial $u(x)$ of degree n can be expressed as

$$u(x) = c_n x^n + \ldots + c_1 x + c_0,$$

while an $(n+1)$-digit number $N = (d_n \ldots d_1 d_0)_b$ in base (radix) b notation represents the quantity

$$N = d_n b^n + \ldots + d_1 b + d_0.$$

The representations are similar, with the variable x in the polynomial playing a role analogous to the base b of the number. Polynomial arithmetic modulo b is similar to base-b arithmetic on numbers, with one principal difference. In polynomial arithmetic, the powers of the variable x serve as place holders and each coefficient acts independently of its neighbors. When we perform arithmetic on numbers, however, there is a notion of "carry" from one digit to the next.

In fact, arithmetic on polynomials modulo b is virtually identical to arithmetic on numbers represented in the base b. For example, compare the multiplication of the binary numbers $(1110)_2$ and $(1101)_2$ with the multiplication of the polynomials $x^3 + x^2 + x$ and $x^3 + x^2 + 1$ modulo 2.

Binary numbers	Polynomials mod 2
1110	1110
× 1101	× 1101
1110	1110
1110	1110
1110	1110
10110110	1000110

The product of the polynomials is obtained by suppressing all carries when adding and is $x^6 + x^2 + x$. Had we multiplied the polynomials over the integers instead of modulo 2, the result would have been $x^6 + 2x^5 + 2x^4 + 2x^3 + x^2 + x$. Notice that the carries are suppressed in this case as well, and the powers of x act merely as place holders.

For most of this chapter, we focus on polynomials in one variable. We show how to represent such polynomials and then present algorithms for the basic arithmetic operations. As with the integers, the standard procedures for addition, subtraction, and multiplication are easily expressed. Division of polynomials with integer coefficients presents some interesting issues, and we show how computer algebra systems perform this operation efficiently.

Much of this chapter focuses on methods for computing the greatest common divisor of two polynomials. Polynomial gcd computation arises in many contexts, some quite surprising, as we shall see in later chapters. Consequently, it is very important to develop efficient computational procedures.

Polynomials of several variables are discussed in the final section. We describe methods for their representation and show how to adapt Euclid's algorithm to determine greatest common divisors.

4.1 Arithmetic on Polynomials

4.1.1 Dense and Sparse Representations

A *polynomial* of a single variable x, also called a *univariate* polynomial, can be written in the form

$$u(x) = c_n x^n + \ldots + c_1 x + c_0, \tag{4.1}$$

where the coefficients c_i are elements of some algebraic domain (or system). Examples of algebraic domains include the integers (denoted $\mathbb{Z}$), the rational numbers (denoted $\mathbb{Q}$), and the integers modulo a prime number p (denoted $\mathbb{Z}_p$). If $c_n \neq 0$, then the polynomial in (4.1) is said to be of *degree* n and c_n is called the *leading coefficient*. A *monic* polynomial is one for which $c_n = 1$.

There are two principal ways to represent a polynomial of a single variable. In a *dense representation*, the coefficients $c_n, c_{n-1}, \ldots, c_0$ are stored in a list that is sorted, either in increasing or decreasing order of exponent. For example, the coefficients of $x^5 - 15x^3 + 27x$ might be stored in the list

$$[1, 0, -15, 0, 27, 0],$$

with the coefficients in decreasing order of the corresponding power.

In a *sparse representation*, only the non-zero coefficients are stored. Each non-zero term, $c_i x^i$, in the polynomial is stored as a pair (i, c_i) and these pairs are placed in a list that is sorted in either increasing or decreasing order of exponent. The polynomial $x^5 - 15x^3 + 27x$ might be represented by the list

$$[\,(5, 1), (3, -15), (1, 27)\,].$$

Here again the coefficients are stored by decreasing power.

The sparse representation corresponds to normal mathematical notation in that we write $3x^2 - 1$, not $3x^2 + 0x - 1$. Moreover, the polynomials encountered in many applications tend to have a large number of zero coefficients. The sparse representation is clearly more storage efficient for polynomials like $x^{100} - 1$. For polynomials with relatively few non-zero coefficients, however, the storage requirement is about twice that of the dense representation since two pieces of information are stored for each non-zero term. Unfortunately, the algorithms to manipulate polynomials in the sparse representation tend to be more complex to code, although they typically execute far more efficiently for polynomials with relatively few non-zero terms.

Two important implementation issues still need to be addressed: the representation of the variable and the data structure used to store the coefficient information. In many contexts, the variable associated with a polynomial of one variable may be assumed, in others it must be stored explicitly. Since the powers of the variable serve primarily as place holders when performing polynomial arithmetic, the arithmetic itself can be carried out without actually knowing the name of the variable.

Both linked lists and dynamic arrays can be used in conjunction with either the dense or the sparse representation. In most cases, computer algebra systems employ links to the coefficients, rather than storing the values themselves in the basic structure, since the coefficients can be arbitrarily large "big numbers." In the final section of this chapter, we consider the issue of data structures for polynomials in greater depth.

4.1.2 Addition and Multiplication

In view of the obvious analogy with big number arithmetic, we need to discuss polynomial addition, subtraction, and multiplication only briefly.

Two polynomials are added by adding the coefficients of the corresponding terms. A procedure for adding two densely represented polynomials is given in Algorithm 4.1. The same procedure can be used for subtraction if the signs of coefficients of the subtrahend are reversed at the outset. If the procedure

```
procedure AddPolynomial(u, v)
{Input: polynomials u = u_m x^m + ... + u_1 x + u_0,
                     v = v_n x^n + ... + v_1 x + v_0, where m ≥ n }
{Output: polynomial w = u + v }
    for i := m downto n + 1 by − 1 do
        w_i := u_i
    end do
    for i := n downto 0 by − 1 do
        w_i := u_i + v_i
    end do
    return(w)
end procedure
```

ALGORITHM 4.1: Addition of dense polynomials

is applied to polynomials having degrees m and n, the running time is clearly proportional to the larger degree, $O(\max(m, n))$.

When two polynomials of equal degree are added, their sum need not be of the same degree. As a consequence, an implementation of Algorithm 4.1 requires eliminating any leading zeros that may be produced when the high order coefficients of the input polynomials cancel.

We next turn our attention to polynomial multiplication. Suppose we wish to multiply two polynomials, u and v, of degrees m and n,

$$u(x) = u_m x^m + ... + u_1 x + u_0,$$
$$v(x) = v_n x^n + ... + v_1 x + v_0.$$

Their product is a polynomial w of degree $m + n$,

$$w(x) = w_{m+n} x^{m+n} + ... + w_1 x + w_0,$$

whose coefficients w_k are given by

$$w_k = \sum_{i=0}^{k} u_{k-i} v_i = u_k v_0 + u_{k-1} v_1 + ... + u_1 v_{k-1} + u_0 v_k. \tag{4.2}$$

Notice that w_k is the sum of all products of the form $u_i v_j$ for which $i + j = k$. For instance, $w_0 = u_0 v_0$, $w_1 = u_1 v_0 + u_0 v_1$, $w_2 = u_2 v_0 + u_1 v_1 + u_0 v_2$, etc.

The standard procedure for polynomial multiplication is to apply formula (4.2) directly to compute each coefficient. This technique is presented formally as Algorithm 4.2. The procedure performs $(m + 1)(n + 1)$ multiplications of coefficients and mn scalar additions, so the work done is $O(mn)$.

procedure $MultiplyPolynomial(u, v)$
{Input: polynomials $u = u_m x^m + \ldots + u_1 x + u_0$,
$\qquad\qquad v = v_n x^n + \ldots + v_1 x + v_0$ }
{Output: polynomial $w = u \cdot v = w_{m+n} x^{m+n} + \ldots + w_1 x + w_0$ }
$\quad$ $w := 0$
$\quad$ **for** $i := m$ **downto** 0 **by** -1 **do**
$\qquad$ **for** $j := n$ **downto** 0 **by** -1 **do**
$\qquad\quad$ $w_{i+j} := w_{i+j} + u_i \cdot v_j$
$\qquad$ **end do**
$\quad$ **end do**
$\quad$ **return**(w)
end procedure

ALGORITHM 4.2: Multiplication of dense polynomials

4.1.3 Karatsuba's Algorithm

The traditional way to multiply two polynomials of degree n involves $O(n^2)$ work. We can develop an asymptotically faster algorithm that is similar to the one presented in Section 3.2.3 for integer multiplication. The basic idea is to observe that the product of two linear polynomials, $u(x) = u_1 x + u_0$ and $v(x) = v_1 x + v_0$, can be found with three multiplications instead of the usual four. The multiplications are $m_1 = (u_1 + u_0) \cdot (v_1 + v_0)$, $m_2 = u_1 \cdot v_1$, and $m_3 = u_0 \cdot v_0$. The desired product, w, can be expressed

$$w(x) = w_2 x^2 + w_1 x + w_0 = u_1 v_1 x^2 + (u_1 v_0 + u_0 v_1) x + u_0 v_0$$
$$= m_2 x^2 + (m_1 - m_2 - m_3) x + m_3.$$

(Our job is to compute the coefficients w_2, w_1, w_0 from u_1, u_0, v_1, v_0. The powers of x in the formula serve merely as placeholders.)

Although our three multiplication scheme uses more additions than the standard method, it can serve as the basis for a procedure to multiply polynomials with a substantially smaller total number of operations as the size of the polynomials increases.

Suppose we want to multiply two cubics,

$$p(x) = p_3 x^3 + p_2 x^2 + p_1 x + p_0,$$
$$q(x) = q_3 x^3 + q_2 x^2 + q_1 x + q_0.$$

We can split the polynomials into their upper and lower halves, expressing them as $p(x) = s(x)\,x^2 + t(x)$ and $q(x) = u(x)\,x^2 + v(x)$, where

$$s(x) = p_3\,x + p_2, \qquad\qquad u(x) = q_3\,x + q_2,$$
$$t(x) = p_1\,x + p_0, \qquad\qquad v(x) = q_1\,x + q_0.$$

The product may be written as

$$p(x) \cdot q(x) = s\,u\,x^4 + (s\,v + t\,u)\,x^2 + t\,v.$$

If we use the classical method to obtain each of the four products of linear polynomials, we perform $4^2 = 16$ scalar multiplications. However, we can take advantage of our three multiplication scheme by forming the products $m_1 = (s + t) \cdot (u + v)$, $m_2 = s \cdot u$, and $m_3 = t \cdot v$. Since each of these three products of linear polynomials can be found by again using the three multiplication scheme, only $3 \cdot 3 = 9$ scalar multiplications are performed overall. The desired result may be written

$$p(x) \cdot q(x) = m_2\,x^4 + (m_1 - m_2 - m_3)\,x^2 + m_3.$$

Now let's apply the three multiplication scheme to polynomials of arbitrary size using an algorithm design paradigm known as *divide-and-conquer*. Assume, solely for ease of description, that the number of terms is a power of two. It also simplifies our discussion to work with polynomials having n terms (or coefficients) (i.e., of degree $n - 1$). We begin by dividing the terms of the input polynomials into upper and lower halves, expressing these polynomials as $p(x) = s(x)\,x^{n/2} + t(x)$ and $q(x) = u(x)\,x^{n/2} + v(x)$. We next form three products of $n/2$-term polynomials: $m_1 = (s + t) \cdot (u + v)$, $m_2 = s \cdot u$, and $m_3 = t \cdot v$. To calculate each product, we apply the algorithm recursively until $n = 1$. The desired product is ultimately formed as

$$p(x) \cdot q(x) = m_2\,x^n + (m_1 - m_2 - m_3)\,x^{n/2} + m_3.$$

How efficient is this procedure? Let $M(n)$ denote the number of scalar multiplications performed in taking the product of two n-term polynomials. Since, in computing the product of the original n-term polynomials, we form three products of $n/2$-term polynomials, we have $M(n) = 3M(n/2)$. This recurrence can be solved by back-substitution (telescoping) as follows:

$$M(n) = 3\,M(n/2) = 3^2\,M(n/4) = \ldots = 3^k\,M(n/2^k).$$

The process stops when $2^k = n$ or, equivalently, $k = \log_2 n$, at which point we use the initial condition $M(1) = 1$ to discover

$$M(n) = 3^{\log_2 n} = n^{\log_2 3} \approx n^{1.59}.$$

So our divide-and-conquer algorithm uses about $n^{1.59}$ scalar multiplications to the classical method's n^2.

What about the overall running time, $T(n)$? It is governed by the recurrence relation

$$T(n) = 3\,T(n/2) + c\,n \tag{4.3}$$

since, in forming the product of two n-term polynomials, we apply the algorithm recursively to form three products of polynomials of half the original size. The cn term on the right captures the number of additions and subtractions that must also be performed, but this work is linear in the size of the polynomials being processed. The solution to the recurrence (4.3) is also $O(n^{\log_2 3}) \approx O(n^{1.59})$, as shown in Section 3.2.3. So, perhaps surprisingly, even the number of additions is smaller than the number in the classical procedure for sufficiently large polynomials.

The algorithm just presented is due to Anatolii Karatsuba and Yu. Ofman (1962). It uses a number of scalar arithmetic operations—both multiplications and additions—that is comparable to the standard method when n is about 16, and its relative performance improves as n increases. Can we do even better? While the fast Fourier transform yields an $O(n \log n)$ algorithm for polynomial multiplication, it is infeasible to implement in a computer algebra system performing exact arithmetic on coefficients that are large integers or rational numbers.

4.1.4 Sparse Polynomial Multiplication

Up to now, all our algorithms have manipulated polynomials in a dense representation. It is instructive to consider at least one manipulation of sparse polynomials, and multiplication is both the obvious and the best choice.

Suppose we want to find the product of the polynomials

$$u(x) = \sum_{k=1}^{m} u_k\, x^{i_k} \quad \text{and} \quad v(x) = \sum_{k=1}^{n} v_k\, x^{j_k},$$

represented by the lists

$$U = [\,(i_1, u_1), (i_2, u_2), \ldots, (i_m, u_m)\,],$$
$$V = [\,(j_1, v_1), (j_2, v_2), \ldots, (j_n, v_n)\,],$$

where the coefficients in both lists are stored in decreasing order of exponent (the first component of each pair). The product,

$$w(x) = \sum_{k=1}^{p} w_k\, x^{l_k},$$

is also to be stored in a list, with the terms ordered by decreasing order of exponent.

We compute the product $w = u \cdot v$ as follows:

1. Construct lists S_k for $1 \leq k \leq n$, where the r-th item is $(i_r + j_k, u_r \cdot v_k)$ for $1 \leq r \leq m$. The list S_k represents the product of the k-th term of v with the entire polynomial u, and the item $(i_r + j_k, u_r \cdot v_k)$ corresponds to the product $u_r x^{i_r} \cdot v_k s^{j_k} = u_r v_k x^{i_r + j_k}$.

2. Merge the lists S_1, S_2 and S_3, S_4, etc. constructed at Step 1, combining terms with identical exponents and maintaining the terms in decreasing order of exponent. Then merge these lists in pairs, again combining like terms and ordering by exponent. Continue until only one list remains.

To examine the running time of the procedure, assume without loss of generality that $m \geq n$. The first step is clearly $O(mn)$. The second step is repeated $\lceil \log_2 n \rceil$ times, and the total work each time is $O(mn)$ since, in the worst case, no terms combine. Therefore, the overall running time is $O(mn \log n)$.

Example 4.1. Find the product of $u(x) = 3x^{15} - 5x^{10} + 6x^5 - 4$ and $v(x) = 7x^{15} - 2x^{10} - 3x^5 + 1$ with the sparse polynomial multiplication algorithm.

Solution. The lists corresponding to the polynomials $u(x)$ and $v(x)$ are

$$U = [\,(15,3),\ (10,-5),\ (5,6),\ (0,-4)\,],$$
$$V = [\,(15,7),\ (10,-2),\ (5,-3),\ (0,1)\,].$$

We first construct the lists S_k corresponding to the products of each term of $v(x)$ with the entire polynomial $u(x)$,

$$S_1 = [\,(30,21),\ (25,-35),\ (20,42),\ (15,-28)\,],$$
$$S_2 = [\,(25,-6),\ (20,10),\ (15,-12),\ (10,8)\,],$$
$$S_3 = [\,(20,-9),\ (15,15),\ (10,-18),\ (5,12)\,],$$
$$S_4 = [\,(15,3),\ (10,-5),\ (5,6),\ (0,-4)\,].$$

Next, we merge the lists S_1, S_2 and S_3, S_4, combining corresponding terms,

$$T_1 = [\,(30,21),\ (25,-41),\ (20,52),\ (15,-40),\ (10,8)],$$
$$T_2 = [\,(20,-9),\ (15,18),\ (10,-23),\ (5,18),\ (0,-4)\,].$$

Finally, we merge these lists to obtain the one corresponding to the product of $w = u \cdot v$,

$$W = [\,(30,21),\ (25,-41),\ (20,43),\ (15,-22),\ (10,-15),\ (5,18),\ (0,-4)\,].$$

So $w(x) = 21x^{30} - 41x^{25} + 43x^{20} - 22x^{15} - 15x^{10} + 18x^5 - 4$. □

We close this section with an example showing how the time complexities of the classical and sparse methods can affect the way polynomial arithmetic should be performed. Suppose we want to compute p^4, where p is a polynomial with n terms in both the dense and sparse representations.

We first consider the dense representation under the assumption that the classical multiplication procedure is used. If we calculate p^4 by repeated squaring, then the time for the first squaring is an^2, and the result is a polynomial with $2n$ terms. Thus the time for the second squaring is $a(2n)^2 = 4an^2$, and the overall running time is $5an^2$. Now suppose we compute p^4 as $(((p \cdot p) \cdot p) \cdot p)$. Again the time for the first product is an^2. The second product requires time $a(2n)n = 2an^2$, and the third $a(3n)n = 3an^2$. Overall, the running time is $6an^2$ and so, as expected, repeated squaring is faster.

Next, consider the sparse representation. If we calculate p^4 by repeated squaring, the time for the first squaring is $bn^2 \log n$. Assuming that few terms combine, the time for the second squaring is $b(n^2)(n^2) \log n^2 = 2bn^4 \log n$, and the total execution time is $b(2n^4 + n^2) \log n$. If we compute p^4 as $(((p \cdot p) \cdot p) \cdot p)$, the times are $bn^2 \log n$ for the first multiplication, $b(n^2)n \log n$ for the second, and $b(n^3)n \log n$ for the third. So the total running time is $b(n^4 + n^3 + n^2) \log n$, which beats repeated squaring! The effect is even more pronounced when we compute p^{2^k} for some integer $k > 2$.

4.1.5 Division

We now turn attention to polynomial *long division*. Suppose the dividend, u, is a polynomial of degree m and the divisor, v, is a non-zero polynomial of degree n,

$$u(x) = u_m x^m + \ldots + u_1 x + u_0,$$
$$v(x) = v_n x^n + \ldots + v_1 x + v_0.$$

We are asked to compute their quotient, q, a polynomial of degree $m - n$,

$$q(x) = q_{m-n} x^{m-n} + \ldots + q_0,$$

and the remainder, r, a polynomial such that $\text{degree}(r) < \text{degree}(v)$,

$$r(x) = r_{n-1} x^{n-1} + \ldots + r_0.$$

The four polynomials satisfy the relationship

$$u(x) = q(x) v(x) + r(x).$$

For simplicity we assume $\text{degree}(u) \geq \text{degree}(v)$, otherwise we merely obtain a zero quotient, $q(x) = 0$, and a remainder $r(x) = v(x)$.

We review the procedure for polynomial long division that we learned in school with an example.

$$\begin{array}{r}
2\,x - 1 \\
3\,x^2 - x + 2 \ \overline{)\ 6\,x^3 - 5\,x^2 + 9\,x + 3} \\
\underline{6\,x^3 - 2\,x^2 + 4\,x} \\
-3\,x^2 + 5\,x + 3 \\
\underline{-3\,x^2 + \ \ x - 2} \\
4\,x + 5
\end{array}$$

FIGURE 4.1: Long division of polynomials.

Example 4.2. The division of $u(x) = 6x^3 - 5x^2 + 9x + 3$ by $v(x) = 3x^2 - x + 2$ is depicted in Figure 4.1. We begin by dividing the highest term of the dividend, $6x^3$, by the leading term of the divisor, $3x^2$. This gives $2x$, which is the first term of the quotient.

We next multiply this term by the entire divisor v, yielding $6x^3 - 2x^2 + 4x$, then subtract the product from the corresponding terms in u, $6x^3 - 5x^2 + 9x$. This eliminates the x^3 term from u and leaves a remainder of $-3x^2 + 5x$. We "bring down" the next term from u (alternatively, we can bring down all of the remaining terms of u) and add this to the remainder, giving $-3x^2 + 5x + 3$. Our problem now reduces to dividing this polynomial by v.

We again divide the highest term of this polynomial, $-3x^2$ by the leading term of the divisor, $3x^2$. This gives -1, which is the next (and, in this case, last) term of the quotient. Multiplying this term by the divisor v gives $-3x^2 + x - 2$, which we subtract from the remainder at the previous step, yielding $4x + 5$. This is the final remainder, and so the quotient is $q(x) = 2x - 1$ and the remainder is $r(x) = 4x + 5$.   □

The procedure for polynomial long division is given formally as Algorithm 4.3. The remainder r is initialized to the dividend u. The degree of the quotient is the difference between the degree of u and that of the divisor v. At each iteration of the outer loop, we compute the next term, $q_i x^i$, of the quotient. (These terms are calculated in decreasing order of power.) We then multiply this term by v in the inner loop, subtracting the product from r.

If degree$(u) = m$ and degree$(v) = n$, the outer loop is executed $m - n + 1$ times. The inner loop is always executed $n + 1$ times, once for each term in the divisor. As a result, the running time of the entire procedure is $O(m\,n)$.

4.1.6   Pseudo-division

When we add, subtract, or multiply polynomials with integer coefficients we obtain a polynomial with integer coefficients. When we divide polynomials with integer coefficients, however, the quotient and remainder are, in general, polynomials with rational, rather than integer, coefficients. A simple example is the division of $u(x) = x^2 + 1$ by $v(x) = 2x - 1$, which produces the quotient $q(x) = \frac{1}{2}x + \frac{1}{4}$ and remainder $r(x) = \frac{5}{4}$.

procedure *DividePolynomial*(u, v)
{Input: polynomials $u = u_m x^m + \ldots + u_1 x + u_0$,
$\qquad\qquad\qquad v = v_n x^n + \ldots + v_1 x + v_0$, where $v \neq 0$ }
{Output: quotient polynomial $q = q_{m-n} x^{m-n} + \ldots + q_1 x + q_0$,
$\qquad$ remainder polynomial $r = r_{n-1} x^{n-1} + \ldots + r_1 x + r_0$,
$\qquad\qquad$ where $u = q\,v + r$ and degree(r) < degree(v) }
$\quad r := u; \quad q := 0$
$\quad$**for** $i := m - n$ **downto** 0 **by** -1 **do**
$\qquad q_i := r_{n+i}/v_n$
$\qquad$**for** $j := n + i$ **downto** i **by** -1 **do**
$\qquad\qquad r_j := r_j - q_i \cdot v_{j-i}$
$\qquad$**end do**
$\quad$**end do**
$\quad$**return**(q, r)
end procedure

ALGORITHM 4.3: Division of dense polynomials

What happens when we divide polynomials with rational coefficients? In this case, the quotient and remainder are both polynomials with rational coefficients. Similarly if we divide two polynomials with coefficients in $\mathbb{Z}_p$, the integers modulo a prime p, we obtain a quotient and remainder that are polynomials whose coefficients are in $\mathbb{Z}_p$. The rational numbers $\mathbb{Q}$ and the integers modulo a prime are examples of a type of algebraic domain called a *field*. The integers $\mathbb{Z}$ do not form a field, however, since integers (other than one) do not have a multiplicative inverse that is an integer. The multiplicative inverse of 3, for example, is the rational number $\frac{1}{3}$ since $3 \cdot \frac{1}{3} = 1$.

An example of the division of two polynomials with integer coefficients that produces a quotient and remainder with rational coefficients is depicted in Figure 4.2. Not only do the quotient and remainder involve rational numbers, but a large number of fractions are generated throughout the course of the computation. Every time a computer algebra system encounters a rational number, whether as a final result or as part of an intermediate computation, it reduces the fraction to lowest terms. This can result in a large number of integer gcd calculations and an even larger number of integer divisions as Euclid's algorithm is iterated to compute each gcd.

There is a way to divide polynomials with integer coefficients and keep the computation within the domain of the integers. The process is known as *pseudo-division*. Looking again at the division in Figure 4.2, we observe that the denominators of all the coefficients in both the quotient $q(x)$ and the remainder $r(x)$—as well as those of all the fractions appearing in the

$$\frac{2}{3}x + \frac{7}{9}$$

$$3x^2 - 2x + 1 \overline{\smash{)}\,2x^3 + x^2 + x + 3}$$

$$2x^3 - \tfrac{4}{3}x^2 + \tfrac{2}{3}x$$

$$\overline{\tfrac{7}{3}x^2 + \tfrac{1}{3}x + 3}$$

$$\tfrac{7}{3}x^2 - \tfrac{14}{9}x + \tfrac{7}{9}$$

$$\overline{\tfrac{17}{9}x + \tfrac{20}{9}}$$

FIGURE 4.2: Division producing polynomials with rational coefficients.

intermediate computations—are powers of the leading coefficient of the divsor $v(x)$ (i.e., $3^0 = 1, 3^1 = 3, 3^2 = 9$),

$$\underbrace{2x^3 + x^2 + x + 3}_{u(x)} = \underbrace{\left(\tfrac{2}{3}x + \tfrac{7}{9}\right)}_{q(x)} \underbrace{\left(3x^2 - 2x + 1\right)}_{v(x)} + \underbrace{\left(\tfrac{17}{9}x + \tfrac{20}{9}\right)}_{r(x)}.$$

As a consequence, if we multiply the dividend $u(x)$ by $3^2 = 9$, the entire computation stays in the domain of the integers,

$$9 \cdot (2x^3 + x^2 + x + 3) = (6x + 7)(3x^2 - 2x + 1) + (17x + 20).$$

To obtain $q(x)$ and $r(x)$, we just divide each of the coefficients of the pseudo-quotient and pseudo-remainder by 9.

When performing pseudo-division, what factor do we use in general? Multiply the dividend $u(x)$ by $\alpha = (v_n)^{\Delta+1}$, where $\Delta = \text{degree}(u) - \text{degree}(v)$ and v_n is the leading coefficient of the divisor $v(x)$. To obtain the actual quotient $q(x)$ and remainder $r(x)$, merely divide the pseudo-quotient and pseudo-remainder by α.

We can even take advantage of pseudo-division to divide polynomials with rational coefficients, where it may seem at first that intermediate computation with fractions is unavoidable. For example, to divide

$$u(x) = \tfrac{3}{2}x^3 - \tfrac{4}{3}x^2 + 2 = \tfrac{1}{6}(9x^3 - 8x^2 + 12) = \tfrac{1}{6}u'(x), \qquad (4.4)$$

$$v(x) = \tfrac{1}{5}x^2 - \tfrac{2}{3}x + 1 = \tfrac{1}{15}(3x^2 - 10x + 15) = \tfrac{1}{15}v'(x), \qquad (4.5)$$

we begin by removing the least common multiples of the denominators of the fractions. Next, we pseudo-divide the polynomials u' and v' with integer coefficients on the right of (4.4) and (4.5). From the pseudo-quotient and pseudo-remainder, we determine the actual quotient q' and remainder r' when u' is divided by v',

$$q'(x) = 3x + \tfrac{22}{3},$$

$$r'(x) = \tfrac{85}{3}x - 98.$$

Since $u' = 6u$ and $v' = 15v$, the desired quotient $q(x) = \frac{15}{6}q' = \frac{15}{2}x + \frac{55}{3}$. To determine the corresponding remainder r, observe that

$$\frac{r}{r'} = \frac{u - qv}{u' - q'v'} = \frac{u - qv}{6u - (\frac{6}{15}q)(15v)} = \frac{1}{6},$$

so $r(x) = \frac{1}{6}r' = \frac{85}{18}x - \frac{49}{3}$. This method can save a great deal of intermediate computation when the sizes of the coefficients are large or there are many non-zero terms in the polynomials being divided.

4.2   Polynomial GCDs

With division under our belts, we turn our attention to the problem of computing greatest common divisors. We concentrate on polynomials with integer coefficients and those with rational coefficients. Polynomial gcd computation turns up in a number of places—many unexpected—in computer algebra systems. Of course, gcds are necessary in performing simple arithmetic operations on rational functions (ratios of polynomials) but, as we shall see in subsequent chapters, they also arise in algebraic simplification, factorization, and integration.

4.2.1   Adapting Euclid's Algorithm

Euclid's algorithm and the extended Euclidean algorithm work for polynomials in much the same way they work for integers, but some minor adjustments are necessary. Recall that Euclid's method for non-negative integers m, n is based on the observation that

$$\gcd(m, n) = \begin{cases} \gcd(n, m \bmod n) & \text{if } m \neq 0, \\ n & \text{if } m = 0, \end{cases}$$

where $m \geq n$,

There are obvious analogies to polynomials. The mod operation corresponds to taking the remainder when dividing two integers. So for polynomials u and v, we also divide and take the remainder. When performing the division, we want degree$(u) \geq$ degree(v). This is the analog of the condition that $m \geq n$ for integers.

What is our halting criterion? For integers, we keep reducing their sizes by dividing and taking remainders until the smaller number eventually becomes zero. When dividing polynomials, we won't always reach a final remainder of zero. Instead the degrees of the polynomials decrease until the one of lesser degree becomes a constant. If the constant is zero, then the polynomials have a non-trivial common divisor. Otherwise, the polynomials have no common factors and their gcd is one.

So as a "first cut," our rule for computing the gcd of two polynomials is

$$
\gcd(u, v) = \begin{cases}
\gcd(v, \operatorname{rem}(u, v)) & \text{if degree}(v) \neq 0, \\
u & \text{if } v = 0, \\
1 & \text{if degree}(v) = 0 \text{ and } v \neq 0,
\end{cases}
$$

where $\text{degree}(u) \geq \text{degree}(v)$. Let's try an example to see how it works.

Example 4.3. Apply Euclid's algorithm to find $\gcd(x^2 - 1, 2x^2 + 4x + 2)$.

Solution. The steps in the computation are

$$
\begin{aligned}
\gcd(x^2 - 1, 2x^2 + 4x + 2) &= \gcd(2x^2 + 4x + 2, -2x - 2) \\
&= \gcd(-2x - 2, 0) \\
&= -2x - 2
\end{aligned}
$$

and the divisions are shown in Figure 4.3. So Euclid's algorithm calculates the gcd to be $-2x - 2$.

This is surprising! We should expect the answer $x + 1$ since the inputs factor as

$$
x^2 - 1 = (x + 1)(x - 1),
$$
$$
2x^2 + 4x + 2 = 2(x + 1)^2.
$$

What happened? If we allow ourselves to use rational coefficients, then we can also express the input polynomials as

$$
x^2 - 1 = (-2x - 2)(-\tfrac{1}{2}x + \tfrac{1}{2}) = \tfrac{1}{2}(-2x - 2)(-x + 1),
$$
$$
2x^2 + 4x + 2 = (-2x - 2)(-x - 1).
$$

So the answer obtained by Euclid's algorithm is correct, it's just not what we expect or want. □

The difficulty in the previous example arises because we can factor the input polynomials in different ways. We want our factorizations to be "unique." Recall that division of polynomials with integer coefficients produces, in general, polynomials with rational coefficients. Unfortunately, factorizations of polynomials over the rational numbers $\mathbb{Q}$ are not unique. There is, however, only one way to factor a polynomial over the integers $\mathbb{Z}$.

The solution to our dilemma is due to Carl Friedrich Gauss (1777–1855). Suppose we are dealing with a polynomial u having integer coefficients. We can split u into two parts. The *content* of u, denoted $\text{cont}(u)$, is defined to be the integer that is the greatest common divisor of the coefficients. The *primitive part*, written $\text{pp}(u)$, is what remains of u when we take out the

$$
\begin{array}{r}
\frac{1}{2} \\[2pt]
2\,x^2 + 4x + 2 \; \overline{\smash{\big)}\; x^2 \qquad\quad -1} \\
\underline{x^2 + 2\,x + 1} \\
-2\,x - 2
\end{array}
\qquad\qquad
\begin{array}{r}
-x - 1 \\[2pt]
-2\,x - 2 \; \overline{\smash{\big)}\; 2\,x^2 + 4x + 2} \\
\underline{2\,x^2 + 2\,x} \\
2\,x + 2 \\
\underline{2\,x + 2} \\
0
\end{array}
$$

(a) First iteration   (b) Second iteration

FIGURE 4.3: Divisions in gcd computation of Example 4.3.

content. This decomposition of u into its content and primitive part is unique up to a unit factor of ± 1. It is conventional to choose the unit to make the leading coefficient of the primitive part positive.

With these definitions, we compute the gcd of two polynomials, u and v, over the integers $\mathbb{Z}$ as

$$\gcd(u, v) = \gcd(\operatorname{cont}(u), \operatorname{cont}(v)) \cdot \gcd(\operatorname{pp}(u), \operatorname{pp}(v)).$$

The integer version of Euclid's algorithm developed in Chapter 3 can be used to compute the unique gcd of the contents. To find the gcd of the primitive parts, we pseudo-divide the polynomials at each iteration and use the primitive part of the remainder at the next step. This works since, by construction, the primitive parts of u and v have no common integer factors. The entire procedure is called the *primitive Euclidean algorithm* and is given in Algorithm 4.4.

Example 4.4. Compute the gcd of $u(x) = 54x^3 - 54x^2 + 84x - 48$ and $v(x) = -12x^3 - 28x^2 + 72x - 32$ with the primitive Euclidean algorithm.

Solution. We begin by splitting the polynomials into their content and primitive parts,

$$
\begin{aligned}
\operatorname{cont}(u) &= 6, \quad \operatorname{unit}(u) = 1, \quad \operatorname{pp}(u) = 9\,x^3 - 9\,x^2 + 14\,x - 8, \\
\operatorname{cont}(v) &= 4, \quad \operatorname{unit}(v) = -1, \quad \operatorname{pp}(v) = 3\,x^3 + 7\,x^2 - 18\,x + 8.
\end{aligned}
$$

Next, we iterate Algorithm 4.4 to determine the gcd of the primitive parts.

$u(x)$	$v(x)$	$r(x)$
$9\,x^3 - 9\,x^2 + 14\,x - 8$	$3\,x^3 + 7\,x^2 - 18\,x + 8$	$-90\,x^2 + 204\,x - 96$
$3\,x^3 + 7\,x^2 - 18\,x + 8$	$15\,x^2 - 34\,x + 16$	$2268\,x - 1512$
$15\,x^2 - 34\,x + 16$	$3\,x - 2$	0

Therefore, $\gcd(\operatorname{pp}(u), \operatorname{pp}(v)) = 3x - 2$ and we have

$$\gcd(u, v) = \gcd(6, 4) \cdot (3\,x - 2) = 6\,x - 4.$$

procedure $PrimitivePolyGCD(p, q)$
{Input: polynomials p, q with integer coefficients,
                where degree$(p) \geq$ degree(q) }
{Output: gcd(p, q) over the integers $\mathbb{Z}$ }
    $c := \gcd(\text{cont}(p), \text{cont}(q))$ { determine gcd of contents }
    $u := \text{pp}(p);$ $v := \text{pp}(q)$ { work with primitive parts }
    while degree$(v) > 0$ **do**
      $r := \text{pseudorem}(u, v)$ { compute pseudo-remainder }
      $u := v;$ $v := \text{pp}(r)$ { take primitive part of r }
    end do
    if $v = 0$ **then return**$(c \cdot u)$
              else return(c)
    end if
end procedure

ALGORITHM 4.4: Primitive Euclidean algorithm for polynomials

To check, we give the factorizations of u and v over the integers,

$$u(x) = (2 \cdot 3)\,(3\,x - 2)\,(3\,x^2 - x + 4),$$
$$v(x) = -(2^2)\,(x + 4)\,(x - 1)\,(3\,x - 2). \qquad \square$$

There is a unique way to express an integer as a product of prime numbers to powers over $\mathbb{Z}$. For example, $60 = 2^2 \cdot 3 \cdot 5$ and, except for reordering the factors, this prime decomposition is unique. Therefore, common integer factors should be included when computing the gcd of polynomials over $\mathbb{Z}$.

The situation is very different with respect to the rational numbers $\mathbb{Q}$. Every non-zero rational number m can be regarded as a *unit*. That is, for any non-zero rational number n, there is a multiplicative inverse n_m^{-1} such that $m = n \cdot n_m^{-1}$. For instance, $-1/2 = (-3/5) \cdot (5/6) = 7 \cdot (-1/14)$ and so forth. Even integers can be written this way (e.g., $60 = 30 \cdot 2 = 120 \cdot (1/2)$) when regarded as rational numbers. As a consequence, we can express a rational number as a product of (reduced) rational numbers in an unlimited number of ways. This is the reason our first "first cut" version of Euclid's algorithm for polynomials had a problem with non-unique factorizations.

We are now ready to present a way to compute unique gcds of polynomials with rational coefficients. If such a polynomial is monic (i.e., has leading coefficient one), then it can be factored uniquely as a product of monic polynomials. This is true because a product of monic polynomials is clearly monic.

For a polynomial u that is not monic, we can factor out its leading coefficient, lcoeff(u). What remains is a monic polynomial that we shall take to be the principal part of u, pp$(u) = 1/\text{lcoeff}(u) \cdot u$. From our discussion above,

procedure $MonicPolyGCD(p, q)$
{Input: polynomials p, q with rational coefficients,
                     where degree(p) $\geq$ degree(q) }
{Output: monic gcd(p, q) over the rational numbers $\mathbb{Q}$ }
    $u := 1/\text{lcoeff}(p) \cdot p$ { divide by leading coefficient }
    $v := 1/\text{lcoeff}(q) \cdot q$ { work with monic polynomials }
    while degree(v) > 0 **do**
        $r := \text{rem}(u, v)$ { compute remainder }
        $u := v$; $v := 1/\text{lcoeff}(r) \cdot r$ { make v monic }
    end do
    if $v = 0$ **then return**(u)
                 else return(1)
    end if
end procedure

ALGORITHM 4.5: Monic Euclidean algorithm for polynomials

lcoeff(u) is a unit in $\mathbb{Q}$, and we have unit(u) = lcoeff(u). Again from this perspective, the content of u is one, cont(u) = 1.

So to compute a unique gcd of polynomials u and v over $\mathbb{Q}$, we begin by normalizing u and v to be monic by factoring out their leading coefficients. Next we iterate Euclid's procedure, dividing polynomials and taking remainders. Since by our convention gcd(u, v) is monic, we can normalize the remainder at each iteration to be monic by factoring out its leading coefficient. This procedure is called the *monic Euclidean algorithm* and is presented as Algorithm 4.5.

Example 4.5. Compute the gcd of $u(x) = 54x^3 - 54x^2 + 84x - 48$ and $v(x) = -12x^3 - 28x^2 + 72x - 32$ with the monic Euclidean algorithm.

Solution. Splitting off the leading coefficients of the polynomials, we obtain the monic representations of the inputs,

$$\text{cont}(u) = 1, \quad \text{unit}(u) = 54, \quad \text{pp}(u) = x^3 - x^2 + \tfrac{14}{9}x - \tfrac{8}{9},$$
$$\text{cont}(v) = 1, \quad \text{unit}(v) = -12, \quad \text{pp}(v) = x^3 + \tfrac{7}{3}x^2 - 6x + \tfrac{8}{3}.$$

We list the values of u, v, and r at each iteration of Algorithm 4.5. Notice that the remainder r at each step is multiplied by the reciprocal of its leading coefficient to give the monic version of v used at the next iteration.

$u(x)$	$v(x)$	$r(x)$
$x^3 - x^2 + \tfrac{14}{9}x - \tfrac{8}{9}$	$x^3 + \tfrac{7}{3}x^2 - 6x + \tfrac{8}{3}$	$-\tfrac{10}{3}x^2 + \tfrac{68}{9}x - \tfrac{32}{9}$
$x^3 + \tfrac{7}{3}x^2 - 6x + \tfrac{8}{3}$	$x^2 - \tfrac{34}{15}x + \tfrac{16}{15}$	$\tfrac{84}{25}x - \tfrac{56}{25}$
$x^2 - \tfrac{34}{15}x + \tfrac{16}{15}$	$x - \tfrac{2}{3}$	0

Therefore, the unique monic representation of $\gcd(u,v) = x - \frac{2}{3}$. To check our answer, we give the factorization of the inputs into irreducible monic polynomials over $\mathbb{Q}$,

$$u(x) = (54)\left(x - \tfrac{2}{3}\right)\left(x^2 - \tfrac{1}{3}x + \tfrac{4}{3}\right),$$
$$v(x) = (-12)(x+4)(x-1)\left(x - \tfrac{2}{3}\right). \qquad \square$$

4.2.2   Polynomial Remainder Sequences

In this section, we explore and compare several strategies for determining polynomial gcds. Since polynomial gcd computation is ubiquitous in computer algebra, we are highly motivated to seek efficient techniques.

In our discussion, we present the sequence of remainders produced by applying each method examined to the polynomials

$$u(x) = x^8 + x^6 - 3\,x^4 - 3\,x^3 + 8\,x^2 + 2\,x - 5,$$
$$v(x) = 3\,x^6 + 5\,x^4 - 4\,x^2 - 9\,x + 21,$$

whose greatest common divisor is one. This famous example is due to W. S. Brown (1971) and appears in virtually every treatment of the subject. It illustrates a common phenomenon in computer algebra known as *intermediate expression swell*. The inputs are polynomials of relatively low degree and all of the coefficients are small; moreover, the final result is simply one. Yet we shall see that the intermediate computation involved to produce the answer can involve some very large numbers.

We observe at the outset that the remainders produced by each of the sequences are proportional to one another when applied to the same pair of polynomials. (Why?) Therefore, we can use any of the methods to compute the greatest common divisor over either the integers $\mathbb{Z}$ or the rational numbers $\mathbb{Q}$ provided we multiply by an appropriate normalization constant at the end of the calculation.

Rational Remainder Sequence. Perhaps the simplest strategy of all is to apply our "first cut" version of Euclid's algorithm, working in the domain of the rational numbers $\mathbb{Q}$. We note that the gcd obtained by this method can be converted to a unique, monic representation at the end of the procedure merely by multiplying the gcd computed by the reciprocal of its leading coefficient.

For the polynomials u and v we are considering, the sequence of remainders calculated is

$$r_1(x) = -\tfrac{5}{9}x^4 + \tfrac{1}{9}x^2 - \tfrac{1}{3},$$
$$r_2(x) = -\tfrac{117}{25}x^2 - 9x + \tfrac{441}{25},$$
$$r_3(x) = \tfrac{233150}{19773}x - \tfrac{102500}{6591},$$
$$r_4(x) = -\tfrac{1288744821}{543589225}.$$

One obvious disadvantage of this method is the rapid growth in the size of the rational coefficients. Another is the integer gcd calculation required to reduce the fractions arising during the course of each polynomial division. □

Monic Remainder Sequence. This is the approach presented in Algorithm 4.5. We again work over $\mathbb{Q}$ but normalize the remainders after each division to have a leading coefficient of one. The normalization process has the effect of reducing the size of the coefficients of the divisor used at the next iteration, as does the fact that the leading coefficient of each divisor is one.

The sequence of remainders produced for our example is

$$r_1(x) = -\tfrac{5}{9}\,x^4 + \tfrac{1}{9}\,x^2 - \tfrac{1}{3},$$
$$r_2(x) = -\tfrac{39}{25}\,x^2 - 3\,x + \tfrac{147}{25},$$
$$r_3(x) = -\tfrac{46630}{2197}\,x + \tfrac{61500}{2197},$$
$$r_4(x) = \tfrac{11014913}{21743569}.$$

Comparing these remainders with those obtained in the rational sequence, we see that this strategy represents a modest improvement. The coefficients are somewhat smaller and, as an added benefit, we compute the gcds of smaller integers as we divide. Unfortunately, we must still perform many integer gcd calculations for each division. □

This is a problem for any polynomial gcd algorithm that works over $\mathbb{Q}$. As a result, rational methods tend to be noticeably slower than all-integer methods. Consequently, we concentrate on all-integer techniques for the rest of this section.

Euclidean Remainder Sequence. In this method, we keep the computation entirely in the domain of the integers $\mathbb{Z}$ by using "pseudo-division." Moreover, to avoid integer gcd computation completely, the content is not removed from the remainder at each step. The remainders generated for our u and v are

$$r_1(x) = -15\,x^4 + 3\,x^2 - 9,$$
$$r_2(x) = 15795\,x^2 + 30375\,x - 59535,$$
$$r_3(x) = 1254542875143750\,x - 1654608338437500,$$
$$r_4(x) = 12593333879550074310093114199 2187500.$$

The only advantage of this approach over the rational sequences is the entire absence of integer gcd calculation—but at an enormous cost. The coefficients, which are proportional to one another in all of the methods, have grown to gargantuan size! □

Primitive Remainder Sequence. This is the method of Algorithm 4.4. It works over $\mathbb{Z}$ using pseudo-division, but removes the content from the remainders at each step. For our example, the remainders after the content has been removed are

$$r_1(x) = 5\,x^4 - x^2 + 3,$$
$$r_2(x) = 13\,x^2 + 25\,x - 49,$$
$$r_3(x) = 4663\,x - 6150,$$
$$r_4(x) = 1.$$

This method keeps the growth of the coefficients to the absolute minimum. Unfortunately, it requires a substantial amount of integer gcd computation. While it is a reasonable choice to implement for the univariate case, its cost is prohibitive for multivariate polynomials where considerably more integer gcd calculation is involved. $\qquad\square$

Before presenting two additional methods, it is useful to generalize the notion of a polynomial remainder sequence and present a formal definition. A *polynomial remainder sequence*, or *PRS*, for $u(x)$ and $v(x)$, where degree$(u) \geq$ degree(v), is a sequence of polynomials $r_{-1}, r_0, r_1, r_2, \ldots, r_m$ satisfying

1. $r_{-1}(x) = u(x)$, $r_0(x) = v(x)$,

2. $\alpha_i\, r_{i-2} = q_i(x)\, r_{i-1}(x) + \beta_i\, r_i(x)$,

3. degree$(r_m) < 1$.[†]

We can place each of the remainder sequences discussed above into this framework. For the rational PRS, we have simply that $\alpha_i = 1$ and $\beta_i = 1$. For the monic PRS, we again have $\alpha_i = 1$, but we take

$$\beta_i = \text{lcoeff}\left(\text{rem}(r_{i-2}, r_{i-1})\right)^{-1}$$

to force the leading coefficient of r_i to be one (i.e., lcoeff$(r_i) = 1$). The Euclidean PRS satisfies

$$\alpha_i = \text{lcoeff}(r_{i-1})^{\Delta_i + 1}, \quad \text{where } \Delta_i = \text{degree}(r_{i-2}) - \text{degree}(r_{i-1})$$

and $\beta_i = 1$. The rule that gives α_i is the same for the primitive remainder sequence but, in this case,

$$\beta_i = \text{cont}(\text{pseudorem}(r_{i-2}(x), r_{i-1}(x)))$$

and so all the remainders r_i are primitive (i.e., their contents are one).

In devising efficient polynomial remainder sequences, our goal is to choose α_i and β_i to make the coefficients of the remainders as small as possible without requiring a great deal of computation. In this regard, the Euclidean PRS

[†]The convention generally used in abstract algebra is to take degree$(0) = -\infty$ and degree$(a) = 0$ for any non-zero element a of $\mathbb{Z}$ or $\mathbb{Q}$.

and the primitive PRS represent opposite extremes. The Euclidean scheme requires no gcd calculation but produces huge coefficients. The primitive scheme calculates integer gcds to remove the content from the remainder at every iteration but produces the smallest coefficients possible. Our next two remainder sequences represent a compromise between calculation and coefficient size.

Reduced Remainder Sequence. While the coefficients produced by this sequence are larger than those in the primitive sequence, no integer gcd calculation is required at all. In this regard, it is like the Euclidean sequence but the coefficients produced are much smaller. The rules for generating the sequence are

$$\alpha_i = \text{lcoeff}(r_{i-1})^{\Delta_i+1}; \quad \beta_1 = 1, \quad \beta_i = \alpha_{i-1} \text{ for } 2 \le i \le m,$$

where, as usual, $\Delta_i = \text{degree}(r_{i-2}) - \text{degree}(r_{i-1})$.

The remainders calculated in our example are

$$r_1(x) = -15\,x^4 + 3\,x^2 - 9,$$
$$r_2(x) = 585\,x^2 + 1125\,x - 2205,$$
$$r_3(x) = -18885150\,x + 24907500,$$
$$r_4(x) = 527933700.$$

If the degrees of the polynomials processed at each iteration differ by at most one (i.e., $\Delta_i \le 1$), then the coefficients produced by the reduced PRS grow approximately linearly in the degree of the inputs. Unfortunately if $\Delta_i > 1$, the growth in the coefficients can be exponential. □

Subresultant Remainder Sequence. This clever scheme was discovered by G. E. Collins (1967) and later improved by W. S. Brown (1971). As with the reduced PRS, it requires no integer gcd calculation but here the growth rate of the coefficients is always guaranteed to be close to linear. Due to this efficiency, the subresultant PRS is often implemented in computer algebra systems.

The rule giving the α_i is the same as for the other all-integer methods, $\alpha_i = \text{lcoeff}(r_{i-1})^{\Delta_i+1}$. The sequence β_i satisfies

$$\beta_1 = (-1)^{\Delta_1+1}, \quad \beta_i = -\text{lcoeff}(r_{i-2}) \cdot \psi_i^{\Delta_i} \text{ for } 2 \le i \le m,$$

where

$$\psi_1 = -1, \quad \psi_i = (-\text{lcoeff}(r_{i-2}))^{\Delta_{i-1}} \cdot \psi_{i-1}^{1-\Delta_{i-1}}.$$

The remainders produced for our example are

$$r_1(x) = 15\,x^4 - 3\,x^2 + 9,$$
$$r_2(x) = 65\,x^2 + 125\,x - 245,$$
$$r_3(x) = 9326\,x - 12300,$$
$$r_4(x) = 260708.$$
□

Many other techniques for computing polynomials gcds have been proposed. These generally fall into one of two categories: heuristics and modular approaches. Working with polynomials modulo a prime p bounds the growth of the coefficients since they always lie in the range $[0, \ldots, p-1]$. The techniques for "lifting" the coefficients from $\mathbb{Z}_p$ to the integers $\mathbb{Z}$ are similar to those we study in Chapter 6 on factorization.

4.2.3 Extended Euclidean Algorithm

The extended Euclidean algorithm is an adaptation of Euclid's algorithm that expresses the gcd of two polynomials u and v as a linear combination of these polynomials,

$$\gcd(u(x), v(x)) = s(x)\, u(x) + t(x)\, v(x).$$

We observe that even when the gcd is computed and expressed over the integers $\mathbb{Z}$, the polynomials s and t have, in general, rational coefficients. (Why?) Consequently, the extended Euclidean algorithm for polynomials is usually carried out over the rationals $\mathbb{Q}$.

When performing the extended Euclidean algorithm, we need both the quotient and the remainder as we divide polynomials. In calculating the sequence of remainders r_i, we simultaneously compute sequences s_i and t_i satisfying

$$r_i(x) = s_i(x)\, u(x) + t_i(x)\, v(x).$$

The quotient q_i and remainder r_i at each step are given by the division

$$\alpha_i\, r_{i-2} = q_i(x)\, r_{i-1}(x) + \beta_i\, r_i(x),$$

where α_i and β_i depend on the remainder sequence used. The corresponding s_i and t_i are calculated as

$$s_i(x) = s_{i-2}(x) - s_{i-1}(x)\, q_i(x),$$
$$t_i(x) = t_{i-2}(x) - t_{i-1}(x)\, q_i(x).$$

Since we are working over $\mathbb{Q}$, it is customary to begin by removing units from u and v. This gives monic polynomials for r_{-1} and r_0. With this convention, the initial conditions for the remainder sequences are

$$r_{-1}(x) = u(x)/\text{lcoeff}(u), \quad s_{-1}(x) = 1/\text{lcoeff}(u), \quad t_{-1}(x) = 0,$$
$$r_0(x) = v(x)/\text{lcoeff}(v), \quad s_0(x) = 0, \quad t_0(x) = 1/\text{lcoeff}(v),$$

where $\deg(u) \geq \deg(v)$. We illustrate the entire procedure with an example.

Example 4.6. Apply the extended Euclidean algorithm to express the gcd of $u(x) = 54x^3 - 54x^2 + 84x - 48$ and $v(x) = -12x^3 - 28x^2 + 72x - 32$ as a linear combination of these polynomials.

Solution. The polynomials u and v are from Example 4.5. We begin by removing units from u and v, yielding polynomials for r_{-1} and r_0 whose leading coefficients are one.

The values of r_i, q_i, s_i, t_i obtained with the rational PRS are as follows:

i	$r_i(x)$	$q_i(x)$	$s_i(x)$	$t_i(x)$
-1	$x^3 - x^2 + \frac{14}{9}x - \frac{8}{9}$		$\frac{1}{54}$	0
0	$x^3 + \frac{7}{3}x^2 - 6x + \frac{8}{3}$		0	$-\frac{1}{12}$
1	$-\frac{10}{3}x^2 + \frac{68}{9}x - \frac{32}{9}$	1	$\frac{1}{54}$	$\frac{1}{12}$
2	$\frac{84}{25}x - \frac{56}{25}$	$-\frac{3}{10}x - \frac{69}{50}$	$\frac{1}{180}x + \frac{23}{900}$	$\frac{1}{40}x + \frac{19}{600}$
3	0	$-\frac{125}{126}x + \frac{100}{63}$		

So the gcd computed is $\frac{84}{25}x - \frac{56}{25}$. To give a unique monic representation for the gcd, we multiply both it and the corresponding coefficient polynomials s_2 and t_2 by $\frac{25}{84}$,

$$\gcd(u,v) = x - \frac{2}{3} = \underbrace{\left(\frac{5}{3024}x + \frac{23}{3024}\right)}_{s(x)} u(x) + \underbrace{\left(\frac{5}{672}x + \frac{19}{2016}\right)}_{t(x)} v(x).$$

Working instead with the monic PRS, we obtain the following:

i	$r_i(x)$	$q_i(x)$	$s_i(x)$	$t_i(x)$
-1	$x^3 - x^2 + \frac{14}{9}x - \frac{8}{9}$		$\frac{1}{54}$	0
0	$x^3 + \frac{7}{3}x^2 - 6x + \frac{8}{3}$		0	$-\frac{1}{12}$
1	$x^2 - \frac{34}{15}x + \frac{16}{15}$	1	$-\frac{1}{180}$	$-\frac{1}{40}$
2	$x - \frac{2}{3}$	$x + \frac{23}{5}$	$\frac{5}{3024}x + \frac{23}{3024}$	$\frac{5}{672}x + \frac{19}{2016}$
3	0	$x - \frac{8}{5}$		

The values of r_1, s_1, and t_1 are scaled by a factor of $-\frac{3}{10}$ from those appearing in the rational PRS. This factor is, of course, the reciprocal of the leading coefficient of the remainder before normalization, β_1. Likewise, the values of r_2, s_2, and t_2 are scaled by a factor of $\beta_2 = \frac{25}{84}$. The desired $s(x)$ and $t(x)$ appear in the same row as the greatest common divisor, $x - \frac{2}{3}$. □

4.3 Multivariate Polynomials

Throughout most of this book, we are concerned with mathematical expressions of a single variable. However in this section, we focus on polynomials of several variables. What are our reasons for doing so?

Computer algebra systems typically do not have a special data structure for polynomials in one variable. These are treated, instead, as a special case of multivariate polynomials. So to understand how computer algebra systems actually store and represent polynomials, we need to examine multivariate polynomials.

Moreover, when we study expressions containing radicals and functions such as exponentials and logarithms in the next chapter, we describe how computer algebra systems manipulate these constructs by treating each one essentially as a special symbol that acts like a new variable. This gives rise to expressions that appear structurally like multivariate polynomials and rational functions (ratios of polynomials), and that are manipulated in a similar way.

In subsequent chapters, we also encounter situations where, in the course of solving a problem involving only a single variable, we require the computation of greatest common divisors of polynomials in several variables. Consequently, it important to know how this is done.

4.3.1   Expanded and Recursive Forms

Two standard forms are commonly used to express polynomials in several variables. In the first, referred to as *expanded* or *distributed form*, the polynomial is written as a sum of products. This results in an expression with two structural levels. An example of a polynomial in expanded (distributed) form is

$$u(x, y, z) = 5\,x^2\,y^2 - 4\,x^2\,y\,z^2 + 2\,x^2\,z^4 - 3\,x + z^3 - 1.$$

The second way to express a multivariate polynomial is *recursive form*. Here, an ordering is placed on the variables. In our case, the ordering might be $x > y > z$. The polynomial is then written in terms the main variable x with coefficients that are polynomials in y and z. These are in turn expressed as polynomials in y with coefficients that are polynomials in z. The recursive representation for the polynomial u under this ordering is

$$u(x, y, z) = (5\,y^2 + (-4\,z^2)\,y + 2\,z^4)\,x^2 + (-3)\,x + (z^3 - 1).$$

We define the *degree of a multivariate polynomial* in one of its variables to be the highest power of that variable appearing in the expanded form. For example, the degree of our polynomial u in x is 2, the degree in y is 2, and the degree in z is 4. The maximum number of terms that can appear in the expanded form of u is $\prod_{i=1}^{r}(\text{degree}(u, x_i) + 1)$, where r is the number of variables. (Why?) Since the actual number of non-zero terms is generally much less, computer algebra systems invariably adopt a sparse representation for multivariate polynomials.

4.3.2   Data Structures for Polynomials

Many data structures can be devised to store multivariate polynomials. We consider three possibilities here. First, we describe a linked representation suitable for use with the recursive form. Second, we show how dynamic arrays can be employed in conjunction with the expanded form. Finally, we discuss the descriptor block representation used in the Altran system.

Linked list representation. In recursive form, we have at each level a polynomial in one variable with coefficients that are either polynomials in some of the other variables or constants. Correspondingly at each level in the linked representation, we have a header node of the form

var_link	first_link

The first field points to the principal variable at this level, and the second to a node representing the first term of the polynomial. The terms are generally stored in either decreasing or increasing order of exponent. Each term is represented by a node of the form

One field points to the header node for the coefficient polynomial, another to the node representing the next term. This field is *NULL* for the last term. Either the exponent itself, in the case of single precision integers, or a pointer to the exponent is stored in the remaining field.

A linked structure for the recursive form polynomial

$$u(x, y, z) = (5\,y^2 + (-4\,z^2)\,y + 2\,z^4)\,x^2 + (-3)\,x + (z^3 - 1)$$

is given in Figure 4.4. Linked structures are commonly used in Lisp-based computer algebra systems like REDUCE and Derive.

Dynamic array representation. Recall that in expanded or distributed form, a multivariate polynomial is regarded as a sum of products. Computer algebra systems employing dynamic arrays typically have a data structure for sums and an analogous structure for products. A typical representation for a sum is

type/length	coefficient$_1$	term$_1$	. . . .	coefficient$_n$	term$_n$

The first field is a descriptor indicating that the structure is a sum. The length can be given either as the number of coefficient-term pairs (n), the number of coefficient and term fields ($2n$), or the total number of fields including the descriptor itself ($2n + 1$). Similarly, a product can be represented as

type/length	exponent$_1$	factor$_1$	. . . .	exponent$_n$	factor$_n$

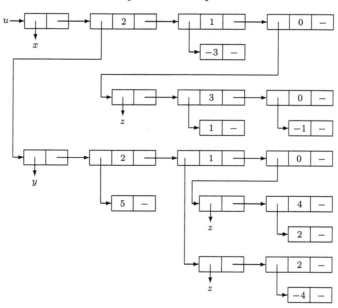

FIGURE 4.4: Linked list representation of a multivariate polynomial.

In the case of a multivariate polynomial in expanded form, the coefficient of each term is an integer or rational number. The term itself is considered a product of factors consisting of variables raised to integer powers. For example, the term $5x^2y^2$ has coefficient 5 and exponent-factor pairs $[2, x]$ and $[2, y]$. Pointers are typically used for the coefficients, rather than the values themselves, since these might be large integers or fractions. Similarly, the names of variables can be longer than the number of characters that can be packed into a computer word so, again, pointers are generally employed.

Dynamic arrays are used in most modern computer algebra systems including Maple and Mathematica. The dynamic array representation of

$$u(x, y, z) = 5\,x^2\,y^2 - 4\,x^2\,y\,z^2 + 2\,x^2\,z^4 - 3\,x + z^3 - 1 \qquad (4.6)$$

is illustrated in Figure 4.5.

Descriptor block representation. The Altran system was developed in the mid-1960s, when computers had extremely limited memory capacities. As a result, its data structures give an interesting historical perspective of how a high degree of computational functionality can be achieved with minimal storage.

The system allocated space dynamically for all data in a large array called the workspace. The basic unit of storage allocation, called a block, contained data of only one type (e.g., pointers, single precision integers, big integers). All blocks were of the same size.

Altran employed an expanded form representation for multivariate polynomials. The terms were ordered lexicographically according to their exponents.

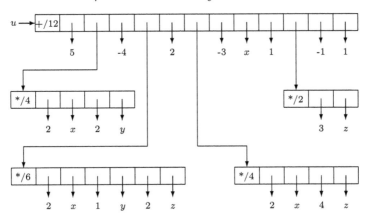

FIGURE 4.5: Dynamic array representation of a multivariate polynomial.

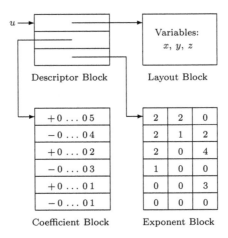

FIGURE 4.6: Descriptor block representation of a multivariate polynomial.

The corresponding data structure consisted of four components. The first was a header, called the descriptor block, that contained pointers to the other components. The layout block enforced an explicit ordering on the variables and indicated how to extract the exponents of a term from a single machine word. The coefficients for each non-zero term were stored in the coefficient block. Recall from Section 3.1 that Altran used fixed length arrays to store large integers. Again for each non-zero term, a tuple of exponents (one for each variable) was stored in the exponent block.

Figure 4.6 shows how the polynomial $u(x, y, z)$ in (4.6) would be represented in Altran.

4.3.3 Multivariate GCD Computation

When dividing two multivariate polynomials, we need to do so with respect to one of the variables. Consider, for example, the polynomials

$$u(x, y) = x\,y - 1, \qquad v(x, y) = x - y.$$

If we divide u by v, performing the division with respect to x, we obtain a quotient q_x and remainder r_x given by

$$q_x(u, v) = y, \qquad r_x(u, v) = y^2 - 1.$$

On the other hand, division with respect to y produces

$$q_y(u, v) = -x, \qquad r_y(u, v) = x^2 - 1.$$

These results are easily verified by expanding $q_x v + r_x$ and $q_y v + r_y$, and checking that both equal u.

With this observation, we describe how to extend the primitive Euclidean procedure given in Algorithm 4.4 to polynomials of two variables, say x and y. Since divisions must be performed with respect to one of the variables, the input polynomials are first placed in recursive form. Taking x as the principal variable, the input polynomials $u(x, y)$ and $v(x, y)$ are considered polynomials in x whose coefficients are polynomials in y. Consequently, $\text{cont}(u)$ and $\text{cont}(v)$ are, in general, polynomials in y that can be found with univariate gcd computation. These contents are then divided out of the coefficients of u and v leaving behind the primitive parts, $\text{pp}(u)$ and $\text{pp}(v)$.

The Euclidean algorithm is then iterated, performing the divisions with respect to the main variable x. If the primitive remainder sequence is used, the content of the remainder (a polynomial in y) is calculated at each step. This again involves only univariate gcd computation. The content is then divided out of the remainder, leaving the primitive part. This polynomial is the divisor at the next iteration.

The procedure is easily generalized in a recursive fashion to deal with more than two variables. With three variables, for instance, the contents of the input polynomials are polynomials in the two non-principal variables, whose gcd can be calculated using the method just described.

We present an example of the two variable case to illustrate the operation of the technique.

Example 4.7. Find the gcd of the multivariate polynomials

$$u(x, y) = 30\,x^3\,y - 60\,x^2\,y^2 + 15\,x^2 - 45\,x\,y + 30\,y^2,$$
$$v(x, y) = -75\,x^2\,y + 125\,x^2 + 135\,x\,y^2 - 225\,x\,y + 30\,y^3 - 50\,y^2$$

computing over the integers $\mathbb{Z}$.

Solution. The basic framework of Algorithm 4.4 can be used to perform the computation, provided we take note of several special considerations for the multivariate case. We begin by expressing u and v in recursive form. Choosing x as the main variable, we have

$$u(x, y) = (30\,y)\,x^3 - (60\,y^2 - 15)\,x^2 - (45\,y)\,x + (30\,y^2),$$
$$v(x, y) = -(75\,y - 125)\,x^2 + (135\,y^2 - 225\,y)\,x + (30\,y^3 - 50\,y^2).$$

Next, we split both u and v into their content and primitive part with respect to the main variable. Since we regard u and v as polynomials in x with coefficients that are polynomials in y, we must apply Algorithm 4.4 recursively to calculate the gcd of univariate polynomials in y to determine the contents with respect to x. For u we find

$$\mathrm{unit}(u(x, y)) = 1,$$
$$\mathrm{cont}(u, x) = \gcd(30\,y,\ 60\,y^2 - 15,\ 45\,y,\ 30\,y^2) = 15,$$
$$\mathrm{pp}(u, x) = (2\,y)\,x^3 - (4\,y^2 - 1)\,x^2 - (3\,y)\,x + (2\,y^2),$$

and for v we obtain

$$\mathrm{unit}(v(x, y)) = -1,$$
$$\mathrm{cont}(v, x) = \gcd(75\,y - 125,\ 135\,y^2 - 225\,y,\ 30\,y^3 - 50\,y^2) = 15\,y - 25,$$
$$\mathrm{pp}(v, x) = 5\,x^2 - (9\,y)\,x - (2\,y^2).$$

Now we iterate Euclid's procedure to determine the gcd of the primitive parts. The remainders are calculated with pseudo-division, regarding both the dividend and divisor as polynomials in the main variable x. The content of each remainder is removed to obtain the divisor at the next iteration. For the remainder $r(x, y) = (2y^3 - 30y)x - 4y^4 + 60y^2$ at the first iteration, we have $\mathrm{cont}(r, x) = 2y^3 - 30y$ and so $\mathrm{pp}(r, x) = x - 2y$.

$u(x, y)$	$v(x, y)$	$r(x, y)$
$2y\,x^3 - (4y^2 - 1)x^2 - 3y\,x + 2y^2$	$5x^2 - 9y\,x - 2y^2$	$(2y^3 - 30y)x - 4y^4 + 60y^2$
$5x^2 - 9y\,x - 2y^2$	$x - 2y$	0

Finally, the greatest common divisor of u and v is computed as

$$\gcd(u, v) = \gcd(\mathrm{cont}(u, x), \mathrm{cont}(v, x)) \cdot \gcd(\mathrm{pp}(u, x), \mathrm{pp}(v, x))$$
$$= 5\,(x - 2\,y) = 5\,x - 10\,y. \qquad \square$$

4.4   Exercises

1. How does your favorite computer algebra systems represent each of these numbers internally?

    (a) the integers 12345678 and -12345678

    (b) the rational number $12345653/12345701$

    (c) the value of 35!

    (d) the value of the rational number $2^{100}/3^{100}$

2. (a) Draw how a *user* views the structure of the following expressions in your favorite computer algebra system. Assume the *user* wants to traverse the "expression tree."

    (i) $x^2\,y^3/z^4$ (ii) $-\sin(x^2)$ (iii) $15\,x\,y - 21\,x - 15\,y^2 + 21\,y$

    (iv) the recursive form of the polynomial in part (iii)

    (b) Now draw the *internal representation* of these expressions.

3. Trace the operation of the sparse multiplication algorithm described in Section 4.1.4 when applied to find the product of the following polynomials.

    (a) $u(x) = x^8 + 16\,x^4 + 2, \quad v(x) = x^8 - 16\,x^4 + 2$

    (b) $u(x) = x^2 - \frac{1}{6}\,x^6 + \frac{1}{12}\,x^{12}, \quad v(x) = x^2 + \frac{1}{6}\,x^6 + \frac{1}{12}\,x^{12}$

    (c) $u(x) = 5\,x^5 + 4\,x^4 + 3\,x^3 + 2\,x^2 + x + 1, \quad v(x) = 5\,x^5 - 4\,x^4 + 3\,x^3$
    $- 2\,x^2 + x - 1$

4. Use pseudo-division to divide $u(x)$ by $v(x)$ for each of the pairs of polynomials given below. Compute the pseudo-quotient and pseudo-remainder, then scale to find the actual quotient and remainder.

    (a) $u(x) = 6\,x^3 - 5\,x^2 + 9\,x + 3, \quad v(x) = 3\,x^3 - x + 2$

    (b) $u(x) = x^4 - 7\,x^2 + 7, \quad v(x) = 3\,x^2 - 7$

    (c) $u(x) = x^3 - 6\,x^2 + 8\,x + 40, \quad v(x) = 4\,x^2 - 12\,x + 8$

    (d) $u(x) = x^6 + x^5 - x^4 + 2\,x^3 + 3\,x^2 - x + 2, \quad v(x) = 2\,x^3 + 2\,x^2 - x + 3$

    (e) $u(x) = \frac{1}{5}\,x^3 + \frac{1}{10}\,x^2 + \frac{1}{15}\,x + \frac{1}{20}, \quad v(x) = \frac{1}{3}\,x^2 - \frac{1}{6}\,x + \frac{1}{9}$

    (f) $u(x) = x^4 - \frac{1}{3}\,x^3 + \frac{1}{2}\,x^2 - x + 1, \quad v(x) = \frac{1}{2}\,x^2 - \frac{1}{3}\,x + \frac{1}{4}$

5. Apply the primitive Euclidean algorithm (Algorithm 4.4) to determine the gcd of the following pairs of polynomials over the integers.

    (a) $u(x) = x^8 - 1, \quad v(x) = x^5 - 1$

    (b) $u(x) = 4\,x^3 - 16\,x^2 + 4\,x + 24, \quad v(x) = 2\,x^3 - 2\,x^2 - 28\,x + 48$

    (c) $u(x) = 48\,x^3 - 84\,x^2 + 42\,x - 36, \quad v(x) = -4\,x^3 - 10\,x^2 + 44\,x - 30$

    (d) $u(x) = x^7 - 4\,x^5 - x^2 + 4, \quad v(x) = x^5 - 4\,x^3 - x^2 + 4$

    (e) $u(x) = 8\,x^8 - 2, \quad v(x) = 5\,x^5 - 2$

    (f) $u(x) = 3\,x^6 + x^5 - x^4 - 2\,x^3 + 3\,x^2 - x + 1, \quad v(x) = 2\,x^3 + 2\,x^2 - x + 3$

6. Apply the monic Euclidean algorithm (Algorithm 4.5) to determine the gcd of the pairs of polynomials in Exercise 5 over the integers.

7. Calculate the sequence of remainders when each of the methods described in Section 4.2.2 is applied to the polynomials

$$u(x) = 3\,x^6 + 5\,x^4 + 9\,x^2 + 4\,x + 8,$$
$$v(x) = 6\,x^3 + 4\,x^2 - 9\,x - 6,$$

whose gcd is one.

    (a) rational remainder sequence   (b) monic remainder sequence

    (c) Euclidean remainder sequence   (d) primitive remainder sequence

    (e) reduced remainder sequence   (f) subresultant remainder sequence

8. Repeat Exercise 7 for the following pairs of polynomials.

    (a) $u(x) = -5x^4 + x^2 - 3,\quad v(x) = 13\,x^2 + 25\,x - 49$

    (b) $u(x) = x^7 - x^5 - 8\,x^2 + 3,\quad v(x) = 6\,x^6 - 5\,x^4 - 16\,x$

9. Apply the extended Euclidean algorithm to find the greatest common divisor of the following polynomials and to express the gcd as a linear combination of the polynomials. Use the monic PRS and write the remainder at each step as a linear combination of the initial polynomials.

    (a) $u(x) = x^2 - 1,\quad v(x) = 2\,x^2 + 4\,x + 2$

    (b) $u(x) = x^8 - 1,\quad v(x) = x^5 - 1$

    (c) $u(x) = x^5 + 1,\quad x^3 - 1$

    (d) $u(x) = 4\,x^4 + 13\,x^3 + 15\,x^2 + 7\,x + 1,\quad v(x) = 2\,x^3 + x^2 - 4\,x - 3$

    (e) $u(x) = x^5 + 1,\quad v(x) = x^4 + 4\,x^3 + 6\,x^2 + 4\,x + 1$

    (f) $u(x) = 48\,x^3 - 84\,x^2 + 42\,x - 36,\quad v(x) = -4\,x^3 - 10\,x^2 + 44\,x - 30$

10. Develop a procedure to apply the extended Euclidean algorithm to a pair of univariate polynomials. Present your procedure as pseudocode using Algorithms 3.6 and 4.5 as a guide. (Programming Project 4 asks that you implement such a procedure in your favorite computer algebra system.)

11. Draw the (a) dynamic array, (b) linked list, and (c) descriptor block representations of the multivariate polynomial

$$2\,x^3 + 4\,x^2\,y + x\,y^5 + y^2 - 6.$$

12. Find the recursive form of the polynomial in Exercise 11, then draw its (a) dynamic array and (b) linked list representations.

13. Repeat Exercise 11 for the polynomial

$$x^3\,y^3 - 9\,x^3\,y\,z^2 - 4\,x\,y^3\,z^2 + 36\,x\,y\,z^4.$$

14. Find the recursive form of the polynomial in Exercise 13, then draw its (a) dynamic array and (b) linked list representations.

15. Compute the gcd of the following pairs of multivariate polynomials over the integers.

(a) $u(x, y) = x^2 + x + y + 1, \quad v(x, y) = x^2 y^2 + x$

(b) $u(x, y) = 4 x^2 - 4 y^2, \quad v(x, y) = 8 x^2 - 16 x y + 8 y^2$

(c) $u(x, y) = 4 x^3 y - 4 x y^3 + 4 x y, \quad v(x, y) = 6 x^4 y + 12 x^3 y^2 + 6 x^2 y^3$

(d) $u(x, y) = 6 x^2 - 3 x y - 3 y^2, \quad v(x, y) = 15 x y - 21 x - 15 y^2 + 21 y$

(e) $u(x, y, z) = x^4 - x^2 y^2 - x^2 z^2 + y^2 z^2, \quad v(x, y, z) = x^3 - x^2 z - x y^2 + y^2 z$

Programming Projects

1. **(Polynomial Arithmetic)**   Write a set of procedures that implements the standard arithmetic operations on polynomials of a single variable with rational coefficients:

- `AddPolynomial(<poly1>,<poly2>)` returns the sum,
- `SubPolynomial(<poly1>,<poly2>)` returns the difference,
- `MultPolynomial(<poly1>,<poly2>)` returns the product,
- `DivPolynomial(<poly1>,<poly2>)` returns a list with the quotient and remainder, `[<quotient>,<remainder>]`,
- `GCDPolynomial(<poly1>,<poly2>)`, where deg(*poly1*) $\geq$ deg(*poly2*), returns the greatest common divisor.

Begin by writing a utility procedure called `CoeffList` that coverts a polynomial represented in your computer algebra system into a list of its coefficients. For example, `CoeffList(1/6*x^3 - 1/2*x^2 + x - 1)` should return the list `[1/6,-1/2,1,-1]`.

Your arithmetic procedures should operate on, and return, lists in `CoeffList` format. Any leading zero coefficients should be suppressed. You can use your computer algebra system's facilities for arithmetic on rational numbers, but not for polynomial manipulation.

Following the convention typically used by computer algebra systems, the procedure `GCDPolynomial` should return a polynomial with integer coefficients if both its arguments have integer coefficients, otherwise return a monic polynomial with rational coefficients.

2. **(Sparse Polynomial Multiplication)**   Write a procedure, `MultSparsePoly`, that forms the product of two sparse univariate polynomials with rational coefficients. Terms with zero coefficients will not appear in either the input or output lists.

You will need a utility routine, `SparseCoeffList`, that converts a polynomial represented in your computer algebra system into its sparse representation. For example, `SparseCoeffList(x^2 - 1/6*x^6 + 1/12*x^12)` should return the list `[[12,1/12],[6,-1/6],[2,1]]`.

As in Programming Project 1, you can use the facilities of your computer algebra system for arithmetic on rational numbers but not those for manipulating polynomials.

3. **(GCDs and the Primitive Remainder Sequence)** Write a procedure to compute the gcd of two univariate polynomials with integer coefficients using the primitive polynomial remainder sequence (primitive PRS).

$$\texttt{GCDPrimitive(<poly1>,<poly2>)},$$

where $\deg(poly1) \geq \deg(poly2)$, returns $\gcd(poly1, poly2)$.

You will need to write utility routines `ContPoly` and `PrimPartPoly` that find the content (i.e., gcd of the coefficients) and primitive (i.e., content removed) part of a polynomial. You will also need to write a procedure to pseudo-divide two polynomials with integer coefficients,

`PseudoDiv(<poly1>,<poly2>)` returns a list `[<psquo>,<psrem>,<scale>]`.

The scale factor, `<scale>`, is the integer that when divided into the pseudo-quotient, `<psquo>`, and pseudo-remainder, `<psrem>`, produces the actual quotient and remainder polynomials.

All polynomials should be in the `CoeffList` format described in Programming Project 1. You should rely on your computer algebra system only for arithmetic operations ($+$, $-$, $*$, quotient, remainder, gcd) on integers.

4. **(Extended Euclidean Algorithm)** Write a procedure that implements the extended Euclidean algorithm for polynomials with rational coefficients.

$$\texttt{EEAPoly(<poly1>,<poly2>)},$$

where $\deg(poly1) \geq \deg(poly2)$, returns a list `[<gcdpoly>,<spoly>,<tpoly>]`, for which

`gcd(poly1,poly2) = gcdpoly = spoly*poly1 + tpoly*poly2.`

Use the monic PRS and follow the convention that a monic polynomial should be returned for the gcd of two polynomials with rational coefficients. To aid in debugging, you may wish to show the values of `r`, `q`, `s`, `t` at each iteration of your procedure.

Chapter 5

Algebraic Simplification

> *Everything should be made as simple as possible, but not simpler.*
>
> Albert Einstein (1879–1955)
> Quoted in *Reader's Digest*, July 1977
>
> *Just because something doesn't do what you planned it to do doesn't mean it's useless.*
>
> Thomas Alva Edison (1847–1931)

Joel Moses (1971) begins his classic essay on algebraic simplification by stating,

> Simplification is the most pervasive process in algebraic manipulation. It is also the most controversial. Much of the controversy is due to the difference between the desires of a user and those of a system designer.

The difficulty arises from the fact that mathematical expressions can generally be written in many different forms. For example,

$$\left(\frac{\cos x}{\sin x}\right)^2, \quad \frac{\cos^2 x}{\sin^2 x}, \quad \frac{\cos^2 x}{1 - \cos^2 x}, \quad \cot^2 x, \quad \frac{1}{\tan^2 x}, \quad \csc^2 x - 1$$

are just a few ways to represent the same trigonometric function.

An issue that designers of computer algebra systems face is how to represent and manipulate mathematical quantities. They prefer representations that are easy to implement and have efficient algorithms. In the case of trigonometric functions, for example, designers might prefer to use only sines and cosines. The corresponding issue that users face is how to coax a system to report results in the form they desire or can best comprehend. Of course, not all users are likely to have the same preferences. As a result, computer algebra systems need to provide systematic methods to transform one representation of an expression into another mathematically equivalent one. How to do this is the principal issue we investigate in this chapter.

A second, very important theme also emerges. In the previous chapter, we focused exclusively on polynomials and rational functions. As we progress through this chapter, we consider much broader classes of mathematical expressions. In particular, we discuss how computer algebra systems represent and manipulate algebraic quantities, like $\sqrt{2}$ and $\sqrt[3]{x^2 - 1}$, and transcendental quantities including logarithms, exponentials, and trigonometric functions.

5.1   Issues in Simplification

Computer algebra systems perform certain types of transformations automatically whenever an expression is processed by the algebra engine or kernel. These include performing arithmetic on integers and rational numbers, collecting like terms with rational coefficients in sums (e.g., $x + 2x \rightarrow 3x$, $x + \frac{1}{2}x \rightarrow \frac{3}{2}x$), and combining like factors raised to rational powers in products (e.g., $x^2\, x \rightarrow x^3$, $x\, x^{1/2} \rightarrow x^{3/2}$). The process of applying such basic transformations is referred to as *evaluation* or *automatic simplification*.

Unless such transformations are made automatically, the results produced by computer algebra systems quickly become unwieldy and an undue burden is placed on the user to request them. For example, consider the steps involved in computing a simple derivative,

$$\frac{d}{dx}\left(\frac{x+1}{x} + \ln x\right) = \frac{(x+1)'}{x} - \frac{(x+1)\cdot x'}{x^2} + \frac{x'}{x}$$
$$= \frac{1+0}{x} - \frac{(x+1)\cdot 1}{x^2} + \frac{1}{x}$$
$$= \frac{1}{x} - \frac{x+1}{x^2} + \frac{1}{x}$$
$$= \frac{2}{x} - \frac{x+1}{x^2}.$$

Most users would be quite unhappy to see any but the final line in the derivation reported as the result.

Another issue in the design and implementation of a computer algebra system is the use of canonical forms. Consider computing the same derivative, only this time applying the quotient rule instead of the product rule to the first term,

$$\frac{d}{dx}\left(\frac{x+1}{x} + \ln x\right) = \frac{(x+1)'\cdot x - (x+1)\cdot x'}{x^2} + \frac{x'}{x}$$
$$= \frac{(1+0)\cdot x - (x+1)\cdot 1}{x^2} + \frac{1}{x}$$
$$= \frac{x - x - 1}{x^2} + \frac{1}{x}$$
$$= -\frac{1}{x^2} + \frac{1}{x}.$$

Although this expression is mathematically equivalent to the one given above, the form of the result is different. A *canonical* simplification always produces the same unique form for mathematically equivalent expressions.

This brings us to an important point. The term "simplification" is a bit of a misnomer. What is "simplest" depends on context. Applying a particular "simplification" may actually produce a more complicated expression. Perhaps a better term for the topic of this chapter is "algebraic transformation."

For example, the expression $x + 1$ is clearly simpler and easier to comprehend than $(x^2 - 1)/(x - 1)$,[†] but $(x^{100} - 1)/(x - 1)$ is simpler and far more compact than $x^{99} + x^{98} + x^{97} + \ldots + x + 1$. Similarly, $x^{100} - 1$, which is in expanded form, is simpler than its factored form,

$$(x - 1)(x + 1) \left(x^2 + 1\right) \left(x^4 - x^3 + x^2 - x + 1\right) \left(x^4 + x^3 + x^2 + x + 1\right)$$
$$\left(x^8 - x^6 + x^4 - x^2 + 1\right) \left(x^{20} - x^{15} + x^{10} - x^5 + 1\right)$$
$$\left(x^{20} + x^{15} + x^{10} + x^5 + 1\right) \left(x^{40} - x^{30} + x^{20} - x^{10} + 1\right).$$

Conversely, the expression $(x + 1)^{100} - 1$ is simpler than the 100 term sum, having coefficients with as many as 30 digits, that results from expanding the power. Even the expression $\frac{1}{4} \left(4 x^3\right) x^4 / ((x^4)^3 + 1)$ might be considered simpler than $x^7/(x^{12} + 1)$ in the context of integration since it suggests the substitution $u = x^4$, which results in the transformation

$$\int \frac{x^7}{x^{12} + 1}\, dx = \int \frac{\frac{1}{4} u}{u^3 + 1}\, du.$$

One transformation we would always like to make is to recognize zero. This is called the *zero-equivalence* problem, and it is important for two reasons. First, users obviously want results expressed in the simplest possible form. Second, the expressions generated by computer algebra systems during the course of a computation can be extremely large, a phenomenon known as *intermediate expression swell*. Vast amounts of computer resources—time and space—can be wasted manipulating and storing unsimplified expressions.

Let's consider three rather small instances of the zero-equivalence problem and examine what is required to solve each one. No special knowledge is necessary to transform the multivariate polynomial

$$12 x^2 y + 20 x - 3 x y - 5 - (4 x - 1)(3 x y + 5)$$

to zero since we can just expand the product and collect terms. A slightly more difficult example is the algebraic expression

$$\left(\sqrt[3]{2} + \sqrt[3]{4}\right)^3 - 6 \left(\sqrt[3]{2} + \sqrt[3]{4}\right) - 6.$$

Here it does not suffice to expand products and combine powers. It is also necessary to recognize and apply the relationship between $\sqrt[3]{2}$ and $\sqrt[3]{4}$ to eliminate one of the radicals. A final example is

$$\ln \left(\tan \left(\frac{x}{2} + \frac{\pi}{4}\right)\right) - \operatorname{arcsinh}(\tan(x)).$$

[†]One might disagree about the equivalence of these expressions since $x + 1$ does not reflect the singularity in $(x^2 - 1)/(x - 1)$ at $x = 1$. From an algebraic perspective, however, the two expressions are equal since their difference is $0/(x - 1) = 0$.

This expression is zero over the domain $-\pi/2 < x < \pi/2$, but how do we even begin to show this? Unfortunately, the problem of recognizing zero is, in general, an unsolvable one. We explore this issue and its implications in the next section.

Moses (1971) classifies simplifiers by the extent to which they insist on transforming expressions to a certain form. Using his terminology, most of today's systems take a *catholic*[‡] approach in that they provide an extensive set of simplification commands. This results in a larger system with more commands for a user to learn. On the other hand, users can almost always transform a result to the form they desire. An advantage from the system designer's perspective is that efficient algorithms can be implemented to perform each specific transformation.

A *radical* simplifier insists on making a particular type of transformation. Derive, for example, always removes the greatest common divisor from the numerator and denominator of a rational expression. The transformation

$$\frac{x^3 - 1}{x^2 - 1} \quad \to \quad \frac{x^2 + x + 1}{x + 1}$$

is performed automatically whenever the expression on the left is encountered. One advantage of the radical approach is that canonical forms can often be handled very efficiently since the internal representations of both the input and output are known in advance. Another is that fewer simplification algorithms are generally required. The disadvantage, of course, is that users have little or no control over the form of the output.

At the opposite extreme are *conservative* simplifiers. These provide very little automatic machinery, and instead give users a facility to implement their own transformations. Two of the earliest computer algebra systems, Fenichel's Famous and Formula Algol, took this approach.

Modern systems often provide a pattern matching mechanism for users to construct their own simplifiers, however these are seldom required in practice because of the rich set of built-in simplification primitives. In Maple, for instance, the command

```
match(3*x^2 - 2*x + 1 = A*x^2 + B*x + C, x, 'p')
```

matches the quadratic form on the left of the = sign and assigns the set $\{A = 3, B = -2, C = 1\}$ to the variable p. In a similar vein, the following Mathematica transformation rules implement the linearity relations for expanding integrals.

form	transformation rule
$\int (y + z)\, dx$	`int[y_ + z_, x_] := int[y, x] + int[z, x]`
$\int c\, y\, dx$	`int[c_ y_, x_] := c int[y, x] /; FreeQ[c, x]`

A technical issue, called the *substitution* or *replacement problem*, arises

---

[‡]The term catholic (with a small "c") means all-encompassing or wide-reaching.

when applying transformations to mathematical expressions. By way of example, consider the problem of substituting a for $x\,y^2$ in the expression $x^2\,y^3$. What do we get? Three answers are possible. The first is to return $x^2\,y^3$ unaltered, as Maple's `subs` command does, since the subexpression $x\,y^2$ does not appear explicitly. A second solution to make an algebraic substitution for the quantity $x\,y^2$ and return $a\,x\,y$, which is what Maple's built-in `algsubs` function produces. However, since $a^2 = x^2\,y^4$, an equally valid algebraic substitution results in a third possibility, a^2/y.

5.2 Normal and Canonical Forms

Earlier we used the term "canonical form" to denote the idea that mathematically equivalent quantities are represented the same way. We also introduced the zero-equivalence problem and indicated that it was unsolvable for most types of mathematical expressions. These two concepts are intimately related. In this section, we examine the notion of canonical simplification more formally. We also indicate some classes of expression for which the zero-equivalence problem is solvable and some for which instances of the problem are known to be unsolvable.

Let $\mathcal{E}$ denote a class of mathematical expressions. Examples include multivariate polynomials with rational coefficients and elementary expressions of a single variable. The latter correspond roughly to the functions studied in introductory calculus. We use the symbol $\sim$ to denote that two expressions are mathematically equivalent although the forms used to express them may be different.

A *normal* (zero-equivalence) *form* is a transformation $f : \mathcal{E} \to \mathcal{E}$ satisfying the following properties for all expressions $e \in \mathcal{E}$:

1. $f(e) \sim e$,
2. if $e \sim 0$, then $f(e) \equiv 0$.

A *canonical form* is a mapping $f : \mathcal{E} \to \mathcal{E}$ that is normal and satisfies the additional property:

3. if $e_1 \sim e_2$, then $f(e_1) \equiv f(e_2)$.

Intuitively, a normal form for an expression is a mathematically equivalent expression with the property that any expression equal to zero is transformed to zero. The existence of a normal form algorithm implies that we can test whether two expressions are equal by determining whether their difference is zero. A canonical form algorithm always transforms two mathematically equivalent expressions to the *same* expression. While the notions of normal and canonical forms are indeed different, the following surprising theorem shows that any class of expressions $\mathcal{E}$ with a normal form also has a canonical form!

Theorem 5.1 (Normal/Canonical Form Theorem). *A class of expressions $\mathcal{E}$ possesses a normal form if and only if $\mathcal{E}$ has a canonical simplification algorithm.*

Proof. It is obvious from the definitions that the existence of a canonical form implies a normal form. We show that if $\mathcal{E}$ has a normal form, then it also has a canonical form.

$\mathcal{E}$ consists of finite length expressions formed using some finite number of operations. Therefore, we can impose a lexicographical (alphanumerical) ordering over all the expressions in $\mathcal{E}$. As a result, an algorithm can be constructed to generate all the elements of $\mathcal{E}$ sequentially according to this lexicographical ordering.

To obtain the canonical form $f(e)$ for any expression e, we can generate the elements of $\mathcal{E}$ in order and apply the zero-equivalence (normal form) algorithm to the difference between each generated element and e. We let the canonical form be the first element of $\mathcal{E}$ in the lexicographical ordering that is equivalent to e. $\qquad\square$

Like many similar existence proofs in the theory of computation, this one may not be particularly satisfying. The canonical form constructed need not be "simpler" than the original expression in any mathematical sense. Moreover, the process of generating and testing the expressions in $\mathcal{E}$ one after another is extraordinarily inefficient and, for all but the smallest expressions, computationally intractable. Nonetheless, the proof is valid and we have the surprising result that the existence of a normal simplifier implies the existence of a canonical simplifier for the same class of expressions.

The next theorem shows that once we get only a little beyond the rational functions, the zero-equivalence problem is not, in general, solvable. Consequently, the class of expressions having normal (canonical) simplification algorithms is very small!

Theorem 5.2 (Richardson's Theorem). *Let $\mathcal{E}$ be the class of expressions in a single variable x generated by:*
  1. *the rational numbers and the two real numbers π and $\ln 2$,*
  2. *the operations addition/subtraction, multiplication, composition, and*
  3. *the sine, exponential, and absolute value functions.*
Then if e is an expression in $\mathcal{E}$, the problem of determining whether there exists a value for x such that $e(x) = 0$ is undecidable.

The theorem is due to Daniel Richardson (1966). Its proof, which is well beyond our scope, is based on the unsolvability of Hilbert's tenth problem, known as the Diophantine problem. This problem is concerned with whether an algorithm exists to determine if polynomials in several variables with integer coefficients have solutions that can be expressed as integers. An example of a second order Diophantine equation is

$$42\,x^2 + 8\,x\,y + 15\,y^2 + 23\,x + 17\,y - 4915 = 0,$$

whose only solution is $x = -11$, $y = -1$.

In view of Richardson's Theorem we might well ask, "What useful classes of expressions have canonical forms?" We next present three such classes.

Univariate Polynomials

The most familiar canonical representation for polynomials in one variable is *expanded form*. For polynomials with integer coefficients, this representation is computed as follows:

    1. Multiply all products of polynomials.
    2. Collect terms of the same degree.
    3. Order the terms by degree.

Omitting the last step results in a normal, rather than canonical, form. We illustrate the procedure with a simple example.

$$
\begin{aligned}
p(x) &= 2\,(x^2 + x - 1)(x^3 + 1) - (x^2 - x + 1)(x^3 - 1) \\
&= (2\,x^5 + 2\,x^2 + 2\,x^4 + 2\,x - 2\,x^3 - 2) - (x^5 - x^2 - x^4 + x + x^3 - 1) \\
&= x^5 + 3\,x^2 + 3\,x^4 + x - 3\,x^3 - 1 \\
&= x^5 + 3\,x^4 - 3\,x^3 + 3\,x^2 + x - 1
\end{aligned}
$$

A second canonical representation for univariate polynomials is *factored form*. This is more expensive to compute than the expanded form. To obtain a canonical, rather than a normal form, a rule is needed to order the factors. One method is to arrange them by degree, say from lowest to highest. Factors of equal degree can be ordered by the values of their coefficients working, say, from highest to lowest degree terms. With this ordering, the factored form of the polynomial p given above is

$$
p(x) = (x^2 - x + 1)(x^3 + 4\,x^2 - 1)
$$

The problem of factoring univariate polynomials is the subject of Chapter 6.

Yet another canonical form, though one not particularly useful in computer algebra, is *Horner's rule*. This representation produces a formula that uses both the fewest number of additions and the fewest number of multiplications to evaluate a general polynomial. For p, it produces

$$
p(x) = x\,(x\,(x\,(x\,(x + 3) - 3) + 3) + 1) - 1.
$$

Multivariate Polynomials

Two canonical representations, the expanded and recursive forms, are very useful in computer algebra. To find the *expanded form*, multiply out all products and collect terms of the same degree, as in the univariate case. This gives a normal form. To obtain a canonical representation, assume an ordering on the variables and arrange terms by degree using this order. For example, with the ordering $x > y > z$, the polynomial

$$
q(x, y, z) = 2\,x^2\,y^2 - 3\,x^2\,y\,z^2 + 6\,x^2\,z^4 + 5\,x - 4\,y^3\,z + y^3 - 4\,y^2 - z^3 + 1
$$

is in expanded form. This representation is also referred to as *distributed form*.

An alternative canonical representation is *recursive form*. Again the variables must be ordered. A multivariate polynomial in this form is written as a polynomial in the first variable with coefficients that are polynomials in the other variables. These coefficient polynomials are then, in turn, represented in recursive form. The recursive form for q, again assuming the ordering $x > y > z$, is

$$q(x, y, z) = (2\,y^2 - 3\,z^2\,y + 6\,z^4)\,x^2 + 5\,x + (-4\,z + 1)\,y^3 - 4\,y^2 - z^3 + 1.$$

It is historically interesting to note that some of the early computer algebra systems took diametrically opposite approaches to the issue of representing multivariate polynomials. Brown's Altran system used an expanded representation, while Collins' PM and SAC-I systems stored polynomials recursively.

A polynomial of degree d in v variables has $(d+1)^v$ coefficients (zero and non-zero) when fully expanded. Consequently, in order to conserve memory, a sparse representation is invariably used to store multivariate polynomials in computer algebra systems. Without this approach, computations would quickly become unwieldy and the results incomprehensible.

As with univariate polynomials, polynomials in several variables can also be expressed canonically in factored form. While users can request this transformation, computer algebra systems never employ it as the standard internal representation since it can be costly to find.

Rational Functions

For simplicity, we direct our attention to the case where both the numerator and denominator are polynomials in only one variable. We first consider the situation where the coefficients are integers. One common simplification is to transform the ratio of the polynomials to *expanded canonical form* with the following procedure:

1. Expand the polynomials in the numerator and denominator.
2. Remove the gcd of these polynomials.
3. Order the terms of the numerator and denominator polynomials by degree.
4. Make the leading coefficients of both polynomials positive.

Here's a simple example,

$$
\begin{aligned}
r(x) &= \frac{6\,x^2 + 2\,x - 4}{-4\,x^2 + 2\,x + 6} \\
&= \frac{2\,(x+1)\,(3\,x - 2)}{2\,(x+1)\,(-2\,x + 3)} \\
&= \frac{3\,x - 2}{-2\,x + 3} \\
&= -\frac{3\,x - 2}{2\,x - 3}.
\end{aligned}
$$

When the coefficients are rational numbers, one of two strategies is used to guarantee a unique representation for expressions like the one on the left in

$$-\frac{-\frac{1}{3}x + \frac{3}{2}}{\frac{2}{5}x - 1} = -\frac{10\,x - 45}{12\,x - 30} = -\frac{\frac{5}{6}x - \frac{15}{4}}{x - \frac{5}{2}}. \tag{5.1}$$

The first is to use only integer coefficients, giving the middle representation in (5.1). This form is obtained by finding the least common multiple of the denominators of the coefficients. In our example, $\mathrm{lcm}(2, 3, 5) = 30$ so we multiply both the numerator and denominator polynomials by 30. The second approach is to force the leading coefficient of the denominator to be one and use rational coefficients, producing the expression on the right in (5.1).

Another canonical form for rational functions is the *partial fraction decomposition*. The steps involved to ensure uniqueness are:

1. Expand the polynomials in the numerator and denominator.
2. Remove the gcd of these polynomials.
3. Use long division to split the rational function into a polynomial part plus a proper rational function.
4. Order the terms of the polynomial part by degree.
5. Do a square-free partial fraction decomposition of the rational part. (See Sections 6.4.2 and 7.4.)
6. For each term in the decomposition, order the numerator and denominator polynomials by degree, then make the leading coefficients of both positive.

Again, we give a simple example,

$$\frac{x^4}{x^3 - x^2 - x + 1} = x + 1 + \frac{2\,x^2 - 1}{x^3 - x^2 - x + 1}$$

$$= x + 1 + \frac{2\,x^2 - 1}{(x - 1)^2\,(x + 1)}$$

$$= x + 1 + \frac{1}{4}\frac{7\,x - 5}{(x - 1)^2} + \frac{1}{4}\frac{1}{x + 1}.$$

5.3 Simplification of Radicals

5.3.1 Algebraic Numbers and Functions

We now turn our attention to algebraic numbers and functions. The term "algebraic" is used here to denote the solutions to a polynomial equation with integer coefficients. For example, $\sqrt{2}$ is an *algebraic number* since it is a solution (one of two) to $\alpha^2 - 2 = 0$. Likewise, $\sqrt[3]{x^2 - 1}$ is an *algebraic function* since it is a solution to $\beta^3 - x^2 + 1 = 0$. In what follows, we consider algebraic numbers but the same techniques can generally be used to manipulate and simplify algebraic functions.

Note that the algebraic number $-\sqrt{2}$ is defined by the same polynomial equation as $\sqrt{2}$, $\alpha^2 - 2 = 0$. In general, the equation defining the radical $\sqrt[n]{k}$, $\alpha^n - k = 0$, has n solutions. For positive integers k and $n > 1$, the radical $\sqrt[n]{k}$ is usually intended to refer the unique positive real solution to the equation. The term *surd* is sometimes used to denote this quantity.

The familiar formula

$$\frac{-b \pm \sqrt{b^2 - 4\,a\,c}}{2\,a}$$

gives the solutions to a general polynomial of degree-2 (quadratic), $ax^2 + bx + c$, in terms of radicals. More complicated formulas exist to express the solutions to polynomials of degree-3 (cubics) and polynomials of degree-4 (quartics) in terms of radicals. In 1826, Niels Henrik Abel demonstrated the surprising result that the solutions to polynomials of degree-5 (quintics) or more cannot, in general, be expressed in terms of radicals. When we think of algebraic numbers, we think of radicals. Clearly, every radical is an algebraic number but Abel's result shows that the reverse is not true.

Theorem 5.3 (Abel's Theorem). *The solutions to polynomials of degree five or more cannot always be expressed in terms of radicals.*

The proof uses abstract algebra (group theory, in particular) and is beyond our scope. The interested reader is referred to the lively, well-written book by Maxfield and Maxfield (1971) for an introduction to abstract algebra that culminates with Galois theory and a proof of Abel's Theorem.

Can we give an example of a polynomial whose roots cannot be expressed in terms of radicals? The following result due to Y. Bhalotra and S. Chowla (1942) provides a large source of examples.

Theorem 5.4. *Let $x^5 + a\,x^3 + b\,x^2 + c\,x + d$ be a polynomial that is irreducible (does not factor) over the integers. If a and b are even and c and d are odd, then the quintic equation*

$$x^5 + a\,x^3 + b\,x^2 + c\,x + d = 0$$

is not solvable in terms of radicals.

In particular, an irreducible quintic trinomial of the form $x^5 + cx + d$ is not solvable by radicals when c and d are both odd. An example is $x^5 - x + 1$.

For the remainder of our discussion, it is useful to distinguish a hierarchy of algebraic expressions:

1. simple radicals: e.g., $\sqrt{2}$ and $\sqrt[3]{x^2 - 1}$,

2. nested radicals: e.g., $\sqrt{1 + \sqrt{2}}$ and $\sqrt[3]{\sqrt{5} + \sqrt{x}}$,

3. general algebraic expressions: e.g., the algebraic number α defined by $\alpha^5 - \alpha + 1 = 0$.

Each class in this hierarchy is properly contained in the next.

Representation of Algebraic Numbers

It is far more difficult to manipulate algebraic numbers than ordinary numbers. If we hope to obtain a canonical form for expressions containing algebraic numbers, we must avoid the use of interdependent quantities like $\sqrt{2}$ and $\sqrt{8}$ since $\sqrt{8} = 2\sqrt{2}$. Similarly, we can use only two of $\sqrt{2}$, $\sqrt{3}$, and $\sqrt{6}$ since $\sqrt{6} = \sqrt{2}\sqrt{3}$. The algebraic numbers $\alpha_1, \alpha_2, \ldots, \alpha_k$ are said to be *algebraically dependent* if there exists a polynomial h with integer coefficients such that $h(\alpha_1, \alpha_2, \ldots, \alpha_k) = 0$.

Example 5.1. Show that the radicals in each of the following sets are algebraically dependent.

(a) $\{\sqrt{2}, \sqrt{8}\}$. Setting $\alpha_1 = \sqrt{2}$ and $\alpha_2 = \sqrt{8}$, we see that

$$h(\alpha_1, \alpha_2) = 2\,\alpha_1 - \alpha_2 = 0.$$

(b) $\{\sqrt{2}, \sqrt{3}, \sqrt{6}\}$. With $\alpha_1 = \sqrt{2}$, $\alpha_2 = \sqrt{3}$ and $\alpha_3 = \sqrt{6}$, we have

$$h(\alpha_1, \alpha_2, \alpha_3) = \alpha_1\,\alpha_2 - \alpha_3 = 0.$$

(c) $\{\sqrt{2}, \sqrt{3}, \sqrt{14 - 4\sqrt{6}}\}$. The polynomial equation expressing the dependency between $\alpha_1 = \sqrt{2}$, $\alpha_2 = \sqrt{3}$, and $\alpha_3 = \sqrt{14 - 4\sqrt{6}}$ is

$$h(\alpha_1, \alpha_2, \alpha_3) = 2\,\alpha_2 - \alpha_1 - \alpha_3 = 0.$$

This relation can be verified by squaring both sides of the equation

$$2\sqrt{3} - \sqrt{2} = \sqrt{14 - 4\sqrt{6}}. \qquad \square$$

Not only must we avoid the use of dependent algebraic quantities, we also cannot use radicals like $\sqrt{1}$, $\sqrt{4}$, $\sqrt{9}$, etc. that simplify to integers. Observe that the defining polynomial of each of these radicals factors. For instance, the defining polynomial for $\sqrt{9}$ is $\alpha^2 - 9 = (\alpha - 3)(\alpha + 3)$. In general, we must avoid any radical whose defining polynomial factors over the integers. For example, $\sqrt[4]{4}$ should not be used as a substitute for $\sqrt{2}$ since $\alpha^4 - 4 = (\alpha^2 - 2)(\alpha^2 + 2)$. More subtly, we should avoid $\sqrt[4]{-4} = (-1)^{3/4}\sqrt{2}$ since $\alpha^4 + 4$ factors as $(\alpha^2 - 2\alpha + 2)(\alpha^2 + 2\alpha + 2)$.

The defining polynomial for an algebraic number is usually referred to as its *minimal polynomial*. It is easy to find the minimal polynomial for a simple radical like $\sqrt{2}$, but expressions like $\sqrt{2} + \sqrt{3}$ are algebraic numbers themselves. In the next example, we explore this idea.

Example 5.2. Find the minimal polynomial for $\alpha = \sqrt{2} + \sqrt{3}$.

Solution. Rearranging the formula for α and squaring to eliminate $\sqrt{3}$, we have

$$\left(\alpha - \sqrt{2}\right)^2 = \left(\sqrt{3}\right)^2$$
$$\alpha^2 - 2\sqrt{2}\,\alpha + 2 = 3.$$

Rearranging and squaring again to eliminate $\sqrt{2}$ gives

$$\left(\alpha^2 - 1\right)^2 = \left(2\sqrt{2}\,\alpha\right)^2$$
$$\alpha^4 - 2\alpha^2 + 1 = 8\alpha^2$$
$$\alpha^4 - 10\alpha^2 + 1 = 0.$$

So the minimal polynomial is $\alpha^4 - 10\alpha^2 + 1$. This polynomial has four solutions,

$$\alpha = \pm\sqrt{2} \pm \sqrt{3},$$

that correspond to the four choices of sign for the radicals. It is the minimal polynomial for all four of these algebraic numbers. $\qquad\square$

With this background, we are ready to tackle one of the zero-equivalence problems presented earlier.

Example 5.3. Simplify $(\sqrt[3]{2} + \sqrt[3]{4})^3 - 6\,(\sqrt[3]{2} + \sqrt[3]{4}) - 6$.

Solution. Letting $\alpha = \sqrt[3]{2}$ and $\beta = \sqrt[3]{4}$, the expression to be simplified is

$$(\alpha + \beta)^3 - 6\,(\alpha + \beta) - 6.$$

Expanding products, we get

$$\alpha^3 + 3\alpha^2\beta + 3\alpha\beta^2 + \beta^3 - 6\alpha - 6\beta - 6,$$

which does not factor over the integers. Substituting back $\alpha = \sqrt[3]{2}$ and $\beta = \sqrt[3]{4}$ gives

$$3 \cdot 2^{2/3} \cdot 4^{1/3} + 3 \cdot 2^{1/3} \cdot 4^{2/3} - 6 \cdot 2^{1/3} - 6 \cdot 4^{1/3}$$

since $\alpha^3 + \beta^3$ cancels with -6 when the expression is evaluated. We have not yet made any progress toward transforming the expression to zero. The problem is that α and β are algebraically dependent.

Recognizing that $\sqrt[3]{4} = \sqrt[3]{2^2} = 2^{2/3} = \alpha^2$, we can write the original expression in terms of just α,

$$(\alpha + \alpha^2)^3 - 6\,(\alpha + \alpha^2) - 6.$$

Expanding products this time results in

$$\alpha^6 + 3\alpha^5 + 3\alpha^4 + \alpha^3 - 6\alpha^2 - 6\alpha - 6,$$

which we note can be factored as

$$(\alpha^3 - 2)(\alpha^3 + 3\alpha^2 + 3\alpha + 3).$$

Substituting $\alpha = \sqrt[3]{2}$ gives 0, which is easily seen from the factored form. $\square$

The previous example illustrates how *explicit algebraic numbers* (radicals) are represented in computer algebra systems. The radicals are replaced by "labels" that act like new variables, transforming the original expression into a polynomial or rational function of one or more variables. In general, the transformation produces a rational function of several variables. The resulting polynomial or rational function is then manipulated using the standard algorithms. In Example 5.3, this involves the expansion of polynomials. The polynomial defining the radical serves as a "side relation" and, at the end of the computation, it is used to substitute the radical back into the transformed expression. Finally, evaluation (automatic simplification) is used to clean up the result.

What about *implicit algebraic numbers*—those that cannot be expressed using radicals? To see how they are represented, let's examine Maple's reply when asked to solve $x^5 - x + 1$.

```
>  solve(x^5 - x + 1);
```

$$RootOf\left(\_Z^5 - \_Z + 1, index = 1\right), \quad RootOf\left(\_Z^5 - \_Z + 1, index = 2\right),$$
$$RootOf\left(\_Z^5 - \_Z + 1, index = 3\right), \quad RootOf\left(\_Z^5 - \_Z + 1, index = 4\right),$$
$$RootOf\left(\_Z^5 - \_Z + 1, index = 5\right)$$

```
>  evalf(%);
```

$$.7648844336 + .3524715460\,I, \quad -.1812324445 + 1.083954101\,I,$$
$$- 1.167303978, \quad -.1812324445 - 1.083954101\,I,$$
$$.7648844336 - .3524715460\,I$$

The `RootOf` procedure plays the role of a placeholder representing each of the roots. The index parameter numbers the roots and orders them by increasing polar angle in the complex plane, as can be seen from the numerical values. Mathematica's reply to `Solve[x^5-x+1==0]` is similar. It uses a procedure called `Root`, whose role is identical to that of Maple's `RootOf`, and reports the real solution first.

5.3.2 Rationalizing Denominators

In elementary algebra, we are taught a simple technique to "rationalize denominators" of expressions involving radicals. The idea is to eliminate all the radicals in the denominator. For example,

$$\frac{1}{\sqrt{2}-1} = \frac{1}{\sqrt{2}-1} \cdot \frac{\sqrt{2}+1}{\sqrt{2}+1} = \frac{\sqrt{2}+1}{2-1} = \sqrt{2}+1.$$

The technique we learned in school works well for small examples, but is neither sufficiently algorithmic nor sufficiently powerful to deal with transformations such as

$$\frac{1}{\sqrt{2}+\sqrt{3}+\sqrt{5}} = \frac{1}{4}\sqrt{2} + \frac{1}{6}\sqrt{3} - \frac{1}{12}\sqrt{2}\sqrt{3}\sqrt{5},$$

$$\frac{1}{5^{3/2}-5^{4/3}} = \frac{1}{100}5^{5/6} + \frac{1}{100}5^{2/3} + \frac{1}{20}5^{1/2} + \frac{1}{20}5^{1/3} + \frac{1}{20}5^{1/6} + \frac{1}{20}.$$

Rationalizing the denominator does not necessarily produce an expression that is "simpler" than the original one, as the last two examples illustrate. Nevertheless, it is imperative if we desire to transform expressions like

$$\sqrt{3}+\sqrt{2} - \frac{1}{\sqrt{3}-\sqrt{2}}$$

to zero. The obvious benefit of rationalizing the denominator is that it produces a normal form.

We can rationalize the denominator of an algebraic number with the aid of the extended Euclidean algorithm. The numerator n and denominator d can be regarded as polynomials of the radicals appearing in the expression. We eliminate, one at a time, each algebraically independent radical α appearing in d as follows. Regarding both n and d as polynomials in α, we apply the extended Euclidean algorithm to find the gcd of d and $p(\alpha)$, the minimal polynomial defining α,

$$\gcd(d(\alpha), p(\alpha)) = 1 = s(\alpha)\, d(\alpha) + t(\alpha)\, p(\alpha). \tag{5.2}$$

The gcd is 1 since p, by construction, is irreducible (i.e., it does not factor). Next, we multiply both the numerator n and denominator d by s. Since $p(\alpha) = 0$, Equation (5.2) tells us that $s\,d = 1$ and so $n/d = (n\,s)/(d\,s) = n\,s$. This procedure eliminates α from the denominator.

We first illustrate the procedure with a simple example in which we compare our algorithm with the "hand method" learned in algebra.

Example 5.4. Rationalize the denominator of $\dfrac{1}{2\sqrt{5}-1}$.

Solution. Letting $\alpha = \sqrt{5}$, its minimal polynomial is $p(\alpha) = \alpha^2 - 5$. The "hand calculation" proceeds as follows:

$$\frac{1}{2\alpha-1} \cdot \frac{2\alpha+1}{2\alpha+1} = \frac{2\alpha+1}{4\alpha^2-1} = \frac{1}{19}(2\alpha+1).$$

The last step is the result of substituting $\alpha^2 = 5$ in the denominator.

Using the extended Euclidean algorithm from Chapter 4, we begin by computing the gcd of the denominator and $p(\alpha)$,

$$\gcd(2\alpha-1, \alpha^2-5) = 1 = \underbrace{\tfrac{1}{19}(2\alpha+1)}_{s(\alpha)}\underbrace{(2\alpha-1)}_{d(\alpha)} - \underbrace{\tfrac{4}{19}}_{t(\alpha)}\underbrace{(\alpha^2-5)}_{p(\alpha)}.$$

Since the numerator $n = 1$, the result is $n s = s$ and so

$$\frac{1}{2\sqrt{5}-1} = \frac{1}{19}(2\alpha+1) = \frac{1}{19}\left(2\sqrt{5}+1\right). \quad \square$$

The next example is more involved in that there are two radicals in the denominator.

Example 5.5. Rationalize the denominator of $\dfrac{1}{\sqrt{3}-\sqrt[3]{2}}$.

Solution. Our algebraic number is $1/(\alpha-\beta)$, where $\alpha = \sqrt{3}$ and $\beta = \sqrt[3]{2}$. The minimal polynomials of α and β are, respectively, $\alpha^2 - 3$ and $\beta^3 - 2$. To eliminate α from the denominator first, we perform the extended gcd computation,

$$\gcd(\alpha-\beta, \alpha^2-3) = 1 = \underbrace{-\frac{(\alpha+\beta)}{\beta^2-3}}_{s_1}\underbrace{(\alpha-\beta)}_{d_1} + \underbrace{\frac{1}{\beta^2-3}}_{t_1}\underbrace{(\alpha^2-3)}_{p_1}.$$

Multiplying the numerator and denominator of $1/(\alpha-\beta)$ by s_1, we have

$$\frac{1}{\alpha-\beta} = \frac{\alpha+\beta}{(\alpha-\beta)(\alpha+\beta)} = -\frac{\alpha+\beta}{\beta^2-3}. \tag{5.3}$$

But this is $n s_1 = s_1$ since the original numerator $n = 1$.

To eliminate β from the denominator, we perform a second extended gcd calculation,

$$\gcd(\beta^2-3, \beta^3-2) = 1 = \underbrace{-\tfrac{1}{23}(3\beta^2+2\beta+9)}_{s_2}\underbrace{(\beta^2-3)}_{d_2} + \underbrace{\tfrac{1}{23}(3\beta+2)}_{t_2}\underbrace{(\beta^3-2)}_{p_2}.$$

Multiplying the numerator and denominator of the algebraic number in (5.3) by s_2 gives

$$-\frac{\alpha+\beta}{\beta^2-3} = -\frac{(\alpha+\beta)(3\beta^2+2\beta+9)}{(\beta^2-3)(3\beta^2+2\beta+9)} = -\frac{(\alpha+\beta)(3\beta^2+2\beta+9)}{3\beta^4+2\beta^3-6\beta-27}.$$

Finally substituting $\alpha = \sqrt{3}$ and $\beta = \sqrt[3]{2}$, we arrive at the solution,

$$\frac{1}{\sqrt{3}-\sqrt[3]{2}} = \frac{1}{23}(3^{1/2}+2^{1/3})(3\cdot2^{2/3}+2\cdot2^{1/3}+9).$$

Again, observe that this result can be obtained far more simply just by multiplying s_2 with the numerator of (5.3), $-(\alpha+\beta)$. $\quad \square$

Our final example illustrates two points. The first is the need for expressing an algebraic number in terms of independent radicals. The second is the difficulty that arises when an innocuously appearing denominator turns out to be zero.

Example 5.6. Rationalize $\dfrac{1}{2 \cdot 5^{5/2} - 2 \cdot 5^{7/3} - 10 \cdot 5^{3/2} + 10 \cdot 5^{4/3}}$.

Solution. The powers of 5 in the denominator can all be expressed in terms of $\alpha = 5^{1/6}$, whose minimal polynomial is $p(\alpha) = \alpha^6 - 5$. Doing so, the denominator is $d(\alpha) = 2\alpha^{15} - 2\alpha^{14} - 10\alpha^9 + 10\alpha^8$. Performing the extended Euclidean algorithm, we find

$$\gcd(d(\alpha), p(\alpha)) = \alpha^6 - 5$$

and not one, as expected! d factors as $2\alpha^8 (\alpha - 1)(\alpha^6 - 5)$, and the minimal polynomial appears in the factorization of the denominator. This implies the denominator is zero, and the algebraic number is undefined. Consequently, it makes no sense to rationalize the denominator. □

5.3.3 Denesting Radicals

One simplification that most users almost always want to make is to denest radicals. Each of the identities listed below has a doubly nested radical on the left and one or more singly nested radicals on the right.

$$\sqrt{9 + 4\sqrt{2}} = 1 + 2\sqrt{2}, \tag{5.4}$$

$$\sqrt{5 + 2\sqrt{6}} = \sqrt{3} + \sqrt{2}, \tag{5.5}$$

$$\sqrt{5\sqrt{3} + 6\sqrt{2}} = \sqrt[4]{27} + \sqrt[4]{12}, \tag{5.6}$$

$$\sqrt{12 + 2\sqrt{6} + 2\sqrt{10} + 2\sqrt{15}} = \sqrt{2} + \sqrt{3} + \sqrt{5}, \tag{5.7}$$

$$\sqrt{9 + \sqrt[3]{9} + 6\sqrt[3]{3}} = 3 + \sqrt[3]{3}, \tag{5.8}$$

$$\sqrt{\sqrt[3]{5} - \sqrt[3]{4}} = \frac{1}{3} \left(\sqrt[3]{2} + \sqrt[3]{20} - \sqrt[3]{25} \right). \tag{5.9}$$

The famous Indian mathematician Srinivasa Ramanujan (1887–1920) discovered many such identities. Unfortunately, he did not provide an algorithm for producing such denestings.

While some progress has been made in recent years, no general algorithm is known to denest radicals or even to determine whether or not a given radical can be denested. Only a few special cases have been solved satisfactorily. One is nested square roots, which was solved by Borodin, Fagin, Hopcroft and Tompa (1985). The article by Susan Landau (1994) is a rigorous, though accessible, introduction to the subject of denesting radicals.

In this section, we present a technique for simplifying square roots of square roots that is due to David J. Jeffrey and Albert D. Rich (1999). While simpler and easier to implement than the technique of Borodin, et. al., it does not solve all possible cases. Thus it is more properly regarded as a heuristic, rather than an algorithm. Nonetheless, the method is very powerful and solves a large number of cases, including the transformations in Equations (5.4)–(5.7).

Where do nested radicals arise? Often they occur when standard formulas are used to solve a problem. For example, the solutions to the quadratic equation $x^2 + 3x - \sqrt{2}$ are

$$x = \frac{-3 \pm \sqrt{9 + 4\sqrt{2}}}{2},$$

giving rise to the radical on the left side of (5.4). The quadratic cannot be factored over the integers, but can be factored using $\sqrt{2}$ as

$$(x + 2 + \sqrt{2})(x + 1 - \sqrt{2}),$$

suggesting how the right side of (5.4) comes about. The evaluation of trigonometric functions is another source of nested roots. For example,

$$\sin(\pi/8) = \frac{1}{2}\sqrt{2 - \sqrt{2}}, \quad \cos(\pi/8) = \frac{1}{2}\sqrt{2 + \sqrt{2}}.$$

Unfortunately, neither of these radicals can be denested.

Why do we want to denest radicals? The obvious reason is that the right sides of Equations (5.4)–(5.9) appear "simpler" to most people. A second, more subtle reason, is the danger that mathematical simplification may depend on the order in which operations are performed. For example if the expression $\sqrt{(1 - \sqrt{3})^2}$ is evaluated "top down" with the rule $\sqrt{x^2} = |x|$, we obtain

$$\sqrt{(1 - \sqrt{3})^2} = \sqrt{3} - 1.$$

However, if we evaluate the expression "bottom up" by expanding the square inside the radical, we get

$$\sqrt{(1 - \sqrt{3})^2} = \sqrt{4 - 2\sqrt{3}}.$$

Without denesting, we produce a different result—a situation that the designers of computer algebra systems strive to avoid. Another reason for denesting radicals is numerical accuracy. Jeffrey and Rich (1999) state that $\sqrt{199999 - 600\sqrt{111110}} \approx 0.00158$ approximates to zero with 10 decimal digits of precision, whereas the denested form, $100\sqrt{10} - 3\sqrt{11111}$, approximates to 0.00158 with only 7 digits of precision.

We now turn our attention to the problem of simplifying nested square roots. Consider (5.5) and its companion identity,

$$\sqrt{5 + 2\sqrt{6}} = \sqrt{3} + \sqrt{2}, \quad \sqrt{5 - 2\sqrt{6}} = \sqrt{3} - \sqrt{2}.$$

The left sides of these equations are of the form $\sqrt{X \pm Y}$ and the right sides have the form $\sqrt{A} \pm \sqrt{B}$. In fact, each of (5.4)–(5.7) is of the form

$$\sqrt{X + Y} = \sqrt{A} + \sqrt{B} \tag{5.10}$$

and has a companion relation of the form

$$\sqrt{X-Y} = \sqrt{A} - \sqrt{B} \tag{5.11}$$

for some suitable choice of X, Y, A, B. Squaring both sides of (5.10), we see that

$$X + Y = A + B + 2\sqrt{A\,B}.$$

One way we can satisfy this equation to let $X = A + B$ and $Y^2 = 4A\,B$. Multiplying (5.10) and (5.11) reveals that

$$\sqrt{X^2 - Y^2} = A - B.$$

With $A + B = X$ and $A - B = \sqrt{X^2 - Y^2}$, we have

$$2\,A = X + \sqrt{X^2 - Y^2},$$

$$2\,B = X - \sqrt{X^2 - Y^2}.$$

Substituting these values for A and B into (5.10) and (5.11) gives rise to the following theorem due to Jeffrey and Rich (1999).

Theorem 5.5 (Square Root Denesting Theorem). *If X and Y are real numbers with $X > Y > 0$, then*

$$\sqrt{X \pm Y} = \sqrt{\tfrac{1}{2}\left(X + \sqrt{X^2 - Y^2}\right)} \pm \sqrt{\tfrac{1}{2}\left(X - \sqrt{X^2 - Y^2}\right)}. \tag{5.12}$$

In our next example, we use the Square Root Denesting Theorem to derive Equations (5.4)–(5.6).

Example 5.7. Denest the following square roots:

(a) $\sqrt{9 + 4\sqrt{2}}$. Setting $X = 9$ and $Y = 4\sqrt{2}$, we have

$$X \pm \sqrt{X^2 - Y^2} = 9 \pm \sqrt{81 - 32} = 9 \pm \sqrt{49} = 9 \pm 7,$$

and so,

$$\sqrt{9 + 4\sqrt{2}} = \sqrt{\tfrac{1}{2} \cdot 16} + \sqrt{\tfrac{1}{2} \cdot 2} = \sqrt{8} + 1 = 2\sqrt{2} + 1.$$

(b) $\sqrt{5 + 2\sqrt{6}}$. Letting $X = 5$ and $Y = 2\sqrt{6}$, we find

$$X \pm \sqrt{X^2 - Y^2} = 5 \pm \sqrt{25 - 24} = 5 \pm 1,$$

therefore,

$$\sqrt{5 + 2\sqrt{6}} = \sqrt{\tfrac{1}{2} \cdot 6} + \sqrt{\tfrac{1}{2} \cdot 4} = \sqrt{3} + \sqrt{2}.$$

(c) $\sqrt{5\sqrt{3}+6\sqrt{2}}$. With $X = 5\sqrt{3}$ and $Y = 6\sqrt{2}$, we obtain

$$X \pm \sqrt{X^2 - Y^2} = 5\sqrt{3} \pm \sqrt{75-72} = 5\sqrt{3} \pm \sqrt{3},$$

thus,

$$\sqrt{5\sqrt{3}+6\sqrt{2}} = \sqrt{3\sqrt{3}} + \sqrt{2\sqrt{3}}.$$

There are still doubly nested square roots on the right side, but $\sqrt{3\sqrt{3}}$ can be expressed as $\sqrt[4]{27}$ and $\sqrt{2\sqrt{3}}$ is equal to $\sqrt[4]{12}$. Therefore,

$$\sqrt{5\sqrt{3}+6\sqrt{2}} = \sqrt[4]{27} + \sqrt[4]{12}.$$

This example illustrates that fourth roots may be needed to denest square roots. Borodin et. al. (1985) showed that only fourth roots and square roots are required. □

The next example is more involved. When more than two terms appear under the radical to be denested, the question arises how to split the terms into X and Y. The solution demonstrates another point. It may be beneficial to apply Theorem 5.5 more than once when solving a problem.

Example 5.8. Denest $\sqrt{10 + 2\sqrt{6} + 2\sqrt{10} + 2\sqrt{15}}$.

Solution. This time it is not obvious how to split the terms under the radical into X and Y. We discuss this issue shortly. For now, let's choose $X = 10 + 2\sqrt{6}$, $Y = 2\sqrt{10} + 2\sqrt{15}$. This gives

$$X \pm \sqrt{X^2 - Y^2} = 10 + 2\sqrt{6} \pm \sqrt{(124 + 40\sqrt{6}) - (100 + 40\sqrt{6})}$$
$$= 10 + 2\sqrt{6} \pm 2\sqrt{6},$$

and so we have

$$\sqrt{10 + 2\sqrt{6} + 2\sqrt{10} + 2\sqrt{15}} = \sqrt{5 + 2\sqrt{6}} + \sqrt{5}.$$

Although we have not completely denested the original radical, we have made substantial progress as the nested radical that remains has fewer terms than the original one. It is, in fact, the radical of Example 5.7b, and so

$$\sqrt{10 + 2\sqrt{6} + 2\sqrt{10} + 2\sqrt{15}} = \sqrt{2} + \sqrt{3} + \sqrt{5}. \qquad (5.13)$$

This example illustrates that Theorem 5.5 can sometimes be applied repeatedly to solve larger problems. □

Let's revisit the issue of splitting the terms in the denesting (5.13) into X and Y. Since the right side of (5.13) is of the form $a + b + c$, the denesting takes the form

$$\sqrt{a^2 + b^2 + c^2 + 2\,ab + 2\,ac + 2\,bc} = \sqrt{(a+b+c)^2} = a + b + c.$$

From symmetry considerations, it does not matter how we choose X and Y since one of them is $a^2 + b^2 + c^2$ plus either $2ab$, $2ac$, or $2bc$. But for a denesting of the form

$$\sqrt{a^2 + b^2 + c^2 + d^2 + 2\,ab + 2\,ac + 2\,ad + 2\,bc + 2\,bd + 2\,cd} = a + b + c + d,$$

only the partitioning $X = a^2 + b^2 + c^2 + d^2 + 2ab + 2cd$ works. So the structure of the expression under a nested radical must be examined if we are to succeed in using Theorem 5.5.

The majority of nested square roots cannot be denested, and the next two examples show how Theorem 5.5 can fail.

Example 5.9. Apply the Square Root Denesting Theorem to $\sqrt{2 + \sqrt{2}}$.

Solution. With $X = 2$, $Y = \sqrt{2}$, we have

$$X \pm \sqrt{X^2 - Y^2} = 2 \pm \sqrt{2}$$

and so, according to Theorem 5.5,

$$\sqrt{2 + \sqrt{2}} = \sqrt{\tfrac{1}{2}\left(2 + \sqrt{2}\right)} + \sqrt{\tfrac{1}{2}\left(2 - \sqrt{2}\right)} = \frac{\sqrt{2}}{2}\left(\sqrt{2 + \sqrt{2}} + \sqrt{2 - \sqrt{2}}\right).$$

The two nested square roots on the right are as complicated as the original nested radical (one is the original radical in this very simple example), and no denesting is possible. □

Example 5.10. Attempt to denest $\sqrt{10 + 2\sqrt{6} + 2\sqrt{10} + 2\sqrt{14}}$ using Theorem 5.5.

Solution. This is a slight variant of the radical in Example 5.8. Again choosing $X = 10 + 2\sqrt{6}$, we have $Y = 2\sqrt{10} + 2\sqrt{14}$ and

$$X \pm \sqrt{X^2 - Y^2} = 10 + 2\sqrt{6} \pm \sqrt{(124 + 40\sqrt{6}) - (96 + 16\sqrt{35})}$$

$$= 10 + 2\sqrt{6} \pm 2\sqrt{7 + 10\sqrt{6} - 4\sqrt{35}}.$$

Therefore the original radical is equal to

$$\sqrt{5 + \sqrt{6} + \sqrt{7 + 10\sqrt{6} - 4\sqrt{35}}} + \sqrt{5 + \sqrt{6} - \sqrt{7 + 10\sqrt{6} - 4\sqrt{35}}}.$$

We have replaced a doubly nested radical containing three square roots by two triply nested radicals. Moreover if we try to simplify $\sqrt{7 + 10\sqrt{6} - 4\sqrt{35}}$ with the Square Root Denesting Theorem, the number of levels of radicals increases to four. In either event, we obtain an expression with greater nesting depth than the original radical. □

So how can we incorporate a stopping condition into a procedure for denesting square roots? We do this by defining a quantity that attempts to measure the structural complexity of an expression with regard to nested radicals. When the denesting formula (5.12) is applied, we check to see whether the complexity of the expression we obtain is less than that of the original radical. If it is, we continue; if not, we abandon this transformation. One such measure of structural complexity is simply the number of levels of nesting. For another measure, the interested reader is referred to Jeffrey and Rich (1999).

5.4 Transcendental Functions

We are now set to extend our repertoire to the transcendental functions. These include the familiar logarithmic, exponential, and trigonometric functions studied in calculus and pre-calculus mathematics. By way of introduction, observe that the transformation

$$\frac{1}{\sin^2 x} + \frac{1}{\cos^2 x} \rightarrow \frac{\cos^2 x + \sin^2 x}{\sin^2 x \cos^2 x}$$

can be regarded as a rational calculation,

$$\frac{1}{\theta_1^2} + \frac{1}{\theta_2^2} \rightarrow \frac{\theta_2^2 + \theta_1^2}{\theta_1^2 \theta_2^2},$$

where $\theta_1 = \sin x$ and $\theta_2 = \cos x$. Similarly, the simplification

$$\frac{1}{\sin^2 x} + \frac{1}{\cos^2 x} \rightarrow \frac{4}{\sin^2(2\,x)}$$

can be regarded as the same rational transformation with the "side relations"

$$\cos^2 x + \sin^2 x = 1,$$
$$\sin(2\,x) = 2\,\sin x\,\cos x,$$

since

$$\frac{\theta_2^2 + \theta_1^2}{\theta_1^2 \theta_2^2} = \frac{1}{(\theta_1\,\theta_2)^2} = \frac{4}{\sin^2(2\,x)}.$$

This example serves to illustrate the general strategy that computer algebra systems employ to manipulate transcendental functions. Each instance of a transcendental function is replaced by an indeterminate symbol (variable). In many instances, this results in a polynomial or rational function of several variables that can be manipulated using standard techniques like those described in Chapter 4. Additional simplifications are often possible by applying side relations appropriate to a particular transcendental function.

5.4.1   Brown's Normal Form Algorithm

In this section, we consider an algorithm due to William S. Brown (1964) for simplifying *rational exponential expressions*. These expressions are built recursively from the four standard arithmetic operations and by forming exponentials of existing rational exponential expressions. We consider here the case where the expressions contain only a single variable, but it is easy to generalize Brown's method to several variables. An example of a rational exponential expression is

$$\frac{e^{e^x} + 2}{e^x + 3\,e^{2x} + x\,e^{4x+1}}.$$

Brown's algorithm produces a normal form for this class of expressions provided an unproven conjecture that the real numbers e and π are algebraically independent is true. That is, under the assumptions of the conjecture, expressions equivalent to zero are simplified to zero by the procedure, but the method does not produce a canonical form.

The idea behind Brown's normal form algorithm is to process each exponential in turn, starting at the innermost level, and to replace it by a new indeterminate (variable) symbol. This is done so long as the exponential is not related algebraically to any previously considered. In other words, the argument of each exponential cannot be expressed as a linear combination (with rational coefficients) of the previous exponentials. For example, e^x and e^{x^2} are independent but e^{2x} and e^{3x-1} are not since $3x - 1 = \frac{3}{2}(2x) - 1$. If the exponential under consideration is algebraically related to one or more of the exponentials already considered, then the algorithm performs the simplification corresponding to the discovered algebraic relation.

This procedure produces a rational expression of the indeterminates introduced and the original variables. The rational expression can then be simplified using the techniques of Chapter 4. After these rational transformations are performed, the substitutions are reversed by replacing the indeterminates by the corresponding exponentials, thereby returning a transformed rational exponential expression. Rather than give a formal description of the procedure, we discuss the operation of Brown's technique and the issues that arise in its implementation with several examples.

Example 5.11. Simplify the exponential expression $\dfrac{e^x - x}{e^{2x} - 2\,x\,e^x + x^2}$.

Solution. There are two distinct exponentials in this expression, e^x and e^{2x}. We first consider e^x. Letting $\theta_1 = e^x$, the argument of θ_1 is $u_1 = x$. Making the substitution of θ_1 for e^x in the expression to be simplified, we have

$$\frac{\theta_1 - x}{e^{2x} - 2\,x\,\theta_1 + x^2}.$$

We now process the second exponential, $\theta_2 = e^{2x}$. Since its argument $u_2 = 2x = 2u_1$, the two exponentials are not algebraically independent and

$$\theta_2 = e^{2x} = (e^x)^2 = \theta_1^2,$$

so our expression becomes

$$\frac{\theta_1 - x}{\theta_1^2 - 2x\,\theta_1 + x^2}.$$

The next step in the simplification process is to calculate and remove the gcd of the numerator and denominator, $\gcd(\theta_1 - x,\ \theta_1^2 - 2x\,\theta_1 + x^2) = \theta_1 - x$, giving us

$$\frac{1}{\theta_1 - x}.$$

This is a purely rational calculation. Finally we substitute back for θ_1, obtaining $1/(e^x - x)$. $\qquad\square$

Whenever possible, we want to avoid the introduction of algebraic quantities into the exponential expression being simplified. The next example illustrates this important point and shows how the algorithm avoids the problem by revisiting earlier substitutions.

Example 5.12. Simplify the rational exponential expression $\dfrac{e^{2x} - e^x}{e^x}$.

Solution. This expression contains the same two exponentials as the previous example. Since we encounter e^{2x} first, suppose we substitute θ_1 for e^{2x}. The argument to θ_1 is $u_1 = 2x$ and the substitution yields

$$\frac{\theta_1 - e^x}{e^x}.$$

Now we deal with the second exponential, $\theta_2 = e^x$. Its argument is $u_2 = x = \frac{1}{2}u_1$, so the two exponentials are not algebraically independent. Since $u_2 = \frac{1}{2}u_1$, the substitution for θ_2 yields

$$\frac{\theta_1 - \sqrt{\theta_1}}{\sqrt{\theta_1}}.$$

This introduces an algebraic quantity into the expression to be simplified! We want rational expressions whenever possible so we redefine θ_1 as θ_2^2, with the resulting algebraic dependency $u_1 = 2u_2$. This substitution produces a rational expression,

$$\frac{\theta_2^2 - \theta_2}{\theta_2}.$$

Canceling the factors of θ_2 and substituting back for θ_2, the expression simplifies to $\theta_2 - 1 = e^x - 1$. $\qquad\square$

Our final example and the previous one show some of the difficulties a simplifier for rational exponential expressions must surmount. The example also demonstrates that Brown's algorithm does not produce a canonical form. We examine three solutions to a simplification problem, the last of which shows how partial fractions can be useful in eliminating algebraic dependencies.

Example 5.13. Simplify the rational exponential expression

$$\frac{e^{1/(x+x^2)} + e^{1/x}\,e^{-1/(x+1)}}{e^{1/x}}.$$

Solution 1. Suppose we make the substitutions in the order

$$\theta_1 = e^{1/(x+x^2)}, \qquad \theta_2 = e^{1/x}, \qquad \theta_3 = e^{-1/(x+1)},$$

$$u_1 = \frac{1}{x+x^2} \qquad\qquad u_2 = \frac{1}{x} \qquad\qquad u_3 = -\frac{1}{x+1}.$$

Observing that $u_3 = u_1 - u_2$, θ_3 can be written in terms of θ_1 and θ_2 as $\theta_3 = \theta_1/\theta_2$. With these substitutions, the expression simplifies to

$$\frac{\theta_1 + \theta_2\,\theta_1/\theta_2}{\theta_2} = \frac{2\,\theta_1}{\theta_2} = \frac{2\,e^{1/(x+x^2)}}{e^{1/x}}.$$

Solution 2. Now suppose the substitutions are made in the order

$$\theta_1 = e^{1/x}, \qquad \theta_2 = e^{-1/(x+1)}, \qquad \theta_3 = e^{1/(x+x^2)},$$

$$u_1 = \frac{1}{x} \qquad\qquad u_2 = -\frac{1}{x+1} \qquad\qquad u_3 = \frac{1}{x+x^2}.$$

This time we have $u_3 = u_1 + u_2$, so θ_3 is written in terms of θ_1 and θ_2 as $\theta_3 = \theta_1\,\theta_2$. With these substitutions, the expression simplifies to

$$\frac{\theta_1\,\theta_2 + \theta_1\,\theta_2}{\theta_1} = 2\,\theta_2 = 2\,e^{-1/(x+1)}.$$

Since this result is different from Solution 1, we observe that Brown's normal form algorithm does not produce a unique (canonical) form.

Solution 3. Yet another way to solve this problem is to use partial fractions to decompose the exponent $1/(x + x^2)$ before substitution. Since

$$\frac{1}{x+x^2} = \frac{1}{x} - \frac{1}{x+1},$$

the original expression can be rewritten as

$$\frac{e^{1/x}\,e^{-1/(x+1)} + e^{1/x}\,e^{-1/(x+1)}}{e^{1/x}} = \frac{2\,e^{1/x}\,e^{-1/(x+1)}}{e^{1/x}}.$$

With the substitutions $\theta_1 = e^{1/x}$ and $\theta_2 = e^{-1/(x+1)}$, we obtain the simplification

$$\frac{2\,\theta_1\,\theta_2}{\theta_1} = 2\,\theta_2 = 2\,e^{-1/(x+1)},$$

the same result as Solution 2. This approach, however, has the advantage of eliminating the algebraic dependency introduced by the third substitution. □

5.4.2   Structure Theorem

In this section, we show how the ideas behind Brown's algorithm can be extended to deal with other types of functions. Having already discussed exponentials, we focus attention on algebraic quantities (radicals) and logarithms. As with Brown's algorithm, we replace each function in an expression, starting at the innermost level. If the quantity under consideration is independent of those previously processed, it is replaced by a new indeterminate (variable) symbol. Otherwise, the algebraic relationship between the new function and the previous ones is discovered, and the quantity being processed is replaced by an appropriate combination of those already considered.

We begin with a fairly simple example involving both logarithms and an algebraic function.

Example 5.14. Simplify $f = \sqrt{(\ln(x-1) + \ln(x+1))\ln(x^2-1)}$.

Solution. Let $\theta_1 = \ln(x-1)$ and $\theta_2 = \ln(x+1)$. The arguments to these logarithms are $v_1 = x-1$ and $v_2 = x+1$. The argument, v_3, to the third logarithm, $\theta_3 = \ln(x^2-1)$, is easily seen to be the product of the arguments to the other logs, $v_3 = v_1 v_2$. As a result,

$$\theta_3 = \ln(v_1 v_2) = \ln v_1 + \ln v_2$$

by the rule for the logarithm of a product. At this point, we choose to replace $\theta_1 + \theta_2$ by θ_3, rather than the other way around, since it results in a more compact expression. Our original expression, f, is algebraic and can be expressed by the symbol θ_4, whose minimal polynomial satisfies $\theta_4^2 - \theta_3^2 = 0$. Therefore, we have $\theta_4 = \theta_3 = \ln(x^2-1)$.

Alternatively, the original expression f can be regarded as logarithmic rather than algebraic. A logarithm $\theta = \ln v$ obeys the differentiation rule $\theta' = v'/v$. Letting $v = x^2 - 1$, we have

$$f' = \frac{v'}{v} = \frac{(x^2-1)'}{x^2-1},$$

from which we can immediately infer that $f = \ln(x^2 - 1)$.   □

The next very simple example prepares the way for a discussion of the independence of logarithms.

Example 5.15. Simplify the sum of logarithms $\ln x + \ln x^2 + \ln x^4$.

Solution. Letting $\theta_1 = \ln x$, its argument is $v_1 = x$. Now consider the second logarithm, $\theta_2 = \ln x^2$. Since its argument, $v_2 = x^2$, is equal to v_1^2, θ_1 and θ_2 are not independent and $\theta_2 = 2\theta_1$. Similarly, the argument $v_3 = x^4$ of $\theta_3 = \ln x^4$ can be written $v_3 = v_1^4$ and so $\theta_3 = 4\theta_1$. So the sum is

$$\theta_1 + \theta_2 + \theta_3 = \theta_1 + 2\theta_1 + 4\theta_1 = 7\theta_1 = 7\ln x.$$

Now let's start instead with $\theta_3 = \ln x^4$. Then $\theta_2 = \frac{1}{2}\theta_3$ since $v_2 = v_3^{1/2}$. Similarly, $\theta_1 = \frac{1}{4}\theta_3$ since $v_1 = v_3^{1/4}$. Therefore,

$$\theta_1 + \theta_2 + \theta_3 = \frac{1}{4}\theta_3 + \frac{1}{2}\theta_3 + \theta_3 = \frac{7}{4}\theta_3 = \frac{7}{4}\ln x^4,$$

which simplifies to $7\ln x$ by the rule for the logarithm of a power. $\quad\square$

When is a new logarithm independent of any previously defined ones? Suppose the argument v_i to the new logarithm $\theta_i = \ln v_i$ can be expressed in terms of the arguments v_j to previously defined logarithms $\theta_j = \ln v_j$, $1 \le j < i$, as

$$v_i = c \prod_{j=1}^{i-1} v_j^{k_j},$$

where the k_j are rational numbers and c is a constant. Then the new logarithm is *not independent* of those already defined and can be expressed as

$$\theta_i = \ln v_i = \ln c + k_j \sum_{j=1}^{i-1} \ln v_j = \ln c + k_j \sum_{j=1}^{i-1} \theta_j,$$

by the laws for the logarithms of products and powers, $\ln(ab) = \ln a + \ln b$ and $\ln a^n = n\ln a$.

Combining this observation about logarithms with those we made about exponentials in the previous section leads to the Structure Theorem, attributed to Robert Risch (1979) and Maxwell Rosenlicht (1976). This theorem gives the conditions for when a new transcendental quantity, either an exponential or logarithmic function, should be introduced because it is independent of those already appearing in an expression.

Theorem 5.6 (Structure Theorem). *Let $\mathbb{Q}$ be the rational numbers and $\theta_1, \ldots, \theta_n$ be either:*

    a) *algebraic (i.e., $p(\theta_i) = 0$ for some polynomial p with rational coefficients),*

    b) *exponential (i.e., $\theta_i = e^{u_i}$, where $\theta_i'/\theta_i = u_i'$), or*

    c) *logarithmic (i.e., $\theta_i = \ln v_i$, where $\theta_i' = v_i'/v_i$),*

with each θ_i defined in terms of $\mathbb{Q}$, the variable x, and $\theta_1, \ldots \theta_{i-1}$. Then

    1) *$\theta_i = e^{u_i}$ is independent of $\theta_1, \ldots, \theta_{i-1}$ if and only if u_i cannot be expressed as a linear combination*

$$c_0 + \sum_{j=1}^{i-1} c_j u_j,$$

where $c_j \in \mathbb{Q}$, and

2) $\theta_i = \ln v_i$ is independent of $\theta_1, \ldots, \theta_{i-1}$ if and only if v_i cannot be expressed as a product

$$c \prod_{j=1}^{i-1} v_j^{k_j},$$

where $c, k_j \in \mathbb{Q}$.

What does the Structure Theorem mean? The theorem can be stated informally as follows. An exponential is independent of the other exponentials if and only if its argument cannot be written as a linear combination with rational coefficients of the arguments of other exponentials. Using independent exponentials guarantees that each exponential is not a product of powers of the other exponentials. A logarithm is independent of the other logarithms if and only if its argument cannot be written as a product with rational exponents of the arguments of other logarithms. Using independent logarithms guarantees that each logarithm is not a linear combination with rational coefficients of the other logarithms.

We also need to consider the case of expressions containing both exponentials and logarithms. To guarantee that the exponentials and logarithms are independent of one another, we must further insure that the argument to each exponential cannot be expressed as linear combination of the logarithms. We must also guarantee that each logarithm cannot be expressed as a product with rational exponents of the exponentials.

As a consequence of the Structure Theorem, the only possible simplifications for exponentials and logarithms are

$$e^{f+g} = e^f\, e^g,$$

$$\ln(f\, g) = \ln f + \ln g,$$

$$e^{\ln f} = \ln e^f = f.$$

How can we determine whether a new transcendental quantity is independent of those previously defined? This can be done by checking whether a solution exists to a set of linear equations. First consider the case of a new exponential quantity, u_i. To determine whether u_i can be expressed as a linear combination $c_0 + \sum_{j=1}^{i-1} c_j u_j$, we differentiate this expression to obtain a linear system of equations for $c_1, \ldots, c_{i-1}$. Once the c_i are determined, we can find c_0. We illustrate the technique by revisiting one of the sets of substitutions in Example 5.13.

Example 5.16. Show that $\theta_3 = e^{-1/(x+1)}$ is not independent of $\theta_1 = e^{1/(x+x^2)}$ and $\theta_2 = e^{1/x}$.

Solution. We want to show how the argument $u_3 = -1/(x+1)$ of θ_3 can be expressed as a linear combination of the arguments $u_1 = 1/(x+x^2)$ and $u_2 = 1/x$ to θ_1 and θ_2, respectively,

$$u_3 = c_0 + c_1\, u_1 + c_2\, u_2$$

or, equivalently,

$$-\frac{1}{x+1} = c_0 + c_1 \frac{1}{x+x^2} + c_2 \frac{1}{x}. \qquad (5.14)$$

Differentiating both sides of (5.14), we have

$$\frac{1}{(x+1)^2} = -c_1 \frac{2x+1}{(x+x^2)^2} - c_2 \frac{1}{x^2}$$

and, clearing denominators,

$$\begin{aligned}
x^2 &= -c_1 \left(2x+1\right) - c_2 \left(x+1\right)^2, \\
&= -c_2 x^2 - 2\left(c_1 + c_2\right)x - \left(c_1 + c_2\right).
\end{aligned}$$

The solution is $c_1 = 1, c_2 = -1$, implying $c_0 = 0$ in (5.14). So u_3 can be expressed as $u_1 - u_2$. □

The situation for logarithms is only slightly more complicated. Suppose we want to determine whether the argument of a new logarithm, $\theta_i = \ln v_i$, can be expressed as a product $v_i = c \prod_{j=1}^{i-1} v_j^{k_j}$, where c and all k_j are rational numbers. This is equivalent to determining whether there exist k_j such that $v_i^{-1} \prod_{j=1}^{i-1} v_j^{k_j}$ is a rational number or, alternatively, whether the derivative of this product is zero. (Why?) The next example shows how this can be done.

Example 5.17. Let $\theta_1 = \sqrt{x^2+1}$, $\theta_2 = \ln(\theta_1 + x)$, and $\theta_3 = \ln(\theta_1 - x)$. Show that θ_2 and θ_3 are not independent, then use this result to simplify their sum,

$$\ln(\sqrt{x^2+1} + x) + \ln(\sqrt{x^2+1} - x).$$

Solution. The argument of θ_2 is $v_2 = \theta_1 + x$ and that of θ_3 is $v_3 = \theta_1 - x$. To show that θ_3 can be expressed in terms of θ_2, we need to find a rational number k such that $v_3^{-1} v_2^k$ is also a rational number. This is equivalent to determining whether a k exists such that $(v_3^{-1} v_2^k)' = 0$,

$$\frac{\left(\sqrt{x^2+1} + x\right)^k (k+1)}{\sqrt{x^2+1}\left(\sqrt{x^2+1} - x\right)} = 0.$$

The equation is clearly satisfied when $k = -1$, and so $v_3 = v_2^{-1}$. Therefore,

$$\theta_3 = \ln(\theta_1 - x) = \ln(\theta_1 + x)^{-1} = -\ln(\theta_1 + x) = -\theta_2,$$

and the sum $\theta_2 + \theta_3 = 0$. The dependence between θ_2 and θ_3 can also be seen by observing that their product is 1,

$$\theta_2 \theta_3 = (\theta_1 + x)(\theta_1 - x) = \theta_1^2 - x^2 = (x^2 + 1) - x^2 = 1. \qquad □$$

Our final example illustrates the interplay between exponentials and logarithms.

Example 5.18. Simplify $e^{(\ln x)/2}$.

Solution. The transcendental quantities appearing in this expression are $\theta_1 = \ln x$ and $\theta_2 = e^{\theta_1/2}$. If we view the expression as an exponential of a logarithm, we make no progress toward simplification,

$$\theta_2 = e^{\theta_1/2} = e^{(\ln x)/2}.$$

However, viewing it as an algebraic function, we have

$$\theta_2^2 = e^{\theta_1} = x$$

and so $\theta_2 = \sqrt{x}$. $\qquad\square$

5.4.3 Expanded and Combined Forms

We noted in the previous section that, as consequence of the Structure Theorem, the only possible simplifications for exponentials and logarithms are

$$e^{u+v} = e^u e^v, \tag{5.15}$$

$$\ln(u\,v) = \ln u + \ln v, \tag{5.16}$$

$$e^{\ln u} = \ln e^u = u.$$

Often, users of a computer algebra system wish to transform the arguments of exponentials or logarithms according to Equations (5.15) and (5.16). The process of applying these identities from left to right is called *expansion*, and the process of applying them from right to left is called *combination*. Combination is sometimes referred to as *contraction* or *collection*. In this section, we present expanded and combined forms for expressions containing exponentials and logarithms. The mathematical appropriateness of these transformations is considered in Chapter 9.

An expression is in *exponential expanded form* if no argument to an exponential is either (1) a sum or (2) a product that includes an integer or rational number. To place an expression in this form, the transformations

$$e^{u+v} \rightarrow e^u e^v, \tag{5.17}$$

$$e^{ku} \rightarrow (e^u)^k, \qquad k \in \mathbb{Z}, \mathbb{Q} \tag{5.18}$$

where k is either an integer or rational number, are applied to the arguments of each exponential in a top-down manner.

Example 5.19. Expand the following exponentials:

(a) e^{3x+5yz}. At the top level, the argument to the exponential is a sum and so we apply the transformation (5.17),

$$e^{3x+5yz} \rightarrow e^{3x} \, e^{5yz}.$$

The argument to each exponential in the resulting expression is a product that includes an integer. These are removed using (5.18) to complete the expansion,

$$e^{3x} \, e^{5yz} \rightarrow (e^x)^3 \, (e^{yz})^5.$$

(b) $e^{2(x+y+z)}$. The argument to this exponential is a product at the top level, and so the integer factor is removed with (5.18),

$$e^{2(x+y+z)} \rightarrow \left(e^{x+y+z}\right)^2.$$

The argument to the exponential on the right is a sum, which is expanded using (5.17), and the resulting product is distributed over the exponent,

$$\left(e^{x+y+z}\right)^2 \rightarrow (e^x \, e^y \, e^z)^2$$
$$\rightarrow (e^x)^2 \, (e^y)^2 \, (e^z)^2.$$

The final transformation is performed automatically by most computer algebra systems as part of the expression evaluation process. □

Exponential combination applies the transformation rules

$$e^u \, e^v \rightarrow e^{u+v}, \tag{5.19}$$
$$(e^u)^v \rightarrow e^{u\,v} \tag{5.20}$$

to products and powers of exponentials with the goal of producing an expression that meets the following criteria. An expression is in *exponential combined form* if

(1) each product contains at most one exponential factor,
(2) no exponential function is raised to a power,[§] and
(3) each complete subexpression is algebraically expanded.

The next example illustrates the conversion of a pair of expressions to exponential combined form.

[§] Note the difference between an exponential raised to a power, like $(e^u)^v$, and a nested exponential function, like e^{e^u}. Nested exponentials, of necessity, are permitted. The Maple syntax for these expressions, `exp(u)^v` vs. `exp(exp(u))`, should make the distinction clear.

Example 5.20. Combine the exponentials in the following expressions:

(a) $e^x(e^x + e^y)$. This expression is not in algebraically expanded form so the first step is to expand the product,

$$e^x(e^x + e^y) \to (e^x)^2 + e^x\, e^y.$$

Both terms on the right have more than one exponential factor so each is combined by applying (5.19) to yield

$$(e^x)^2 + e^x\, e^y \to e^{2x} + e^{x+y}.$$

(b) $\left(e^{e^x}\right)^{e^y}$. This expression takes the form of a power with an exponential for a base. Applying (5.20), we have

$$\left(e^{e^x}\right)^{e^y} \to e^{e^x\, e^y}.$$

The exponentials in the exponent are then combined using (5.19) to yield the exponential combined form,

$$e^{e^x\, e^y} \to e^{e^{x+y}}. \qquad \square$$

We can define analogous expanded and combined forms for logarithms. An expression is in *logarithmic expanded form* if no argument to a logarithm is a product or power. An expression is converted to log-expanded form by applying the transformations

$$\ln(u\,v) \to \ln u + \ln v,$$
$$\ln(u^w) \to w \ln u$$

to each natural logarithm. For example,

$$\ln(x\,y^2) + \ln\left(\frac{y^{1/2}}{z}\right) \to \ln x + \tfrac{5}{2} \ln y - \ln z$$

converts the expression on the left to log-expanded form.

Similarly, we can define a log-combined form in which the transformations

$$\ln u + \ln v \to \ln(u\,v),$$
$$k \ln u \to \ln u^k, \qquad k \in \mathbb{Z}, \mathbb{Q}$$

where k is either a integer or rational number, are applied. An expression is in *logarithmic combined form* if

(1) a sum has at most one term that is a logarithm, and
(2) no product with a logarithmic factor also has a factor that is either an integer or rational number.

For example,

$$\ln x + \tfrac{5}{2} \ln y - \ln z \to \ln\left(\frac{x\,y^{5/2}}{z}\right)$$

converts the expression on the left to log-combined form.

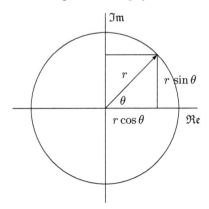

FIGURE 5.1: Rectangular and polar representations of a complex number.

5.4.4   Euler's Formula

In our discussion of the Structure Theorem, we seem to have omitted a very important class of functions—the trigonometric functions. What happened to them? It turns out that the trigonometric functions can be expressed as exponential functions with complex arguments using a formula attributed to Leonhard Euler. In fact, the formula was proved in an obscure form by Roger Cotes in 1714, then rediscovered and popularized by Euler in 1748. The geometrical interpretation of the formula, which we now explore, arose some 50 years later and is due to Caspar Wessel.

Before presenting Euler's Formula, a brief review of complex numbers is in order. Recall that a complex number, $x + iy$ where $i = \sqrt{-1}$, represents a point in the complex plane with rectangular coordinates (x, y). Alternatively, the number can be written using polar coordinates (r, θ). Here r is the distance from the origin to the point, and θ is the angle between the real axis and the line from the origin to the point. The two representations are diagrammed in Figure 5.1. The relation between the rectangular and polar coordinates is given by

$$x = r \cos \theta, \qquad y = r \sin \theta.$$

This is readily apparent from Figure 5.1. Euler's Formula expresses this relation as

$$re^{i\theta} = r \cos \theta + i r \sin \theta.$$

We now state Euler's formula as a theorem and give a proof.

Theorem 5.7 (Euler's Formula). *For any real number θ,*

$$e^{i\theta} = \cos \theta + i \sin \theta, \tag{5.21}$$

where e is the base of the natural logarithm and $i = \sqrt{-1}$.

Proof. We prove the formula by examining the Taylor series expansions of the left and right sides of (5.21). The series expansion of e^z about $z = 0$ is

$$e^z = 1 + z + \frac{z^2}{2!} + \frac{z^3}{3!} + \frac{z^4}{4!} + \frac{z^5}{5!} + \frac{z^6}{6!} + \frac{z^7}{7!} + \cdots .$$

Substituting $z = i\theta$ and recognizing that the powers of the imaginary number i are

$$i^0 = 1, \quad i^1 = i, \quad i^2 = -1, \quad i^3 = -i, \quad i^4 = 1, \quad \dots,$$

the left side of (5.21) is given by

$$e^{i\theta} = 1 - \frac{\theta^2}{2!} + \frac{\theta^4}{4!} - \frac{\theta^6}{6!} + \cdots + i\left(\theta - \frac{\theta^3}{3!} + \frac{\theta^5}{5!} - \frac{\theta^7}{7!} + \cdots\right).$$

The series expansions for the cosine and sine functions are

$$\cos\theta = 1 - \frac{\theta^2}{2!} + \frac{\theta^4}{4!} - \frac{\theta^6}{6!} + \cdots,$$

$$\sin\theta = \theta - \frac{\theta^3}{3!} + \frac{\theta^5}{5!} - \frac{\theta^7}{7!} + \cdots .$$

Multiplying the series expansion of the sine by i and adding the two series gives

$$\cos\theta + i\sin\theta = 1 - \frac{\theta^2}{2!} + \frac{\theta^4}{4!} - \frac{\theta^6}{6!} + \cdots + i\left(\theta - \frac{\theta^3}{3!} + \frac{\theta^5}{5!} - \frac{\theta^7}{7!} + \cdots\right),$$

which is identical to the expansion of $e^{i\theta}$. $\qquad\square$

Let's explore some interesting consequences of Euler's Formula. Substituting $\theta = \pi$ into (5.21) yields

$$e^{i\pi} = \cos\pi + i\sin\pi = -1 \qquad (5.22)$$

since $\cos\pi = -1$ and $\sin\pi = 0$. This remarkable formula, when rewritten as $e^{i\pi} + 1 = 0$, gives a relation between what are perhaps the five most important mathematical constants: $e, i, \pi, 1, 0$. Taking the logarithm of both sides of (5.22) reveals that

$$\ln(-1) = i\pi.$$

What is $\ln(-2)$? By the rule for the logarithm of a product,

$$\ln(-2) = \ln(-1 \cdot 2) = \ln 2 + \ln(-1) = \ln 2 + i\pi.$$

So, contrary to what we are told in calculus, the natural logarithm can take a negative argument! When this happens, the value of the function is complex.

We now show how the trigonometric functions can be expressed in terms of complex exponentials. Euler's Formula provides a way to write e^{ix} and e^{-ix} in terms of the sine and cosine,

$$e^{ix} = \cos x + i \sin x, \tag{5.23}$$

$$e^{-ix} = \cos x - i \sin x. \tag{5.24}$$

By subtracting these equations we obtain the desired representation for the sine, and by adding them we obtain the cosine,

$$\sin x = \frac{1}{2i}\left(e^{ix} - e^{-ix}\right), \tag{5.25}$$

$$\cos x = \frac{1}{2}\left(e^{ix} + e^{-ix}\right). \tag{5.26}$$

The tangent can, of course, be expressed as their ratio,

$$\tan x = \frac{\sin x}{\cos x} = -i\,\frac{e^{ix} - e^{-ix}}{e^{ix} + e^{-ix}} = -i\,\frac{(e^{ix})^2 - 1}{(e^{ix})^2 + 1}.$$

Many familiar trigonometric identities can be derived using Euler's Formula. For example, the Pythagorean identity relating $\cos^2 x$ and $\sin^2 x$ results from multiplying (5.23) and (5.24),

$$e^{ix}\,e^{-ix} = 1 = (\cos x + i \sin x)(\cos x - i \sin x)$$
$$= \cos^2 x + \sin^2 x.$$

The double angle formulas for $\cos(2x)$ and $\sin(2x)$ arise from squaring (5.23),

$$(e^{ix})^2 = (\cos x + i \sin x)^2$$
$$= (\cos^2 x - \sin^2 x) + i\,(2 \sin x \cos x).$$

Since $(e^{ix})^2 = e^{i\,2x} = \cos(2x) + i \sin(2x)$, we obtain

$$\cos(2x) = \cos^2 x - \sin^2 x,$$
$$\sin(2x) = 2 \sin x \cos x$$

by equating the real and imaginary parts. The angle-sum relations are derived in a similar fashion,

$$e^{i(x+y)} = e^{ix}\,e^{iy} = (\cos x + i \sin x)(\cos y + i \sin y)$$
$$= (\cos x \cos y - \sin x \sin y) + i\,(\sin x \cos y + \cos x \sin y).$$

Recognizing that $e^{i(x+y)} = \cos(x+y) + i \sin(x+y)$, we have

$$\cos(x+y) = \cos x \cos y - \sin x \sin y,$$
$$\sin(x+y) = \sin x \cos y + \cos x \sin y.$$

DeMoivre's formula,

$$(\cos x + i \sin x)^n = \cos(n\,x) + i \sin(n\,x),$$

also follows immediately from Euler's Formula.

Just as the trigonometric functions can be expressed as complex exponentials, the inverse trigonometric (arctrig) functions can be written using logarithms with complex arguments:

$$\arcsin x = -i \ln \left(\sqrt{1 - x^2} + ix \right),$$

$$\arccos x = -i \ln \left(x + i\sqrt{1 - x^2} \right),$$

$$\arctan x = \frac{1}{2} i \ln \left(\frac{i + x}{i - x} \right).$$

The hyperbolic functions and their inverses comprise the other major class of transcendental functions considered in calculus. The hyperbolic functions are defined in terms of ordinary (non-complex) exponentials,

$$\sinh x = \tfrac{1}{2} \left(e^x - e^{-x} \right),$$

$$\cosh x = \tfrac{1}{2} \left(e^x + e^{-x} \right),$$

$$\tanh x = \frac{e^{2x} - 1}{e^{2x} + 1},$$

and the inverse hyperbolics can be expressed as logarithms with real arguments,

$$\operatorname{arcsinh} x = \ln \left(x + \sqrt{x^2 + 1} \right),$$

$$\operatorname{arccosh} x = \ln \left(x + \sqrt{x^2 - 1} \right),$$

$$\operatorname{arctanh} x = \frac{1}{2} \ln \left(\frac{1 + x}{1 - x} \right).$$

and the inverse hyperbolics can be expressed as logarithms with real arguments, It is interesting to observe the analogies between the corresponding trigonometric and hyperbolic functions, and their inverses.

5.4.5 Trigonometric Functions

Transforming trigonometric functions to and from complex exponentials in order to manipulate them presents some difficulties. The recombination of complex exponentials into sines and cosines by applying (5.25) and (5.26) from right to left is a highly non-trivial computational task for all but simple expressions.

In this section, we explore how trigonometric functions can be manipulated directly in a computer algebra system. We first consider the problem of expanding trigonometric functions using angle-sum and multi-angle identities. Next, we visit the inverse problem of combining trig functions by applying the same identities in reverse. Finally, we investigate the evaluation of trig functions in terms of algebraic numbers (e.g., $\cos(\pi/6) = \sqrt{3}/2$) whenever this is possible. Throughout this section, we restrict our attention to sines and cosines since the other trigonometric functions can be expressed in terms of them.

Expansion

Trigonometric functions are expanded by repeatedly applying the rules

$$\cos(u + v) \to \cos u \cos v - \sin u \sin v, \tag{5.27}$$

$$\sin(u + v) \to \sin u \cos v + \cos u \sin v. \tag{5.28}$$

The expansions of $\cos(n\,u)$ and $\sin(n\,u)$, where n is a positive integer, are a special case. (Why?) For negative arguments, the identities $\cos(-u) = \cos u$ and $\sin(-u) = -\sin u$ are employed.

The result of applying these rules is an expression that typically contains products of sums. These are expanded algebraically and like terms combined in an effort to work toward a standard representation. A consequence of the Pythagorean relation $\sin^2 u + \cos^2 u = 1$, however, is that trig-expanded expressions typically have multiple representations. To obtain a canonical form, computer algebra systems generally replace powers of sines greater than one by powers of cosines (or less frequently, powers of cosines greater than one by powers of sines).

Example 5.21. Expand the following trigonometric functions:

(a) $\sin(3\,x)$. Expressing $\sin(3\,x)$ as $\sin(2\,x + x)$, repeatedly applying (5.27) and (5.28), and expanding algebraically gives

$$\sin(3\,x) \to \sin(2\,x)\cos x + \cos(2\,x)\sin x$$
$$\to (\sin x \cos x + \cos x \sin x)\cos x + (\cos^2 x - \sin^2 x)\sin x$$
$$\to 3\cos^2 x \sin x - \sin^3 x.$$

This representation is not unique. If we replace $\sin^2 x$ by $1 - \cos^2 x$ in the $\sin^3 x$ term and again expand algebraically, we obtain

$$4\cos^2 x \sin x - \sin x,$$

the canonical representation a computer algebra system would most likely report.

(b) $\sin(2(x+y))$. Applying (5.28) with both u and v equal to $x+y$, we get

$$\sin(2(x+y)) \to 2 \sin(x+y) \cos(x+y).$$

Using both (5.27) and (5.28) to rewrite the result, we obtain

$$2(\sin x \cos y + \cos x \sin y)(\cos x \cos y - \sin x \sin y).$$

Expanding this product algebraically gives

$$2 \cos x \sin x \cos^2 y - 2 \sin^2 x \cos y \sin y$$
$$+ 2 \cos^2 x \cos y \sin y - 2 \cos x \sin x \sin^2 y.$$

Finally we use the Pythagorean identity to replace $\sin^2 x$ and $\sin^2 y$, and expand the resulting expresssion algebraically to yield

$$4 \cos^2 x \cos y \sin y + 4 \cos x \sin x \cos^2 y$$
$$- 2 \cos x \sin x - 2 \cos y \sin y,$$

our canonical representation. □

Redundant recursion can arise when expanding trigonometric functions unless we are careful. Observe that the expansion of $\cos(u+v)$ using (5.27) may require the expansion of both $\cos u$ and $\sin u$. Similarly, the expansion of $\sin(u+v)$ with (5.28) may also require the expansion of both $\cos u$ and $\sin u$. Suppose $u+v$ is a sum of n terms and we regard $n-1$ of these as belonging to u and one to v. Then $2^{n-1} - 1$ rule applications are used to expand a cosine or a sine recursively. A way around this exponential growth is to expand $\cos w$ and $\sin w$ as a pair at each stage. This approach results in only $2(n-1)$ rule applications.

Combination

Trigonometric combination can be regarded as the inverse of trigonometric expansion. Combination is accomplished by repeatedly applying the rules

$$\cos u \cos v \to \tfrac{1}{2} \cos(u+v) + \tfrac{1}{2} \cos(u-v), \tag{5.29}$$

$$\sin u \sin v \to \tfrac{1}{2} \cos(u-v) - \tfrac{1}{2} \cos(u+v), \tag{5.30}$$

$$\cos u \sin v \to \tfrac{1}{2} \sin(u+v) - \tfrac{1}{2} \sin(u-v), \tag{5.31}$$

to pairs of factors occurring in a product of sines and cosines. As the combination process proceeds, the resulting expressions are expanded algebraically, allowing similar terms to add or cancel. Repeated application the transformations permits us to obtain combined forms for $\cos^n u$ and $\sin^n u$.

Example 5.22. Combine the trigonometric functions in the following expressions:

(a) $\sin^2 x + \sin^2(2\,x)$. Applying (5.30) twice gives

$$\sin^2 x + \sin^2(2\,x) \to \tfrac{1}{2}\cos(x - x) - \tfrac{1}{2}\cos(x + x)$$
$$+ \tfrac{1}{2}\cos(2\,x - 2\,x) - \tfrac{1}{2}\cos(2\,x + 2\,x)$$
$$\to 1 - \tfrac{1}{2}\cos(2\,x) - \tfrac{1}{2}\cos(4\,x)$$

since $\cos(0) = 1$.

(b) $3\cos^2 x \sin x - \sin^3 x$. From (5.29), $\cos^2 x = \tfrac{1}{2}\cos(2\,x) + \tfrac{1}{2}$ and so

$$\cos^2 x \sin x \to \left(\tfrac{1}{2}\cos(2\,x) + \tfrac{1}{2}\right)\sin x$$
$$\to \tfrac{1}{2}\cos(2\,x)\sin x + \tfrac{1}{2}\sin x.$$

Applying (5.31) to $\cos(2\,x)\sin x$ gives $\tfrac{1}{2}\sin(3\,x) - \tfrac{1}{2}\sin x$. Therefore

$$\cos^2 x \sin x \to \tfrac{1}{4}\sin(3\,x) + \tfrac{1}{4}\sin x. \tag{5.32}$$

From (5.30), $\sin^2 x = \tfrac{1}{2} - \tfrac{1}{2}\cos(2\,x)$ and

$$\sin^3 x \to \left(\tfrac{1}{2} - \tfrac{1}{2}\cos(2\,x)\right)\sin x$$
$$\to \tfrac{1}{2}\sin x - \tfrac{1}{2}\cos(2\,x)\sin x.$$

Making the same substitution as before for $\cos(2\,x)\sin x$, we have

$$\sin^3 x \to \tfrac{3}{4}\sin x - \tfrac{1}{4}\sin(3\,x). \tag{5.33}$$

Combining (5.32) and (5.33) produces the solution,

$$3\cos^2 x \sin x - \sin^3 x \to \sin(3\,x).$$

These manipulations invert the expansion of $\sin(3\,x)$ given in Example 5.21. $\quad\square$

Evaluation

Computer algebra systems typically evaluate trigonometric functions whose arguments are of the form $k\pi/n$, where k and n are integers and n is small. In this context, a small n usually means $n = 1, 2, 3, 4, 6$ and sometimes also includes $n = 5, 8, 10, 12$. Although the value of a sine or cosine with an argument of the form $k\pi/n$ is always an algebraic number, the representation can be quite involved and is rarely useful for values of n other than those mentioned.

We show by example how to evaluate $\cos(\pi/n)$ for $n > 2$. The value of $\cos(k\pi/n)$ for any integer k can be obtained from that of $\cos(\pi/n)$ with the multi-angle formula and the fact that $\cos(-x) = \cos(x)$.

Example 5.23. Find the value of $\cos(\pi/5)$.

Solution. To evaluate $\cos(\pi/n)$, we first apply the multi-angle formula to expand $\cos(n\,\theta)$ as a trigonometric polynomial in $\cos\theta$. This can always be done without the use of the sine function. In this instance, we have

$$\cos(5\,\theta) = 16\cos^5\theta - 20\cos^3\theta + 5\cos\theta.$$

We next use this expansion to relate the value of $\cos\pi$ to $\cos(\pi/5)$,

$$\cos\pi = -1 = 16\cos^5\frac{\pi}{5} - 20\cos^3\frac{\pi}{5} + 5\cos\frac{\pi}{5}.$$

Therefore, the minimal polynomial for the algebraic number representing $\cos(\pi/5)$ is $16\alpha^5 - 20\alpha^3 + 5\alpha + 1$. The solutions to the polynomial are

$$-1,\ \frac{1}{4}\left(1 + \sqrt{5}\right)\ \text{(double root)},\ \frac{1}{4}\left(1 - \sqrt{5}\right)\ \text{(double root)}.$$

Since $\theta = \pi/n$ always lies in the first quadrant for $n > 2$, the solution of interest is positive and has a value closest to that of $\cos(0) = 1$. In this case, we have $\cos(\pi/5) = \frac{1}{4}(1 + \sqrt{5})$. $\qquad\square$

We can develop a similar procedure for evaluating sines of angles having the form $k\pi/n$. A slight difficulty arises, however, in that the expansion of $\sin(n\,\theta)$ is a product of $\sin\theta$ with a polynomial in $\cos\theta$, whereas the expansion of $\cos(n\,\theta)$ can be expressed simply as a polynomial in $\cos\theta$. While we can easily surmount this problem, an alternative approach is to realize that $\sin(k\pi/n) = \cos((n-2k)\pi/(2n))$, as can be seen by applying the angle-sum formula (5.27) to $\cos(\pi/2 - k\pi/n)$.

5.5   Exercises

1. Perform the partial fraction decomposition of the following rational functions.

   (a) $\dfrac{x^5 - 1}{x^4 - x^2}$ (b) $\dfrac{x^2 + 1}{(x + 2)\,(x - 1)\,(x^2 + x + 1)}$ (c) $\dfrac{x^2 - 1}{(x + 2)\,(x - 1)\,(x^2 + x + 1)}$

2. (a) Express the rational function

   $$\frac{\frac{7}{5} - x}{\frac{3}{2}x^2 - \frac{1}{4}x + \frac{1}{3}}$$

   as a ratio of polynomials with integer coefficients.

   (b) Now express it as a ratio of polynomials where the leading coefficient of the denominator is one.

3. Show that the following sets of radicals are algebraically dependent by finding a polynomial h, as in Example 5.1, that gives the dependency between them.

(a) $\{\sqrt{3}, \sqrt[3]{81}\}$ (b) $\{\sqrt{6}, \sqrt{10}, \sqrt{15}\}$ (c) $\{\sqrt{5} - \sqrt{3}, \sqrt{5} + \sqrt{3}\}$

(d) $\{\sqrt[4]{-4}, \sqrt{2}, i = \sqrt{-1}\}$ (e) $\{\sqrt{3}, \sqrt{5}, 47 - 12\sqrt{15}\}$

4. Find the minimal polynomial for the following algebraic numbers.

(a) $\dfrac{1}{\sqrt{2} + \sqrt{3}}$ (b) $\sqrt{5 + 2\sqrt{6}}$ (c) $(-8)^{1/3}$

(d) $1 + \sqrt{2} + \sqrt{3}$ (e) $\sqrt{2} - \sqrt{3} + \sqrt{5}$ (f) $\left(2^{1/3} - 1\right)^{1/3}$

5. (a) Express $\sin(\pi/12)$ in terms of $\cos(\pi/6)$ using the half-angle formula

$$\sin(x/2) = \sqrt{\frac{1 - \cos x}{2}}.$$

(b) Express $\sin(\pi/12)$ in terms of the sines and cosines of $\pi/4$ and $\pi/6$ using the angle-difference formula

$$\sin(x - y) = \sin x \, \cos y - \cos x \, \sin y.$$

(c) Denest the radical in part (a) to show that the algebraic numbers in parts (a) and (b) are equivalent.

6. Rationalize and simplify the following expressions to radical normal form.

(a) $\dfrac{1}{\sqrt{5} - \sqrt{3}}$ (b) $\dfrac{1}{\sqrt{7} - \sqrt{11} + \sqrt{13}}$ (c) $\dfrac{1}{2\sqrt{2}\sqrt{5} - 3\sqrt{2}\sqrt{3} + \sqrt{3}\sqrt{5}}$

(d) $\dfrac{1}{5^{3/2} - 5^{1/3}}$ (e) $\dfrac{1}{9^{1/3} - 2^{1/3}}$ (f) $\dfrac{\sqrt[3]{2} - \sqrt{3}}{\sqrt{8} - 2\sqrt[3]{6}}$

7. Denest the following expressions involving nested radicals.

(a) $\sqrt{37 + 20\sqrt{3}}$ (b) $\sqrt{107 - 12\sqrt{77}}$ (c) $\sqrt{2\sqrt{19549} + 286}$

(d) $\sqrt{4 + 3\sqrt{2}}$ (e) $\sqrt{3 + 2\sqrt{3}}$ (f) $\sqrt{x + \sqrt{x^2 - 1}}$

(g) $\sqrt{12 + 2\sqrt{6} + 2\sqrt{14} + 2\sqrt{21}}$

8. (a) Show that the Square Root Denesting Theorem fails when applied to each of the following radicals.

$$\sqrt{1 + \sqrt{3}}, \qquad \sqrt{3 + 3\sqrt{3}}, \qquad \sqrt{10 + 6\sqrt{3}}.$$

(b) The radicals in part (a) cannot, in fact, be denested. Show, however, that the sum

$$\sqrt{1 + \sqrt{3}} + \sqrt{3 + 3\sqrt{3}} - \sqrt{10 + 6\sqrt{3}}$$

denests and is equal to zero.

9. The following theorem, based on the work of Ramanujan, states when and how the square root of a sum of two cube roots can be denested.

Theorem 5.8 (Square Root of Cube Roots Denesting Theorem).
$\sqrt{\sqrt[3]{\alpha} + \sqrt[3]{\beta}}$ can be denested if and only the rational number $\frac{\alpha}{\beta}$ is not a perfect cube and the polynomial

$$F(t) = t^4 + 4\,t^3 + 8\frac{\beta}{\alpha}t - 4\frac{\beta}{\alpha}$$

has a rational root s. The denesting is given by

$$\sqrt{\sqrt[3]{\alpha} + \sqrt[3]{\beta}} = \pm\frac{1}{\sqrt{f}}\left(\sqrt[3]{\beta^2} + s\sqrt[3]{\alpha\,\beta} - \frac{s^2\sqrt[3]{\alpha^2}}{2}\right),$$

where $f = \beta - s^3\alpha$.

(a) Apply the theorem with $\alpha = 5$, $\beta = -4$, and $s = -2$ to show that

$$\sqrt{\sqrt[3]{5} - \sqrt[3]{4}} = \frac{1}{3}\left(\sqrt[3]{2} + \sqrt[3]{20} - \sqrt[3]{25}\right).$$

(b) Apply the theorem with $\alpha = 28$, $\beta = -27$, and $s = -3$ to show that

$$\sqrt{\sqrt[3]{28} - \sqrt[3]{27}} = -\frac{1}{3}\left(1 + \sqrt[3]{28} - \sqrt[3]{98}\right).$$

(c) Show that $\sqrt{\sqrt[3]{3} + \sqrt[3]{2}}$ cannot be denested.

10. Show how the following expressions would be represented and simplified using the Structure Theorem.

(a) $e^x\,e^{2x}\,e^{x/2}$ (b) $e^{\frac{1}{x-1}}/e^{\frac{1}{x+1}}$ (c) $\left(e^{e^x}\right)^{e^x}$

(d) $\ln x - \ln x^2 + \ln\sqrt{x}$ (e) $\exp(x^2 + (\ln x)/2)$ (f) $\ln(e^{2x}/e^{\sqrt{x}})$

(g) $\sqrt{\ln(x^2 + 3x + 2)\cdot(\ln(x+1) + \ln(x+2))}$

(h) $\dfrac{e^{1/(x^2-1)} + e^{1/(x+1)}\,e^{-1/(x-1)}}{e^{1/(x+1)}}$

11. Find the real and imaginary parts of these expressions involving $i = \sqrt{-1}$. Write each expression in standard rectangular form, $a + bi$, where a and b are real.

(a) e^i (b) $\ln i$ (c) $\sqrt{i}$ (d) $\sqrt[3]{i}$ (e) i^i

(f) $\cos i$ (g) $\sin i$ (h) $\cosh i$ (i) $\sinh i$ (j) $\arctanh i$

12. Expand fully the following expressions involving exponentials and logarithms.

(a) e^{2x+y} (b) $e^{2(x+y)}$ (c) $e^{x^2}\,e^{2x}$ (d) $(e^x)^2\,e^{2x}$

(e) $\ln(1/x)$ (f) $\ln(x\,y)^3$ (g) $\dfrac{1}{e^{x^2} - e^{2x}}$ (h) $\dfrac{1}{(e^x)^2 - e^{2x}}$

(i) $e^{(x+y)^2}$ (j) $e^{(e^{2x}-(e^x)^2+1)\,(2x+3y)}$ (k) $e^{x+\ln y}$ (l) $e^{x+\ln(y/z)}$

(m) $\ln\sqrt{\dfrac{x\,y^2}{z^3}}$ (n) $\ln\left(\dfrac{x^2}{y}\right)-\ln\left(\dfrac{y^2}{x}\right)$ (o) $e^{\ln\sqrt{x\,y^2/z^3}}$ (p) $e^{\ln(x^2/y)-\ln(y^2/x)}$

13. Combine the exponentials and logarithms in the following expressions.

(a) $e^{8x}e^{-4x}e^{2x}e^{-x}$ (b) $(e^{5x})^2 e^{-3x}$ (c) $e^{x^2}(e^{xy})^2 e^{y^2}$

(d) $e^{x+3\ln y}$ (e) $3\ln\left(\frac{x^2}{y}\right) - 2\ln\left(\frac{y^2}{x}\right)$ (f) $3\ln 2 - 2\ln 3$

(g) $5\ln x + 2\ln(4x) - \ln(8x^5)$ (h) $\frac{1}{2}\ln(81x^{12}) + \ln(2x^2)$

14. Express the following multiple angle trigonometric functions in terms of sines and cosines of x. Replace powers of sines greater than one by powers of cosines.

(a) $\cos(3x)$ (b) $\cos(2(x-y))$ (c) $\cos(4x)$ (d) $\sin(4x)$

(e) $\cos(5x)$ (f) $\sin(5x)$ (g) $\cos(6x)$ (h) $\sin(6x)$

15. Express $\cos(u+v+w)$ and $\sin(u+v+w)$ in terms of the sines and cosines of u, v, and w.

16. Apply trigonometric combination to the following expressions.

(a) $(\cos x + \sin y)\cos x$ (b) $\cos^2 x \sin^2 x$

(c) $2\cos^2 x - 2\cos x \sin x - 1$ (d) $6\cos^2 x - 4\cos x \sin x - 3$

(e) $\cos^4 x + \sin^4 x$ (f) $1 - 2\cos^2 x \sin^2 x$

(g) $\cos^3 x \sin^3 x$ (h) $\cos x \cos y \cos z$

17. Expand the following as a trigonometric polynomial in $\sin x$ and $\cos x$.

(a) $\sin(4x)$ (b) $\cos(4x)$ (c) $\sin(8x)$ (d) $\cos(8x)$

Perform each expansion twice—first using Strategy 1, then Strategy 2—to apply rules (5.27) and (5.28).

Strategy 1. Express $\sin(nx)$ as $\sin((n-1)x + x)$ and $\cos(nx)$ as $\cos((n-1)x + x)$ at every step.

Strategy 2. Express $\sin(nx)$ as $\sin(\frac{n}{2}x + \frac{n}{2}x)$ and $\cos(nx)$ as $\cos(\frac{n}{2}x + \frac{n}{2}x)$ at every step.

With both strategies, use the Pythagorean identity $\sin^2 x + \cos^2 x = 1$ to replace powers of sines greater than one by powers of cosines.

18. (a) Transform the trigonometric expression $\sin(2x)\cos(3x)$ into complex exponential form.

(b) Express $\sin(2x)\cos(3x)$ as a polynomial in $\sin x$ and $\cos x$.

(c) Express $\sin(2x)\cos(3x)$ as a linear combination of sines and cosines of $5x$ and x.

19. (a) Derive expressions for $\operatorname{arcsinh} x$ and $\operatorname{arccosh} x$ in terms of logarithms.

(b) Derive expressions for $\arcsin x$ and $\arccos x$ in terms of complex logarithms.

20. (a) Give the exact value of $\sin(\pi/5)$ in terms of algebraic numbers.

(b) Give the exact values of $\cos(k\pi/5)$ for $k = 1, 2, 3, 4, 5$.

21. Express the exact values of the following cosines in terms of algebraic numbers.

(a) $\cos(\pi/8)$ (b) $\cos(\pi/10)$ (c) $\cos(\pi/12)$

22. Prove the multiangle identity for tangents,

$$\tan(u + v) = \frac{\tan u + \tan v}{1 - \tan u \tan v}.$$

Start by expressing $\tan(u + v)$ as the ratio of the sine and cosine.

23. Use the result of Exercise 22 to express each of the following in terms of $\tan x$.

    (a) $\tan(2x)$ (b) $\tan(3x)$ (c) $\tan(4x)$

24. Verify the following equalities involving the arctangent.

    (a) $\arctan(1/2) + \arctan(1/3) = \dfrac{\pi}{4}$ (b) $4\arctan(1/5) - \arctan(1/239) = \dfrac{\pi}{4}$

    Hint. Use the result of Exercise 22.

25. Verify the identity

$$\tan(4x) = \frac{\sin(x) + \sin(3x) + \sin(5x) + \sin(7x)}{\cos(x) + \cos(3x) + \cos(5x) + \cos(7x)}.$$

    Hint. Express $\tan(4x)$ as $\sin(4x)/\cos(4x)$, cross multiply the numerators and denominators, and apply trigonometric expansion.

26. Show that the following expressions are equal to zero.

    (a) $\sqrt{2} + 1 - \left(7 + 5\sqrt{2}\right)^{1/3}$ (b) $\left(\sqrt{3} + \sqrt{2}\right)^{1/3}\left(\sqrt{3} - \sqrt{2}\right)^{1/3} - 1$

27. Show $\ln(\tan(x/2 + \pi/4)) - \operatorname{arcsinh}(\tan(x)) = 0$ for $-\pi/2 < x < \pi/2$.

Programming Projects

1. **(Trigonometric Expansion)** Write two procedures,

   `ExpandSin(<n>,<theta>)` and `ExpandCos(<n>,<theta>)`

   that return expansions for `sin(<n><theta>)` and `cos(<n><theta>)`, respectively. Assume that `<n>` is an integer (positive or negative) and allow for the fact that `<theta>` may be a sum. When expanding sines and cosines, move powers of sine greater than one to cosines.
   For example, `ExpandCos(-3,x)` should return

   $$4\cos(x)^3 - 3\cos(x)$$

   and `ExpandSin(1,2x-y)` should return

   $$2\cos(x)\sin(x)\cos(y) - 2\cos(x)^2\sin(y) + \sin(y)$$

   Make your code as efficient as possible.

2. **(Rationalizing Denominators)** Write a procedure to rationalize the denominator of an algebraic number,

$$\text{Rationalize}(\texttt{<numer>},\texttt{<denom>},[\texttt{<r-list>}]) \; .$$

`<numer>` represents the numerator of the algebraic number, `<denom>` represents
the denominator. Every radical appearing in the algebraic number is replaced
by a variable, and the variable name and its minimal polynomial appear as a
pair in `r-list`.

Here are two examples of `Rationalize` commands and transformations they
produce.

command	before	after
`Rationalize(1,2*a-1,[[a,a^2-5]])`	$\dfrac{1}{2\sqrt{5}-1}$	$\dfrac{1}{19}\left(2\sqrt{5}+1\right)$
`Rationalize(1,a-b,[[a,a^2-3],[b,b^2-2]])`	$\dfrac{1}{\sqrt{3}+\sqrt{2}}$	$\sqrt{3}-\sqrt{2}$

Chapter 6

Factorization

It has often been said that a person does not really understand some-thing until he can teach it to someone else. Actually a person does not really understand something until he can teach it to a computer.

Donald E. Knuth
"Computer Science and Its Relation to Mathematics,"
American Mathematical Monthly, 81 (1974)

As soon as an Analytical Engine exists, it will necessarily guide the future course of the science. Whenever any result is sought by its aid, the question will then arise—By what course of calculation can these results be arrived at by the machine in the shortest time.

Charles Babbage (1791–1871)
Passages from the Life of a Philosopher, 1864

The methods we learned in algebra to factor polynomials rely on a certain amount of pattern matching, guesswork, and just plain luck. They work well for quadratics, a few special cubics, and perhaps an occasional polynomial of higher degree where one or two roots are obvious. One of the more impressive capabilities of computer algebra systems is their ability to factor polynomials reliably and efficiently. How do these systems solve this problem?

6.1 What is Factoring?

We begin by addressing the question of what it means to factor a polyno-mial. Let's examine the response of a typical computer algebra system, Maple, to a few simple factorization problems.

$$\texttt{factor(x\textasciicircum 2 - 9)} \rightarrow (x-3)\,(x+3) \tag{6.1}$$

$$\texttt{factor(4x\textasciicircum 2 + 16*x + 16)} \rightarrow 4\,(x+2)^2 \tag{6.2}$$

$$\texttt{factor(x\textasciicircum 2 - 9/16)} \rightarrow \tfrac{1}{16}\,(4\,x-3)\,(4\,x+3) \tag{6.3}$$

$$\texttt{factor(x\textasciicircum 2 - 7)} \rightarrow x^2 - 7 \tag{6.4}$$

$$\texttt{factor(x\textasciicircum 2 + 9)} \rightarrow x^2 + 9 \tag{6.5}$$

The first two answers are not surprising. By analogy with (6.1), we might expect $(x - \frac{3}{4})(x + \frac{3}{4})$ to be returned for (6.3). The polynomials in (6.4) and (6.5) are returned unfactored. Why not answer $(x - \sqrt{7})(x + \sqrt{7})$ and $(x - 3\,i)(x + 3\,i)$, respectively?

When factoring polynomials, we need to do so with respect to some algebraic system. Computer algebra systems typically choose to factor over the integers unless directed otherwise. A polynomial that cannot be factored is said to be *irreducible*. For example, we see from (6.4) above that $x^2 - 7$ is irreducible over the integers, but factors as $(x - \sqrt{7})(x + \sqrt{7})$ when algebraic numbers are allowed to express the result. Similarly in (6.5), $x^2 + 9$ is irreducible over the integers, although it factors over the complex numbers as $(x - 3\,i)(x + 3\,i)$.

Factoring polynomials with rational coefficients is equivalent to factoring over the integers. To factor

$$p(x) = \frac{a_n}{b_n}\,x^n + \frac{a_{n-1}}{b_{n-1}}\,x^{n-1} + \cdots + \frac{a_1}{b_1}\,x + \frac{a_0}{b_0},$$

we first compute the least common multiple (lcm) of the b_i and then remove $1/\text{lcm}(b_n, b_{n-1}, \ldots, b_1, b_0)$ from p, leaving a polynomial with integer coefficients. So factoring $x - \frac{9}{16}$ is equivalent to factoring $\frac{1}{16}\,(16\,x^2 - 9)$, explaining Maple's answer to (6.3).

In this chapter, we explore how computer algebra systems factor polynomials in one variable with integer coefficients. Our problem is to take such a polynomial, $p(x)$, and to express it as a product

$$p(x) = c\,p_1(x)^{e_1}\,p_2(x)^{e_2}\,\cdots\,p_r(x)^{e_r},$$

where c is the content (gcd of the coefficients) of $p(x)$, the polynomials $p_i(x)$ are irreducible, and the exponents e_i are positive integers. This factorization is unique (canonical) up to the order of the factors in the product.

6.2   Rational Roots and Linear Factors

We begin by identifying the linear factors, which are given by the rational roots (or zeros) of a polynomial. For example, the rational roots of

$$p(x) = x^4 + 2x^3 + x^2 - 2x - 2$$

are $x = 1$ and $x = -1$ since $p(1) = 0$ and $p(-1) = 0$. Therefore $(x-1)(x+1)$ evenly divides p, leaving the quadratic $x^2 + 2x + 2$ as the quotient. As a result, p factors as

$$p(x) = (x-1)(x+1)(x^2 + 2x + 2).$$

The following theorem identifies the rational roots of a polynomial.

Theorem 6.1 (Rational Root Test). *Suppose*

$$p(x) = c_n x^n + c_{n-1} x^{n-1} + \cdots + c_1 x + c_0$$

is a polynomial with integer coefficients and $x = \frac{a}{b}$ (reduced to lowest terms) is a rational root of p. Then a divides c_0 and b divides c_n.

Proof. To see this is true, if $x = \frac{a}{b}$ is a root of p then $p(\frac{a}{b}) = 0$,

$$c_n \left(\frac{a}{b}\right)^n + c_{n-1}\left(\frac{a}{b}\right)^{n-1} + \cdots + c_1 \left(\frac{a}{b}\right) + c_0 = 0.$$

Multiplying both sides by b^n, we have

$$c_n a^n + c_{n-1} a^{n-1} b + \cdots + c_1 a b^{n-1} + c_0 b^n = 0$$

or

$$\left(c_n a^{n-1} + c_{n-1} a^{n-2} b + \cdots + c_1 b^{n-1}\right) a = -c_0 b^n.$$

Since the left side is evenly divisible by a, the right side must also be evenly divisible by a. Moreover, a divides c_0 because a and b are relatively prime. We can also rearrange $p(\frac{a}{b}) = 0$ as

$$\left(c_{n-1} a^{n-1} + c_{n-2} a^{n-2} b + \cdots + c_0 b^{n-1}\right) b = -c_n a^n,$$

so by similar reasoning, b must divide c_n.   □

Example 6.1. Find the rational roots of $p(x) = 6x^4 - x^3 + 4x^2 - x - 2$.

Solution. The possible values for a are the divisors of -2: $\pm 1, \pm 2$. The possible values for b are the positive divisors of 6: $1, 2, 3, 6$. Therefore, the candidates for rational roots are (in decreasing order of magnitude):

$$\pm 2, \quad \pm 1, \quad \pm \tfrac{2}{3}, \quad \pm \tfrac{1}{2}, \quad \pm \tfrac{1}{3}, \quad \pm \tfrac{1}{6}.$$

Substituting each of these 12 values into p, we find $p(\frac{2}{3}) = p(-\frac{1}{2}) = 0$ and $p(x) \neq 0$ for the other 10 choices. Therefore, $(x-\frac{2}{3})(x+\frac{1}{2}) = \frac{1}{6}(3x-2)(2x+1)$ evenly divides p and

$$p(x) = \left(x - \tfrac{2}{3}\right)\left(x + \tfrac{1}{2}\right)\left(6x^2 + 6\right) = (3x - 2)(2x + 1)(x^2 + 1).   □$$

6.3   Schubert-Kronecker Factorization

Isaac Newton, in his *Arithmetica Universalis* (1707), described a method for finding the linear and quadratic factors of a univariate polynomial. In 1793, the astronomer Friedrich von Schubert showed how to extend this technique to find factors of any degree. Schubert's method was rediscovered by Leopold Kronecker some 90 years later. Their procedure is fairly straightforward to understand and implement. The key ideas behind the Schubert-Kronecker method are:

- If the polynomial p can be factored, then one factor has degree at most $m = \lfloor \frac{\text{degree}(p)}{2} \rfloor$.

- If the polynomial q divides p, then $q(a)$ evenly divides $p(a)$ for any integer a.

- A unique polynomial of degree d (or less) can be interpolated through $d + 1$ points.

Here are the steps in the Schubert-Kronecker method:

1. Choose $m + 1$ integers: $a_0, a_1, \ldots, a_m$. $(0, \pm 1, \pm 2, \ldots$ will work.)

2. Evaluate p at each of these values a_i.

3. Find the set D_i of distinct (+ and −) divisors of $p(a_i)$.

4. To find the degree d factors of p, choose combinations of $d + 1$ points (a_i, b_i) where $b_i \in D_i$ and interpolate a polynomial q through these points. If q evenly divides p, then q is a factor of p.

A more complete description of the procedure is given in Algorithm 6.1. We illustrate the technique with an example.

Example 6.2. Factor the polynomial $p(x) = x^4 + 4$ using the Schubert-Kronecker method.

Solution. If p can be factored, then one factor must be of degree $m = 2$ or less. (Why?) Let's evaluate p at the $m + 1$ points: $a_0 = 0$, $a_1 = 1$, $a_2 = -1$. The values of p at these points and their divisors are

$$p(a_0) = p(0) = 4, \qquad\qquad D_0 = \{\pm 1, \pm 2, \pm 4\},$$
$$p(a_1) = p(1) = 5, \qquad\qquad D_1 = \{\pm 1, \pm 5\},$$
$$p(a_2) = p(-1) = 5, \qquad\qquad D_2 = \{\pm 1, \pm 5\}.$$

The Rational Root Test reveals that p has no linear factors. To find its quadratic factors, there are $6 \cdot 4 \cdot 4 = 96$ triples to test.

procedure $SKFactor(p)$
{Input: primitive polynomial p with integer coefficients}
{Output: factorization of p over the integers, pf}
  $r := p$ {r is the unfactored portion of p}
  $pf := 1$ {pf is the factored portion of p}
  $i := 0$
  while $i \leq \lfloor \text{degree}(r)/2 \rfloor$ **do**
    choose an integer a_i and evaluate $p(a_i)$ {$a_i = 0, \pm 1, \pm 2, \ldots$ works}
    if $p(a_i) = 0$ **then** {a linear factor is found}
      while $x - a_i$ divides r **do**
        $pf := pf \cdot (x - a_i);$ $r := r/(x - a_i)$
      end do {is it a repeated linear factor?}
    else
      $D_i :=$ set of distinct divisors of $p(a_i)$
    end if
  end do
  $m := \lfloor \text{degree}(r)/2 \rfloor$
  for $d := 1$ **to** m **do**
    for every subset of $d + 1$ values b_i, one from each $D_i, 0 \leq i \leq d$ **do**
      interpolate q through $(a_0, b_0), (a_1, b_1), \ldots, (a_d, b_d)$
      while q divides r **do** {q is a factor of p}
        $pf := pf \cdot q;$ $r := r/q$
        if $\text{degree}(r) < 2d$ **then** {we're done!}
          return$(pf \cdot r)$
        end if
      end do {q may be a repeated factor}
    end do
    if $\text{degree}(r) < 2(d + 1)$ **then** {we're done!}
      return$(pf \cdot r)$
    end if
  end do
end procedure

ALGORITHM 6.1: Schubert-Kronecker factorization

One triple that doesn't succeed is $b_0 = 1$, $b_1 = -1$, $b_2 = 5$. Interpolating the quadratic form $q(x) = A\,x^2 + B\,x + C$ through the three points, we obtain three equations in the three unknowns:

$$(0,1): \qquad 1 = A \cdot 0^2 + B \cdot 0 + C,$$
$$(1,-1): \qquad -1 = A \cdot 1^2 + B \cdot 1 + C,$$
$$(-1,5): \qquad 5 = A \cdot (-1)^2 + B \cdot (-1) + C.$$

Solving, we find that $A = 1, B = -3, C = 1$, and the corresponding quadratic $q(x) = x^2 - 3x + 1$ does not evenly divide p.

A triple that works is $b_0 = 2$, $b_1 = 5$, $b_2 = 1$. Here we have

$$(0,2): \qquad 2 = A \cdot 0^2 + B \cdot 0 + C,$$
$$(1,5): \qquad 5 = A \cdot 1^2 + B \cdot 1 + C,$$
$$(-1,1): \qquad 1 = A \cdot (-1)^2 + B \cdot (-1) + C,$$

and $A = 1, B = 2, C = 2$. This time $q(x) = x^2 + 2x + 2$ evenly divides p and the quotient is $x^2 - 2x + 2$. Therefore, the factorization of p is given by $p(x) = (x^2 + 2x + 2)(x^2 - 2x + 2)$. $\qquad\qquad\square$

One way to improve the efficiency of the Schubert-Kronecker method is in the selection of points where the polynomial p to be factored is evaluated. The procedure runs more quickly when the sizes of the sets D_i formed in Step 3 are small. So instead of selecting and using the first points to come along, we can sample the value of p at many more integer values a_i and choose those having the fewest divisors. Points where $p(a_i)$ is prime are particularly nice to use. In fact, if the value of p is prime for sufficiently many points or for sufficiently large a_i, it can be shown that p is irreducible. The following theorem due to M. Ram Murty (2002) provides such an avenue to improve the efficiency of the procedure.

Theorem 6.2. *Let* $p(x) = c_n x^n + c_{n-1}x^{n-1} + \ldots + c_1 x + c_0$ *be a polynomial of degree* n *with integer coefficients and let*

$$M = \max_{0 \le i < n} |c_i / c_n|.$$

If $p(a)$ *is prime for some integer* $a \ge M + 2$, *then* $p(x)$ *is irreducible over the integers.*

Even with this improvement, the running time of the Schubert-Kronecker algorithm is, unfortunately, exponential in the degree of p. As a result, the method is practical only for polynomials of very small degree.

6.4 Simplifying Polynomial Factorization

Before we develop a far more efficient algorithm, we observe that our problem can be greatly simplified by restricting the types of polynomials considered.

6.4.1 Monic Polynomials

We first show how the factorization of a polynomial with an arbitrary leading coefficient can be obtained from that of an "equivalent" monic polynomial—a polynomial whose leading coefficient is one.

Suppose we want to factor a polynomial p whose leading coefficient $c_n \neq 1$. We perform these steps:

1. Form the monic polynomial $q(x)$ by substituting x/c_n for x in $p(x)$ and multiplying by c_n^{n-1}: $q(x) = c_n^{n-1} p(x/c_n)$.

2. Factor $q(x)$.

3. Reverse the substitution and divide $q(c_n x)$ by c_n^{n-1}: $p(x) = q(c_n x)/c_n^{n-1}$.

We give a simple example to illustrate the technique.

Example 6.3. Find the factorization of the polynomial $p(x) = 2x^2 + 3x + 1$ from the factorization of an "equivalent" monic polynomial.

Solution. First, we form the monic polynomial q by substituting $\frac{x}{2}$ for x in p and multiplying by 2 (the leading coefficient of p raised to the power one less than its degree),

$$q(x) = 2\left(2\left(\tfrac{x}{2}\right)^2 + 3\left(\tfrac{x}{2}\right) + 1\right)$$
$$= x^2 + 3x + 2.$$

Next, we factor q: $q(x) = (x+1)(x+2)$. Finally, we reverse the process by substituting $2x$ for x in q and dividing by 2 to uncover the factorization of p,

$$p(x) = \tfrac{1}{2}(2x+1)(2x+2)$$
$$= (2x+1)(x+1). \qquad \square$$

It is advantageous to start with a polynomial p that is *primitive*—one for which content (gcd of the coefficients) is removed. In this case, we can remove the gcd of the coefficients from each factor produced at Step 3 to obtain the final result. The following more involved example illustrates this subtle point.

Example 6.4. Discover the factorization of $p(x) = 112x^4 + 58x^3 - 31x^2 + 107x - 66$ from the factorization an "equivalent" monic polynomial.

Solution. We begin by forming the monic polynomial

$$q(x) = (112)^3 \left(112 \left(\tfrac{x}{112}\right)^4 + 58 \left(\tfrac{x}{112}\right)^3 - 31 \left(\tfrac{x}{112}\right)^2 + 107 \left(\tfrac{x}{112}\right) - 66\right)$$
$$= x^4 + 58x^3 - 3472x^2 + 1342208x - 92725248.$$

This polynomial factors as $q(x) = (x^2+98x-9408)(x^2-40x+9856)$. Reversing the substitution, we obtain $p(x) = q(112x)/(112)^3$,

$$\tfrac{1}{(112)^3} \left((112x)^2 + 98 \cdot 112x - 84 \cdot 112\right) \left((112x)^2 - 40 \cdot 112x + 88 \cdot 112\right).$$

We readily see that 112 can be removed from both of the quadratic factors, leaving
$$p(x) = \tfrac{1}{112} \left(112x^2 + 98x - 84\right)\left(112x^2 - 40x + 88\right).$$

The content (gcd of the coefficients) of what remains of the first factor is 14 and that of the second is 8, and their product cancels the 112 in the denominator,

$$p(x) = \tfrac{1}{112} \cdot 14 \left(8x^2 + 7x - 6\right) \cdot 8 \left(14x^2 - 5x + 11\right)$$
$$= \left(8x^2 + 7x - 6\right)\left(14x^2 - 5x + 11\right).$$

Notice that the final result could have been obtained simply by discarding the content from each of the quadratic factors of $q(112x)$. □

6.4.2 Square-Free Decomposition

In the last section, we saw that a procedure for factoring monic polynomials gives us a technique for factoring polynomials with arbitrary leading coefficients. Another simplification we can employ involves the detection of repeated factors of a polynomial. These are determined quickly by calculating the *square-free decomposition.*

An example of a polynomial in square-free form is

$$p(x) = (x^2 + 1)(x^2 - 1)^4 (x^3 + 3x)^5.$$

Once the square-free form has been found, each of its components can be factored separately to produce the complete factorization,

$$p(x) = (x^2 + 1)(x - 1)^4 (x + 1)^4 x^5 (x^2 + 3)^5.$$

A primitive (content removed) polynomial p is *square-free* if it has no repeated factors. This is equivalent to saying that there is no polynomial q with degree$(q) \geq 1$ such that q^2 divides p. The *square-free decomposition* of p is of the form

$$p(x) = \prod_{i=1}^{r} p_i(x)^i = p_1(x)\, p_2(x)^2 \cdots p_i(x)^i \cdots p_r(x)^r,$$

where each p_i is a square-free polynomial and $\gcd(p_i, p_j) = 1$ for $i \neq j$. For the polynomial $p(x) = (x^2 + 1)(x^2 - 1)^4(x^3 + 3x)^5$, the components of the decomposition are

$$p_1(x) = x^2 + 1, \qquad\qquad p_4(x) = x^2 - 1,$$
$$p_2(x) = 1, \qquad\qquad\quad p_5(x) = x^3 + 3\,x.$$
$$p_3(x) = 1,$$

Notice that p_4 and p_5 are not irreducible.

In determining the square-free decomposition of a primitive polynomial, we use the notion of the derivative p' of a polynomial p:

$$p(x) = c_n x^n + c_{n-1} x^{n-1} + \ldots + c_i x^i + \ldots + c_2 x^2 + c_1 x + c_0,$$
$$p'(x) = n c_n x^{n-1} + (n-1) c_{n-1} x^{n-2} + \ldots + i c_i x^{i-1} + \ldots + 2 c_2 x + c_1.$$

The following theorem provides a test for repeated factors. Its proof is instructive in that it embodies an algorithm for computing the square-free decomposition.

Theorem 6.3 (Repeated Factors Theorem). *The primitive polynomial $p(x)$ has repeated factors if and only if $\gcd(p(x), p'(x)) \neq 1$.*

Proof. p has repeated factors $\Rightarrow \gcd(p, p') \neq 1$. If p has repeated factors, then we can write

$$p(x) = s(x)^2 t(x),$$

where $s(x)$ is the product of all the repeated factors taken once. (Here we include in $t(x)$ the additional factors of the polynomials in $s(x)$ along with the non-repeated factors of p.) Taking derivatives using the product rule,

$$p'(x) = 2\,s(x)\,s'(x)\,t(x) + s(x)^2\,t'(x)$$
$$= s(x)\,(2\,s'(x)\,t(x) + s(x)\,t'(x))\,.$$

Therefore, s is a common factor of p and p' and $\gcd(p, p') \neq 1$.

We prove by contradiction that $\gcd(p, p') \neq 1 \Rightarrow p$ has repeated factors. Suppose $\gcd(p, p') \neq 1$ but that p is square-free. If the factorization of p is

$$p(x) = \prod_{i=1}^{r} p_i(x) = p_1(x)\, p_2(x) \cdots p_i(x) \cdots p_r(x),$$

then, by the product rule, its derivative p' is

$$p'(x) = \sum_{i=1}^{r} p_1(x) \cdots p_{i-1}(x)\, p_i'(x)\, p_{i+1}(x) \cdots p_r(x). \tag{6.6}$$

Since $\gcd(p, p') \neq 1$, at least one factor of p, say p_i, divides both p and p'. This further implies p_i divides every term in (6.6), including $p_1 \cdots p_i' \cdots p_r$. But since $\gcd(p_i, p_j) = 1$ for $i \neq j$, p_i must divide p_i', which is impossible since $\text{degree}(p_i) > \text{degree}(p_i')$. $\qquad\square$

The proof of the theorem suggests a method for computing the square-free decomposition. Suppose

$$p(x) = \prod_{i=1}^{r} p_i(x)^i = p_1(x)\, p_2(x)^2 \cdots p_i(x)^i \cdots p_r(x)^r.$$

Then, by the product rule for differentiation,

$$p'(x) = \sum_{i=1}^{r} p_1(x) \cdots i\, p_i'(x)\, p_i(x)^{i-1} \cdots p_r(x)^r,$$

and their common factors, g, are

$$g(x) = \gcd(p(x),\, p'(x)) = \prod_{i=2}^{r} p_i(x)^{i-1}.$$

The product of the square-free factors without their multiplicities is given by

$$r(x) = p(x)/g(x) = p_1(x)\, p_2(x) \cdots p_r(x) = \prod_{i=1}^{r} p_i(x).$$

To obtain the first square-free factor, p_1, we calculate

$$s(x) = \gcd(g(x),\, r(x)) = p_2(x)\, p_3(x) \cdots p_r(x) = \prod_{i=2}^{r} p_i(x)$$

and divide this out of r,

$$p_1(x) = r(x)/s(x).$$

Finding the second square-free factor, $p_2(x)$, is the same as finding the first square-free factor of g. But observe that the common factors of g and g' can be found simply by dividing g by s,

$$g_{new} = g(x)/s(x) = \gcd(g(x),\, g'(x)) = \prod_{i=3}^{r} p_i(x)^{i-2},$$

and the product of the remaining square-free factors is just s,

$$r_{new} = s(x) = \prod_{i=2}^{r} p_i(x).$$

We can continue iterating this way to find each square-free factor. The entire procedure, due to David Yun (1976), is given in Algorithm 6.2.

Example 6.5. Find the square-free factorization of $p(x) = x^8 - 2x^6 + 2x^2 - 1$.

Solution. We begin by calculating p's derivative, $p'(x) = 8x^7 - 12x^5 + 4x$, and finding their common factors, g,

$$g := \gcd(p(x),\, p'(x)) = x^4 - 2\,x^2 + 1.$$

Dividing g out of p, what remains is the product of the square-free factors,

$$r := p/g = x^4 - 1.$$

procedure *SquareFree(u)*
{Input: monic polynomial u}
{Output: square-free factorization of u, $sqfd$}
$p := u;$ $sqfd := 1;$ $i := 1$
$g := \gcd(p, p');$ $r := p/g$
while $g \neq 1$ **do**
  $s := \gcd(g, r);$ $p := r/s$
  $sqfd := sqfd \cdot p^i;$ $i := i + 1$
  $r := s;$ $g := g/s$
end do
$sqfd := sqfd \cdot r^i$
return$(sqfd)$
end procedure

ALGORITHM 6.2: Square-free decomposition

To obtain the first square-free factor, p_1, we enter the while-loop in Algorithm 6.2 and calculate

$$s := \gcd(g, r) = x^2 - 1,$$

then divide this out of r,

$$p_1 := r/s = x^2 + 1.$$

To get ready for the next iteration of the loop, we find the new r and g,

$$r := s = x^2 - 1,$$
$$g := g/s = x^2 - 1.$$

To obtain the second square-free factor, p_2, we enter the loop again and calculate

$$s := \gcd(g, r) = x^2 - 1,$$

then divide this out of r,

$$p_2 := r/s = 1.$$

Next we update r and s,

$$r := s = x^2 - 1,$$
$$g := g/s = 1.$$

Since $g = 1$, we exit the loop with the value of the third (and last) square-free factor, p_3, in r.

$$p_3 := r = x^2 - 1.$$

FIGURE 6.1: Roundabout factorization.

So the square-free decomposition of p is

$$p(x) = p_1 \, p_2^2 \, p_3^3 = (x^2 + 1)(x^2 - 1)^3. \qquad \square$$

With the simplifications presented in this section and the previous one, our attention can be directed toward the problem of factoring a monic polynomial with no repeated factors.

6.5 Roundabout Factorization

We noted that the Schubert-Kronecker algorithm runs very slowly when implemented on a computer. So how do today's computer algebra systems factor? They employ a "roundabout"[†] technique first suggested by Kurt Hensel in 1908. This method, shown diagrammatically in Figure 6.1, is based on factoring a polynomial modulo a prime number m, and then "lifting" this result to the factorization over the integers.

6.5.1 Factoring Polynomials mod m

By a polynomial modulo some prime m, we mean one whose coefficients are from the set $\{0, 1, 2, \ldots, m - 1\}$. Arithmetic operations are performed on such polynomials by reducing the coefficients modulo m at each step.

Example 6.6. List all polynomials of degree three or less modulo 2. Give the factorization of each.

Solution. A polynomial of degree three or less modulo 2 has the form $c_3 x^3 + c_2 x^2 + c_1 x + c_0$, where the coefficients are chosen from the set $\{0, 1\}$. There are

---

[†]The term "roundabout factorization" was coined by Alkiviadis G. Akritas (1989).

$2^4 = 16$ such polynomials. Two are constants (degree-0): 0, 1; two are linear (degree-1): x, $x + 1$. Among the four quadratics (degree-2), three are obtained by multiplying a pair of linear polynomials and one is irreducible:

$$x^2 = x \cdot x, \quad x^2 + 1 = (x + 1)^2, \quad x^2 + x = x\,(x + 1), \quad x^2 + x + 1.$$

Similarly among the eight cubics, two are obtained by cubing a linear polynomial, four are non-cubes obtained by multiplying a linear polynomial with a quadratic, and two are irreducible:

$$x^3 = x \cdot x \cdot x, \qquad\qquad\qquad\qquad x^3 + x^2 = x^2\,(x + 1),$$
$$x^3 + 1 = (x + 1)\,(x^2 + x + 1), \qquad x^3 + x^2 + 1,$$
$$x^3 + x = x\,(x + 1)^2, \qquad\qquad\qquad x^3 + x^2 + x = x\,(x^2 + x + 1),$$
$$x^3 + x + 1, \qquad\qquad\qquad\qquad\quad x^3 + x^2 + x + 1 = (x + 1)^3.$$

□

Factorization modulo m is clearly a *finite* problem since there are only a *finite* number of polynomials of any given degree and only a *finite* number of combinations to try. In contrast, factorization over the integers appears to be a problem of unbounded dimension, although we will soon see there is a way to bound the size of the coefficients in the factorization.

This is a convenient juncture to introduce some algebraic notation. The set of integers (positive, negative, and zero) is denoted $\mathbb{Z}$. By extension, the algebraic structure consisting of polynomials of a single variable x with integer coefficients is written $\mathbb{Z}[x]$. Similarly, the set of integers modulo a prime m, $\{0, 1, 2, \ldots, m - 1\}$, is denoted $\mathbb{Z}_m$. Correspondingly, the notation $\mathbb{Z}_m[x]$ is used to represent polynomials with coefficients from $\mathbb{Z}_m$.

A polynomial with integer coefficients can be transformed into a polynomial modulo m by taking the remainder when each coefficient is divided by m. For example, the polynomial $p(x) = 112x^4 + 58x^3 - 31x^2 + 107x - 66$ is $2x^4 + 3x^3 + 4x^2 + 2x + 4$ modulo 5. Of course, this transformation cannot be reversed since each number in the set $\{0, 1, 2, 3, 4\}$ represents infinitely many integers.

The key idea behind roundabout factorization is the following theorem.

Theorem 6.4 (Roundabout Factorization). *If $p(x)$ is monic and $p(x) = s(x)\,t(x)$ over the integers, then $p(x) = s(x)\,t(x)$ mod m for every prime m.*

Note, importantly, that the reverse of the theorem is not always true! For example, $x^2 + 1$ is irreducible over the integers but factors mod 2 as $(x + 1)^2$. $x^4 + 1$ is irreducible over the integers but factors modulo *every* prime.

Let's explore some implications of Theorem 6.4 by considering the polynomial p of Example 6.2. Over the integers, its factorization is given by $p(x) = x^4 + 4 = (x^2 - 2x + 2)(x^2 + 2x + 2)$. The value of p and these two factors, s and t, modulo 5 are

$$p(x) \equiv x^4 + 4 \quad (\text{mod } 5),$$
$$s(x) = x^2 - 2x + 2 \equiv x^2 + 3x + 2 \quad (\text{mod } 5),$$
$$t(x) = x^2 + 2x + 2 \equiv x^2 + 2x + 2 \quad (\text{mod } 5).$$

The theorem asserts that if we expand $s(x)\, t(x)$, we get back the polynomial p modulo 5,

$$(x^2 + 3x + 2)\,(x^2 + 2x + 2) \equiv x^4 + 4 \quad (\text{mod } 5).$$

The actual factorization of p mod 5 is

$$p(x) = (x + 1)\,(x + 2)\,(x + 3)\,(x + 4),$$

which you should verify by expanding the product modulo 5. The theorem further implies that the product of one pair of these linear factors must be s (mod 5) and the product of the other pair is t (mod 5),

$$s(x) \equiv (x + 1)\,(x + 2) \ (\text{mod } 5) \equiv x^2 + 3x + 2 \ (\text{mod } 5),$$
$$t(x) \equiv (x + 3)\,(x + 4) \ (\text{mod } 5) \equiv x^2 + 2x + 2 \ (\text{mod } 5).$$

(Can you explain why this is so?)

The next two examples illustrate the usefulness of Theorem 6.4.

Example 6.7. Show that $p(x) = x^6 + 4$ over the integers is irreducible by examining its factorizations modulo 11 and 13.

Solution. The factorizations of p modulo 11 and modulo 13 are

$$p(x) \equiv (x^2 + 5)\,(x^2 + 2x + 5)\,(x^2 + 9x + 5) \quad (\text{mod } 11)$$
$$\equiv (x^3 + 3)\,(x^3 + 10) \quad (\text{mod } 13).$$

Since p has three factors of degree two modulo 11, any factorization over the integer must involve either 1) three factors of degree two or 2) one factor of degree four and one of degree two. Similarly, since p has two factors of degree three modulo 13, its integer factorization must also involve two factors of degree three. These two modular factorizations are incompatible, so p is irreducible over the integers. (This can also be inferred immediately from the fact that p is irreducible modulo 7.) □

So far, we have used the set $\{0, 1, 2, \ldots, m - 1\}$ to represent the integers modulo m. This is called the *non-negative representation* for $\mathbb{Z}_m$. An alternative way to represent $\mathbb{Z}_m$ is with the set $\{-\lfloor m/2 \rfloor, \ldots, -1, 0, 1, \ldots, \lfloor m/2 \rfloor\}$ when $m > 2$. For example, $\{-2, -1, 0, 1, 2\}$ is used for $\mathbb{Z}_5$. This is known as the *symmetric representation*. Observe that modulo 5, -1 is written in quotient-remainder form as $-1 \cdot 5 + 4$, and -2 is expressed as $-1 \cdot 5 + 3$. So the negative numbers $-1, \ldots, -\lfloor m/2 \rfloor$ in the symmetric representation correspond, respectively, to the integers $m - 1, \ldots, \lceil m/2 \rceil$ in the non-negative representation.

Since we are interested in factoring polynomials whose coefficients can be either positive or negative, it is frequently more convenient for us to use the symmetric representation. Suppose we want to factor a polynomial p over the integers and we know that all the coefficients of p and its factorization lie between $-m/2$ and $m/2$, where m is some prime. In transforming a factorization problem over the integers into a modular one, we map non-negative coefficients over $\mathbb{Z}$ onto $0, 1, \ldots, \lfloor m/2 \rfloor$ and negative coefficients onto either $-1, -2, \ldots, -\lfloor m/2 \rfloor$ in the symmetric representation of $\mathbb{Z}_m$, or $m - 1$, $m - 2, \ldots, \lceil m/2 \rceil$ in the non-negative representation. We use this idea in the next example.

Example 6.8. Determine the factorization of $p(x) = x^6 - 1$ from its modulo 3 factorization.

Solution. The modulo 3 factorization of p is

$$p(x) \equiv (x + 1)^3 (x + 2)^3 \pmod{3}.$$

The factor $p_1 = x + 1 \pmod{3}$ corresponds to $x + 1$ in $\mathbb{Z}[x]$, and the factor $p_2 = x + 2 \pmod{3}$ corresponds to $x - 1$. Since both p_1 and p_2 evenly divide p when taken over $\mathbb{Z}[x]$ (i.e., $\mathrm{rem}(p, p_1) = \mathrm{rem}(p, p_2) = 0$ in $\mathbb{Z}[x]$), both are factors of p in $\mathbb{Z}[x]$. After dividing p_1 and p_2 out of p, what remains is $q(x) = x^4 + x^2 + 1$. Since $\mathrm{rem}(q, p_1) = \mathrm{rem}(q, p_2) = 3$ in $\mathbb{Z}[x]$, p_1 and p_2 are not repeated factors of p in $\mathbb{Z}[x]$.

Next, we take the product of the p_i in pairs:

$$p_1^2 = x^2 + 2x + 1 \pmod{3} \equiv x^2 - x + 1 \in \mathbb{Z}[x],$$
$$p_2^2 = x^2 + x + 1 \pmod{3} \equiv x^2 + x + 1 \in \mathbb{Z}[x],$$
$$p_1 p_2 = x^2 + 2 \pmod{3} \equiv x^2 - 1 \in \mathbb{Z}[x],$$

and see if each product divides q (and therefore, also divides p) over $\mathbb{Z}$. We find that $\mathrm{rem}(q, p_1^2) = \mathrm{rem}(q, p_2^2) = 0$ but $\mathrm{rem}(q, p_1 p_2) = 3$, so both p_1^2 and p_2^2 are factors of p in $\mathbb{Z}[x]$. Therefore, the desired factorization is

$$p(x) = (x + 1)(x + 2)(x^2 - x + 1)(x^2 + x + 1). \qquad \square$$

6.5.2   Choosing a Modulus m

Bounding Coefficient Size

We had just enough "wiggle room" in Example 6.8 to deduce the factorization over the integers from the modulo 3 factorization since all the coefficients of the factored polynomial are between -1 and 1. Unfortunately we do not, in general, know the coefficients of the factored polynomial in advance. Moreover the coefficients of the factors can, in general, be larger in magnitude than the coefficients of the polynomial itself. For example,

$$x^4 + 2x^3 - x^2 + 2x + 1 = (x^2 - x + 1)(x^2 + 3x + 1).$$

The next theorem deals with this difficulty by showing how to select a prime modulus m large enough to capture all of the coefficients in the factorization. In other words, all coefficients in the factorization are guaranteed to lie between $-m/2$ and $m/2$. The theorem is due to Maurice Mignotte (1974), building upon earlier work of Edmund Landau and Wilhelm Specht. Its proof is well beyond our scope.

Theorem 6.5 (Landau-Mignotte Bound). *Let* $p(x) = c_n x^n + c_{n-1} x^{n-1} + \ldots + c_1 x + c_0$ *and*

$$\|p\| = \sqrt{c_n^2 + c_{n-1}^2 + \ldots + c_1^2 + c_0^2}.$$

Then if $q(x) = b_d x^d + b_{d-1} x^{d-1} + \ldots + b_1 x + b_0$ *is a factor of* $p(x)$,

$$|b_j| \leq \binom{d-1}{j} \|p\| + \binom{d-1}{j-1} |c_n|.$$

We give an example to illustrate the application and usefulness of the theorem.

Example 6.9. What is the minimum modulus m given by the Landau-Mignotte bound that guarantees the factorization of

$$p(x) = x^8 + x^6 - 3x^4 - 3x^3 + 8x^2 + 2x - 5$$

mod m captures its factorization over the integers?

Solution. If p can be factored, it must have a factor q of degree $d \leq 4$. From the Landau-Mignotte bound,

$$\|p\| = \sqrt{1 + 1 + 9 + 9 + 64 + 4 + 25} = \sqrt{113}.$$

Setting $j = 2$ maximizes the size of the binomials in the theorem, so

$$|b_j| \leq \binom{3}{2} \sqrt{113} + \binom{3}{1} \cdot 1 \approx 34.89.$$

Thus, the coefficients of any factor of p can be at most 34 in magnitude, so any prime $m > 2 \cdot 34 = 68$ will work.   □

The polynomial p in this example is irreducible, and so any prime $m \geq 17$ would suffice. (Why?) As in the example, the bound given by the theorem is generally considerably larger than actually required.

Square-Free Polynomials mod m

Another consideration in choosing our prime modulus m is that the polynomial p we are factoring be square-free modulo m. That is, its mod m factorization should have no repeated factors. A polynomial that is square-free over the integers might not be square-free modulo a certain prime. For example, $p(x) = x^2 + 1$ is irreducible over the integers, but factors as $(x+1)^2$ modulo 2.

Theorem 6.3 still holds modulo a prime. In our example, the derivative $p'(x) = 2x \equiv 0 \pmod{2}$, and so $\gcd(p, p') = x^2 + 1 \pmod{2}$ even though p and p' have no common factors over the integers.

Fortunately, a polynomial that is square-free over the integers can have repeated factors modulo only a *finite* number of primes. So if the first prime we try doesn't work, the second or third almost always yields a modular square-free factorization.

6.5.3   Roundabout Factorization Algorithm

We are now in a position to outline the entire roundabout strategy for factoring a polynomial with integer coefficients.

0. Ensure that we are dealing with a primitive, monic, square-free polynomial in subsequent steps:

   a. Remove the content (gcd of the coefficients) from the polynomial.

   b. Transform to a monic polynomial whose factorization is equivalent.

   c. Perform a square-free decomposition.

Factor each square-free component p found in Step 0 as follows:

1. Choose a prime modulus m such that:

   a. p remains square-free mod m.

   b. m is sufficiently large that the coefficients of the factors of p over the integers lie between $-m/2$ and $m/2$.

2. Factor p mod m.

3. Suppose the factorization found in Step 2 is $p = p_1 p_2 \cdots p_r$.

   a. Take each p_i as a polynomial over the integers and test whether it divides p. Every p_i that does is an irreducible factor of p.

   b. If there are any p_i that do not divide p, form products of pairs, $p_i p_j$ mod m. Test whether these polynomials, taken over the integers, divide p. Every polynomial that does is an irreducible factor of p.

   c. If there are some p_i left over, form products of triples, $p_i p_j p_k$ mod m. Test whether these polynomials, taken over the integers, divide p. Every polynomial that does is an irreducible factor of p.

  d. Continue until every combination or its complement has been
     tested. If there are any p_i left, their product is an irreducible factor
     of p.

A procedure to accomplish Step 2 is described in the next section. In closing our discussion of the roundabout method, we observe that its running time in the worst case can be exponential in the degree of the polynomial p to be factored. If p is of degree d, its factorization modulo a prime m might produce d linear factors. If so and p is either irreducible over the integers, or is the product of two irreducible factors of degree $d/2$, then each subset of factors (or its complement) up to size $d/2$ is tested at Step 3. The number of such subsets is 2^{d-1}. Despite its worst case exponential performance, the roundabout procedure typically provides a considerable improvement in efficiency over the Schubert-Kronecker method.

A polynomial time algorithm for factoring over the integers is described by Lenstra, Lenstra and Lovász (1982), but its $O(n^{12})$ running time makes it impractical for implementation in computer algebra systems.

6.6 Distinct Degree Factorization

At the heart of the roundabout method is the ability to factor a polynomial modulo a prime number m. Two strategies are available: one due to Elwyn R. Berlekamp (1967) and the other to David G. Cantor and Hans Zassenhaus (1981). The latter technique, called *distinct degree factorization*, is considerably simpler to explain and we describe it here.

6.6.1 Finding Distinct Degree Factors

The basis of distinct degree factorization is the following remarkable theorem.

Theorem 6.6 (Distinct Degree Factorization). $x^{m^r} - x$ *is the product of all monic irreducible polynomials in $\mathbb{Z}_m[x]$ whose degree divides r.*

The following example illustrates the meaning of the theorem.

Example 6.10. Find all irreducible polynomials of degrees 1, 2 and 3 in $\mathbb{Z}_2[x]$.

Solution. Setting $m = 2$ and $r = 1$ in Theorem 6.6, we find that

$$x^2 - x \equiv x\,(x + 1) \pmod 2$$

is the product of all irreducible polynomials whose degree is one in $\mathbb{Z}_2[x]$. Substituting $r = 2$ in the theorem, we see that

$$x^4 - x \equiv x\,(x + 1)\,(x^2 + x + 1) \pmod 2$$

is the product of all irreducible polynomials of degrees one and two over $\mathbb{Z}_2[x]$, and so $x^2 + x + 1$ is the only irreducible polynomial of degree two. Similarly with $r = 3$, we have that

$$x^8 - x \equiv x\,(x+1)\,(x^3 + x^2 + 1)\,(x^3 + x + 1) \pmod 2$$

is the product of all irreducible polynomials of degrees one and three. Therefore, there are two irreducible polynomials of degree three in $\mathbb{Z}_2[x]$: $x^3 + x^2 + 1$ and $x^3 + x + 1$. $\quad\square$

It is useful and instructive for the reader to compare the results of this exercise with those of Example 6.6.

How can we use Theorem 6.6 to help factor a polynomial $p(x)$ in $\mathbb{Z}_m[x]$? Setting $r = 1$ in the theorem and taking $\gcd(p(x), x^m - x)$ in $\mathbb{Z}_m[x]$ yields the product s_1 of all linear factors of p mod m. Dividing these out of p and taking the gcd of what remains with $x^{m^2} - x$, gives the product s_2 of all quadratic factors. We proceed in this way, using a gcd calculation at each step to split off the product of all factors of the next degree. The next example puts this idea to work.

Example 6.11. Factor the polynomial $p(x) = x^{15} + 1$ modulo 2.

Solution. Letting $q_d(x) = x^{2^d} - x$, we can determine the linear factors (degree $d = 1$) by finding the greatest common divisor of p and q_1 mod 2,

$$
\begin{aligned}
p_1(x) &= \gcd(p,\, q_1) \\
&= \gcd(x^{15} + 1,\, x^2 - x) \\
&\equiv x + 1 \pmod 2.
\end{aligned}
$$

So p has one linear factor, $p_1(x) = x + 1$. Now divide p_1 out of p, then find the degree $d = 2$ factors the same way,

$$
\begin{aligned}
p_2(x) &= \gcd(p/p_1,\, q_2) \\
&= \gcd(x^{14} + x^{13} + x^{12} + x^{11} + x^{10} + x^9 + x^8 + x^7 + x^6 + x^5 + x^4 + \\
&\qquad x^3 + x^2 + x + 1,\, x^4 - x) \\
&\equiv x^2 + x + 1 \pmod 2.
\end{aligned}
$$

There is one factor of degree two, $p_2(x) = x^2 + x + 1$. Now look for any degree $d = 3$ factors,

$$
\begin{aligned}
p_3(x) &= \gcd(p/(p_1\,p_2),\, q_3) \\
&= \gcd(x^{12} + x^9 + x^6 + x^3 + 1,\, x^8 - x) \\
&\equiv 1 \pmod 2.
\end{aligned}
$$

There are none. Next look for degree $d = 4$ factors,

$$p_4(x) = \gcd(p/(p_1\,p_2\,p_3),\ q_4)$$
$$= \gcd(x^{12} + x^9 + x^6 + x^3 + 1,\ x^{16} - x)$$
$$\equiv x^{12} + x^9 + x^6 + x^3 + 1 \quad (\mathrm{mod}\ 2).$$

Since degree$(p_4) = 3$, there are three degree-4 factors and that's all that's left of p. The problem remains of splitting the three degree-4 factors,

$$p_4(x) \equiv (x^4 + x^3 + x^2 + x + 1)\,(x^4 + x^3 + 1)\,(x^4 + x + 1) \quad (\mathrm{mod}\ 2).$$

This is done with another set of gcd computations that we now discuss.   □

6.6.2   Splitting Equal Degree Factors

Theorem 6.6 allows us to express a polynomial p as $p = p_1 p_2 \cdots p_n$, where p_d is the product of all its degree-d factors in $\mathbb{Z}_m[x]$. If degree$(p_d) = d$, then p_d is an irreducible factor of p. Otherwise, degree(p_d) is a multiple r of d (why?), and the problem remains of splitting p_d into r irreducible factors. The key mathematical idea that allows us to split equal degree factors for an odd modulus m is the following theorem due to Cantor and Zassenhaus (1981).

Theorem 6.7 (Equal Degree Factorization, m odd). *For m odd, $x^{m^d} - x$ can be expressed as*

$$x^{m^d} - x = x\left(x^{(m^d-1)/2} + 1\right)\left(x^{(m^d-1)/2} - 1\right) \bmod m.$$

Note that the factorization given in the theorem is not, in general, the complete factorization of $x^{m^d} - x$ into its irreducible factors mod m. Nonetheless, the theorem is quite useful in that it implies if we substitute a polynomial q for x, then $q(x)^{m^d} - q(x)$ can be written

$$q(x)^{m^d} - q(x) = q(x)\left(q(x)^{(m^d-1)/2} + 1\right)\left(q(x)^{(m^d-1)/2} - 1\right) \bmod m. \quad (6.7)$$

Furthermore by Theorem 6.6, $x^{m^d} - x$ is the product of all monic irreducible polynomials whose degree divides d, and so $q(x)^{m^d} - q(x)$ is a multiple of all monic irreducible polynomials of degree-d.

Our task is to split a polynomial p_d of degree-rd, obtained from the distinct degree decomposition, into its r irreducible factors. From Equation (6.7), it holds that

$$p_d(x) = \gcd(p_d(x),\ q(x)) \cdot \gcd(p_d(x),\ q(x)^{(m^d-1)/2} + 1)$$
$$\cdot \gcd(p_d(x),\ q(x)^{(m^d-1)/2} - 1)$$

for any polynomial q.

It turns out that if q is chosen at random from among the m^{rd} polynomials mod m of degree $< rd$, then $\gcd(p_d(x), q(x)^{(m^d-1)/2} - 1)$ is a *proper* factor of $p_d(x) = p_{d,1} \cdots p_{d,r}$ with probability

$$1 - \frac{1}{2^r}\left(\left(1 + \frac{1}{m^d}\right)^r + \left(1 - \frac{1}{m^d}\right)^r\right).$$

This suggests a method for splitting the factors of p_d. We take random polynomials q and compute $\gcd(p_d, q^{(m^d-1)/2} - 1)$. Since $r \geq 2$, $m \geq 3$, and $d \geq 1$, the probability is at least $4/9 \approx .444\ldots$ that p_d splits! Of course, we may have to do this several times in order to retrieve all of p_d's irreducible factors. Note that this method works well even if we try the simplest possible choices for q, linear polynomials of the form $q(x) = x + s$, although there may not be enough linear polynomials to do the job.

We illustrate the procedure with two examples.

Example 6.12. Split $p(x) = x^5 - 1$ into five linear ($d = 1$) factors over $\mathbb{Z}_{11}[x]$.

Solution. We begin by calculating the degree to which each random q is raised in the gcd computation, $(m^d - 1)/2 = 5$. Let's try to split p using $q(x) = x + 1$,

$$\gcd(x^5 - 1, (x + 1)^5 - 1) \equiv x^2 + 4x + 1 \quad (\text{mod } 11).$$

Success! So we can write $p(x) = (x^2 + 4x + 1)(x^3 + 7x^2 + 4x + 10)$, where the second factor is the result of dividing the gcd into p. Now let's try to split $x^2 + 4x + 1$. With $q(x) = x + 2$, we have

$$\gcd(x^2 + 4x + 1, (x + 2)^5 - 1,) \equiv x + 8 \quad (\text{mod } 11),$$

so $x^2 + 4x + 1 = (x + 8)(x + 7)$. Now let's try to split $x^3 + 7x^2 + 4x + 10$ also using $q(x) = x + 2$,

$$\gcd(x^3 + 7x^2 + 4x + 10, (x + 2)^5 - 1) \equiv x + 10 \quad (\text{mod } 11).$$

Success again—we're on a roll! So $x^3 + 7x^2 + 4x + 10 = (x + 10)(x^2 + 8x + 1)$. We just have to split $x^2 + 8x + 1$. Let's try $q(x) = x + 3$.

$$\gcd(x^2 + 8x + 1, (x + 3)^5 - 1) \equiv x + 2 \quad (\text{mod } 11),$$

so $x^2 + 8x + 1 = (x + 2)(x + 6)$ and we are done,

$$p(x) \equiv (x + 8)(x + 7)(x + 10)(x + 2)(x + 6) \quad (\text{mod } 11). \qquad \square$$

Example 6.13. Split $p(x) = x^{10} + x^5 + 1$ into five quadratic ($d = 2$) factors over $\mathbb{Z}_{11}[x]$.

Solution. Again we begin by calculating the degree to which each random q must be raised, $(m^d - 1)/2 = 60$. Let's try to split p with $q(x) = x + 1$.

$$\gcd(x^{10} + x^5 + 1, (x + 1)^{60} - 1) \equiv x^2 + x + 1 \pmod{11}$$

and we have $p(x) = (x^2 + x + 1)(x^8 + 10x^7 + x^5 + 10x^4 + x^3 + 10x + 1)$. Now let's try using $q(x) = x + 2$ to split the second factor.

$$\gcd(x^8 + 10\,x^7 + x^5 + 10\,x^4 + x^3 + 10\,x + 1, (x + 2)^{60} - 1) \equiv$$
$$x^4 + 2\,x^3 + x^2 + 6\,x + 9 \pmod{11},$$

so $x^8 + 10\,x^7 + x^5 + 10x^4 + x^3 + 10x + 1 = (x^4 + 2x^3 + x^2 + 6x + 9)(x^4 + 8x^3 + 5x^2 + 10x + 5)$. Let's try splitting the first of these quartic factors using $q(x) = x + 3$.

$$\gcd(x^4 + 2\,x^3 + x^2 + 6\,x + 9, (x + 3)^{60} - 1) \equiv 1 \pmod{11},$$

so we've struck out for the first time! If at first you don't succeed, try and try again. This time let's use $q(x) = x + 4$,

$$\gcd(x^4 + 2\,x^3 + x^2 + 6\,x + 9, (x + 4)^{60} - 1) \equiv x^2 + 4\,x + 5 \pmod{11}.$$

Now we've got it, $x^4 + 2x^3 + x^2 + 6x + 9 = (x^2 + 4x + 5)(x^2 + 9x + 4)$. Finally, we need to split the second quartic factor obtained earlier, $x^4 + 8x^3 + 5x^2 + 10x + 5$. Trying $q(x) = x + 3$ we succeed,

$$\gcd(x^4 + 8\,x^3 + 5\,x^2 + 10\,x + 5, (x + 3)^{60} - 1) \equiv x^2 + 3\,x + 9 \pmod{11}.$$

So $x^4 + 8x^3 + 5x^2 + 10x + 5 = (x^2 + 3x + 9)(x^2 + 5x + 3)$, and the complete factorization of p is

$$p(x) \equiv (x^2 + x + 1)\,(x^2 + 4x + 5)\,(x^2 + 9x + 4)\,(x^2 + 3x + 9)\,(x^2 + 5x + 3) \pmod{11}.$$
□

What if our modulus $m = 2$? This is the only prime for which Theorem 6.7 does not hold. We can obtain an analogous result using the notion of the *trace polynomial*, defined as

$$\mathrm{Trace}_d(x) = x + x^2 + x^4 + \ldots + x^{2^{d-1}}.$$

The analog of Theorem 6.7 is the following result, which can be verified easily from the definition of the trace polynomial.

Theorem 6.8 (Equal Degree Factorization, $m = 2$). $x^{2^d} - x$ *can be expressed as*

$$x^{2^d} - x = \text{Trace}_d(x) \cdot (\text{Trace}_d(x) + 1) \text{ mod } 2.$$

As with Theorem 6.7, this result does not give the complete factorization of $x^{2^d} - x$. In fact, Theorem 6.6 tells us that $x^{2^d} - x$ is the product of *all* irreducible polynomials whose degree divides d.

Our job is to split a polynomial p_d of degree-rd, obtained from the distinct degree decomposition, into its r irreducible factors. From Theorem 6.8,

$$p_d(x) = \gcd(p_d(x), \text{Trace}_d(q(x))) \cdot \gcd(p_d(x), \text{Trace}_d(q(x)) + 1),$$

for any polynomial q. It turns out that if we choose a random polynomial q of degree $rd - 1$, $\gcd(p_d(x), \text{Trace}_d(q(x)))$ is a *proper* factor of $p_d(x) = p_1 \cdots p_r$ with probability $1 - 1/2^{r-1}$. So to split the factors of p_d, we take random polynomials q of the proper degree and compute the gcd of p_d with $\text{Trace}_d(q)$.

We illustrate this technique by completing the factorization of $x^{15} + 1$ in $\mathbb{Z}_2[x]$ started in Example 6.11.

Example 6.14. Split $p(x) = x^{12} + x^9 + x^6 + x^3 + 1$ into three degree-4 factors over $\mathbb{Z}_2[x]$.

Solution. Since $r = 3$ and $d = 4$, our first test polynomial q is of degree $rd - 1 = 11$. Let's choose $q(x) = x^{11} + 1$.

$$\text{Trace}_4(q(x)) \equiv x^{88} + x^{44} + x^{22} + x^{11} \pmod{2},$$

$$\gcd(p(x), \text{Trace}_4(q(x))) \equiv x^4 + x^3 + 1 \pmod{2}.$$

So one factor is $x^4 + x^3 + 1$. Dividing this out of p, we now need to split $x^8 + x^7 + x^6 + x^4 + 1$. Our new test polynomial q must be of degree-7. (Why?) Let's try to split the rest of p with $q(x) = x^7 + 1$.

$$\text{Trace}_4(q(x)) \equiv x^{56} + x^{28} + x^{14} + x^7 \pmod{2},$$

$$\gcd(x^8 + x^7 + x^6 + x^4 + 1, \text{Trace}_4(q(x)) \equiv 1 \pmod{2}.$$

We don't succeed. Let's try another polynomial, $q(x) = x^7 + x$.

$$\text{Trace}_4(q(x)) \equiv x^{56} + x^{28} + x^{14} + x^8 + x^7 + x^4 + x^2 + x \pmod{2},$$

$$\gcd(x^8 + x^7 + x^6 + x^4 + 1, \text{Trace}_4(q(x))) \equiv x^4 + x^3 + x^2 + x + 1 \pmod{2}.$$

This time we are successful and obtain the other two factors, $x^4 + x^3 + x^2 + x + 1$ and its cofactor when we divide it into what was left of p, $x^4 + x + 1$. So the desired factorization of p is

$$p(x) \equiv (x^4 + x^3 + 1)(x^4 + x^3 + x^2 + x + 1)(x^4 + x + 1) \pmod{2}. \qquad \square$$

6.7   Hensel Lifting

The Landau-Mignotte bound can yield a very large modulus. This occurs when the degree of the polynomial to be factored is high, or it has one or more large coefficients. Hensel lifting provides a mechanism for us to work with a small, manageable modulus that is usually considerably smaller than the one given by the Landau-Mignotte bound. The idea is to start with the factorization of a polynomial modulo a small prime m. We then find in succession the factorizations mod m^2, m^3, ..., until we reach a modulus that is large enough to give the factorization over the integers.

To illustrate the technique, we describe how to lift the factorization of a polynomial $p(x)$ mod m to its factorization mod m^2. As usual, we assume that p is monic and square-free mod m. Moreover, we use the symmetric representation for all modular polynomials in what follows to accommodate both positive and negative coefficients.

Let $p_1 = p \bmod m$, and suppose its factorization is given by $p_1 = s_1 t_1$ in $\mathbb{Z}_m[x]$. We want to find factors s_2 and t_2 such that $p_2 = s_2 t_2 \bmod m^2$, where $p_2 = p \bmod m^2$, $p_1 \equiv p_2 \pmod{m}$, $s_1 \equiv s_2 \pmod{m}$, and $t_1 \equiv t_2 \pmod{m}$.

Consider the difference between p and $s_1 t_1$. It's this difference that we want to make disappear as we lift p's factorization. The difference is given by

$$d_1 = p - s_1 t_1 \qquad \in \mathbb{Z}[x]$$

$$c_1 = d_1/m \bmod m \quad \in \mathbb{Z}_m[x].$$

In $\mathbb{Z}_{m^2}[x]$, $d_1 \equiv m c_1$. This is the portion of d_1 that disappears as we lift the factorization mod m to one mod m^2. The differences d_1 and c_1 can be calculated from the original polynomial and its mod m factorization.

Notice we can write

$$p_2 = p_1 + m c_1,$$

where $m c_1$ is the part of p_2 that is "visible" modulo m^2 but "invisible" mod m because $m \equiv 0 \bmod m$. Similarly, we can write

$$s_2 = s_1 + m \hat{s}_2,$$

$$t_2 = t_1 + m \hat{t}_2,$$

where $m \hat{s}_2$ and $m \hat{t}_2$ are the parts of s_2 and t_2 that are "visible" modulo m^2 but "invisible" modulo m. Substituting these quantities into the factorization of p in $\mathbb{Z}_{m^2}[x]$, $p_2 = s_2 t_2$, we find

$$p_1 + m c_1 = (s_1 + m \hat{s}_2)(t_1 + m \hat{t}_2)$$
$$= s_1 t_1 + m (s_1 \hat{t}_2 + t_1 \hat{s}_2) + m^2 \hat{s}_2 \hat{t}_2.$$

Since $p_1 = s_1 t_1$ and $m^2 \equiv 0 \bmod m^2$,

$$c_1 = s_1 \hat{t}_2 + t_1 \hat{s}_2. \tag{6.8}$$

How do we find s_2 and t_2? $\hat{s}_2$ and $\hat{t}_2$ are the polynomials with degrees less than those of s_1 and t_1, respectively, that satisfy (6.8). The restriction on the degrees of $\hat{s}_2$ and $\hat{t}_2$ holds because the leading coefficient of each factor of p over $\mathbb{Z}$ is one given that p is monic.

Since s_1 and t_1 are relatively prime modulo m, we can use the extended Euclidean algorithm to find s' and t' such that

$$s' s_1 + t' t_1 = 1 \qquad \in \mathbb{Z}_m[x]. \tag{6.9}$$

Comparing (6.8) and (6.9), we have

$$\hat{s}_2 = \text{rem}(t' c_1, s_1) \pmod{m},$$
$$\hat{t}_2 = \text{rem}(s' c_1, t_1) \pmod{m},$$

where the divisions are performed mod m. Finally we obtain the desired factors s_2 and t_2 of p_2 in $\mathbb{Z}_{m^2}[x]$,

$$s_2 = s_1 + m\,\hat{s}_2,$$
$$t_2 = t_1 + m\,\hat{t}_2.$$

We can repeat this lifting procedure using p's factorization in $\mathbb{Z}_{m^2}[x]$ to determine its factorization in $\mathbb{Z}_{m^3}[x]$, etc. It is instructive to see how to lift the factorization of p from $p_{i-1} = s_{i-1}t_{i-1}$ in $\mathbb{Z}_{m^{i-1}}[x]$ to $p_i = s_i t_i$ in $\mathbb{Z}_{m^i}[x]$. The difference between p and $s_{i-1}t_{i-1}$ is given by

$$d_{i-1} = p - s_{i-1}\,t_{i-1} \qquad \in \mathbb{Z}[x]$$
$$c_{i-1} = d_{i-1}/m^{i-1} \bmod m \qquad \in \mathbb{Z}_m[x].$$

Importantly, m^{i-1} evenly divides d_{i-1}. As before, we can write

$$p_i = p_{i-1} + m^{i-1} c_{i-1},$$
$$s_i = s_{i-1} + m^{i-1} \hat{s}_i,$$
$$t_i = t_{i-1} + m^{i-1} \hat{t}_i,$$

where

$$p_{i-1} + m^{i-1} c_{i-1} = (s_{i-1} + m^{i-1} \hat{s}_i)(t_{i-1} + m^{i-1} \hat{t}_i) \qquad \pmod{m^i}$$
$$= s_{i-1} t_{i-1} + m^{i-1}(s_{i-1} \hat{t}_i + t_{i-1} \hat{s}_i) + m^{2(i-1)} \hat{s}_i \hat{t}_i.$$

Noting that $p_{i-1} = s_{i-1}t_{i-1}$ and $m^{2(i-1)} \equiv 0 \bmod m^i$, we have

$$c_{i-1} = s_{i-1} \hat{t}_i + t_{i-1} \hat{s}_i \pmod{m^i}.$$

But since $s_{i-1} \equiv s_1$ and $t_{i-1} \equiv t_1 \bmod m$, this equation can be expressed as

$$c_{i-1} = s_1 \hat{t}_i + t_1 \hat{s}_i \pmod{m}.$$

procedure *HenselLift*(p, s_1, t_1, m, k)
{Input polynomials: $p \in \mathbb{Z}[x]$, $(s_1, t_1) \in \mathbb{Z}_m[x]$ where $s_1 \cdot t_1 \equiv p \bmod m$}
{Output polynomials: $s, t \in \mathbb{Z}_{m^k}[x]$ where $s \cdot t \equiv p \bmod m^k$}
$\quad (s', t') := \text{ExtendedEuclid}(s_1, t_1) \bmod m;$
$\quad\quad\quad\quad \{\gcd(s_1, t_1) \equiv 1 \equiv s's_1 + t't_1 \bmod m\}$
$\quad s := s_1; \quad t := t_1$
$\quad d := p - s \cdot t \in \mathbb{Z}[x]; \quad c := d/m \in \mathbb{Z}_m[x]$
$\quad$**for** $i := 2$ **to** k **while** $d \neq 0$ **do**
$\quad\quad \hat{s} := \text{rem}(t'c, s) \in \mathbb{Z}_m[x]; \quad \hat{t} := \text{rem}(s'c, t) \in \mathbb{Z}_m[x]$
$\quad\quad s := s + m^{i-1}\,\hat{s}; \quad\quad\quad\quad t := t + m^{i-1}\,\hat{t}$
$\quad\quad d := p - s \cdot t \in \mathbb{Z}[x]; \quad\quad c := d/m^i \in \mathbb{Z}_m[x]$
$\quad$**end do**
$\quad$**return**(s, t)
end procedure

ALGORITHM 6.3: Hensel lifting

Solving this equation for $\hat{s}_i$ and $\hat{t}_i$ involves the same extended Euclidean computation since our arithmetic is performed modulo m. As before, $\hat{s}_i$ and $\hat{t}_i$ are polynomials whose degrees are less than those of s_1 and t_1, respectively, and can be calculated as

$$\hat{s}_i = \text{rem}(t'\,c_{i-1}, s_1) \quad (\bmod\ m),$$
$$\hat{t}_i = \text{rem}(s'\,c_{i-1}, t_1) \quad (\bmod\ m),$$

where the arithmetic is done modulo m.

As we iterate the lifting process, the extended Euclidean computation is always applied to the same two polynomials, s_1 and t_1, and so it only needs to be performed once at the outset! The entire procedure is given in Algorithm 6.3. We present an example to illustrate its operation.

Example 6.15. Lift the factorization of $p(x) = x^3 + 10x^2 - 432x + 5040$ in $\mathbb{Z}_5[x]$ to its factorization over the integers.

Solution. The values of p and its factorization in $\mathbb{Z}_5[x]$ using the symmetric representation are

$$p_1 = x^3 - 2x = x\,(x^2 - 2) \quad (\bmod\ 5).$$

Letting $s_1 = x$ and $t_1 = x^2 - 2$, we apply the extended Euclidean algorithm to find

$$s'\,s_1 + t'\,t_1 = 1 = -2x\,(x) + 2\,(x^2 - 2) \quad (\bmod\ 5),$$

so $s' = -2x$ and $t' = 2$. The difference $d_1 = p - p_1 = 10x^2 - 430x + 5040$ in $\mathbb{Z}[x]$ and $c_1 = d_1/5 \pmod 5 = 2x^2 - x - 2$ in $\mathbb{Z}_5[x]$. Next, we calculate $\hat{s}_2 = \text{rem}(t'c_1, s_1) \pmod 5 = 1$ giving $s_2 = s_1 + m\hat{s}_2 = x + 5$, and $\hat{t}_2 = \text{rem}(s'c_1, t_1) \pmod 5 = x - 1$ giving $t_2 = t_1 + m\hat{t}_2 = x^2 + 5x - 7$. So we have obtained the factorization of p in $\mathbb{Z}_{25}[x]$,

$$p_2 = (x + 5)(x^2 + 5x - 7) \pmod{25}.$$

The lifting process continues and the key values at each iteration are shown in the table.

i	m^i	$\hat{s}_i$	$\hat{t}_i$	s_i	t_i	d_i
1	5	—	—	x	$x^2 - 2$	$10x^2 - 430x + 5040$
2	25	1	$x - 1$	$x + 5$	$x^2 + 5x - 7$	$-450x + 5075$
3	125	1	$-x + 2$	$x + 30$	$x^2 - 20x + 43$	$125x + 3750$
4	625	0	1	$x + 30$	$x^2 - 20x + 168$	0

Observe that at each step, the difference d_i is evenly divisible by the modulus m^i. After three lifting iterations, $d = 0$ and the process terminates with p's factorization over the integers,

$$p(x) = (x + 30)(x^2 - 20x + 168).$$

It is truly amazing that we can deduce the complete factorization of p, one of whose coefficients is greater than 5000, from such little information as its factorization modulo 5, $x(x^2 - 2)$.   □

Up to now, we have considered the case where the factorization of p involves only two factors. What if there are more than two? In that case, we lift the factors one at a time. To lift the first factor s, we let t be the product of the remaining factors. So if there are k factors, we iterate the lifting procedure $k - 1$ times.

Another complication arises when the factor s that we are attempting to lift is not one of the true factors of p over the integers. This occurs when p is irreducible over the integers, or s combines with one or more other factors in the true factorization. In either case, the difference polynomial d is never zero and the iterations may continue indefinitely. For example, we noted earlier that $x^4 + 1$ is irreducible over the integers but factors modulo every prime. Let's examine what happens when we try to lift one of its modular factorizations.

Example 6.16. Apply the Hensel lifting algorithm to the factorization of $p(x) = x^4 + 1$ in $\mathbb{Z}_3[x]$.

Solution. The factorization of p in $\mathbb{Z}_3[x]$ is given by

$$p_1 = (x^2 - x - 1)(x^2 + x - 1) \pmod 3.$$

Letting s_1 be the first factor and t_1 the second, we apply the extended Euclidean algorithm to find

$$s' s_1 + t' t_1 = 1 = (x+1)(x^2 - x - 1) + (-x+1)(x^2 + x - 1) \pmod{3},$$

so $s' = x+1$ and $t' = -x+1$. The first few iterations of the Hensel procedure are given in the table.

i	m^i	$\hat{s}_i$	$\hat{t}_i$	s_i	t_i	d_i
1	3	–	–	$x^2 - x - 1$	$x^2 + x - 1$	$3x^2$
2	9	$-x$	x	$x^2 - 4x - 1$	$x^2 + 4x - 1$	$18x^2$
3	27	x	$-x$	$x^2 + 5x - 1$	$x^2 - 5x - 1$	$27x^2$
4	81	$-x$	x	$x^2 - 22x - 1$	$x^2 + 22x - 1$	$486x^2$
5	243	0	0	$x^2 - 22x - 1$	$x^2 + 22x - 1$	$486x^2$
6	729	x	$-x$	$x^2 + 221x - 1$	$x^2 - 221x - 1$	$48843x^2$

Notice that the differences d_i never decrease and the iterations can continue indefinitely. □

How do we deal with this difficulty? If we know in advance the magnitude B of the largest coefficient either in p or its factorization over the integers, then we can stop iterating when $m^k \geq 2B$. (Why?) We can bound B using the Landau-Mignotte bound (see Theorem 6.5) or by some other means.

The Hensel procedure we have described is referred to as *linear lifting*. Another procedure, called *quadratic lifting*, was developed by Zassenhaus (1969). In quadratic lifting, the exponent doubles at each step. Starting with the factorization mod m, the factorizations mod $m^2, m^4, m^8, \ldots, m^{2^k}$ are computed at successive iterations. Since the cost per iteration is higher, linear lifting is generally preferred in most situations.

6.8 Exercises

1. Find any linear factors of the following polynomials over the integers.
   (a) $x^3 + 14x^2 + 56x + 64$
   (b) $18x^3 - 57x^2 + 53x - 12$
   (c) $6x^3 - 73x^2 - 86x + 273$
   (d) $6x^4 - x^3 + 4x^2 - x - 2$
   (e) $6x^4 - 11x^3 + 8x^2 - 33x - 30$
   (f) $5x^5 - 6x^4 - 24x^3 + 20x^2 + 7x - 2$
   (g) $15x^5 - 11x^4 + 47x^3 + 27x^2 - 38x + 8$

2. Prove that if a monic polynomial with integer coefficients has a rational root, then this root must be an integer.

3. Show that $x^2 + 1$ is irreducible (cannot be factored) over the integers in two ways: (a) with the Rational Root Test and (b) using the Schubert-Kronecker method.

4. Show how each of the following polynomials is factored by the Schubert-Kronecker method.

   (a) $x^4 - 4$ (b) $x^4 - 8x^2 - 9$ (c) $6x^4 - 7x^3 + 5x^2 - 20x + 17$

   (d) $x^6 - 1$ (e) $x^6 + 1$ (f) $x^5 + x + 1$

5. Use the Schubert-Kronecker method to show that $x^4 + x^3 + x^2 + x + 1$ is irreducible over the integers.

6. Convert each of the following to an "equivalent" monic polynomial for factoring. Use a computer algebra system to factor the monic polynomial, then map this result onto the factorization of the original polynomial.

   (a) $6x^2 + 11x - 35$

   (b) $25x^4 - 16$

   (c) $6x^4 - 19x^3 + 24x^2 - 13x + 4$

   (d) $2x^6 + 2x^5 + 2x^4 + 4x^3 + 5x^2 - 3x - 2$

   (e) $8x^5 - 48x^4 + 90x^3 - 90x^2 + 117x - 27$

   (f) $30x^5 + 39x^4 + 35x^3 + 25x^2 + 9x + 2$

7. Calculate the square-free decompositions of the following polynomials over the integers.

   (a) $x^5 - x^4 - 2x^3 + 2x^2 + x - 1$

   (b) $x^{10} + 2x^9 + 2x^8 + 2x^7 + x^6 + x^5 + 2x^4 + x^3 + x^2 + 2x + 1$

   (c) $x^8 - 4x^6 + 16x^2 - 16$

   (d) $x^{12} - 3x^{10} - 2x^9 + 3x^8 + 6x^7 - 6x^5 - 3x^4 + 2x^3 + 3x^2 - 1$

8. Trace the action of Algorithm 6.2 as it obtains each of the following square-free decompositions. $A, B, C,$ and D are all irreducible polynomials.

   (a) $A\ B^2\ C^3\ D^4$

   (b) $A\ B^3\ C^5\ D^7$

   (c) $A^2\ B^4\ C^6\ D^8$

   (d) $A\ B^2\ C^6\ D^{10}$

9. (a) Write out the addition and multiplication tables for the integers modulo 3. What are the additive and multiplicative inverses of each integer?

   (b) Repeat for the integers modulo 5.

10. Consider the polynomials $p(x) = x^{12} + x^9 + x^6 + x^3 + 1$ and $q(x) = x^8 + x^7 + x^5 + x^4 + x^3 + x + 1$. Calculate by hand each of the following in $\mathbb{Z}_2[x]$.

   (a) their sum and difference

   (b) their product

   (c) the quotient and remainder when p is divided by q

   (d) their greatest common divisor

11. List and factor all monic polynomials of degree 3 or less in $\mathbb{Z}_3[x]$.

12. A sieve can be constructed to build a list of irreducible polynomials in $\mathbb{Z}_p[x]$. (See Programming Project 2.3 for the prime number sieve of Eratosthenes.) Begin by listing all of the polynomials of each degree up to the maximum degree of interest. First cross out the multiples of the linear polynomials, then those of the remaining quadratics, those of the remaining cubics, etc.

   (a) Trace the action of this technique by building a list of all irreducible polynomials in $\mathbb{Z}_2[x]$ up to degree 6.

   (b) Repeat for the monic polynomials in $\mathbb{Z}_3[x]$ up to degree 4.

13. (a) Prove that if $p(x)$ takes on prime values infinitely often, then $p(x)$ is irreducible.

   (b) Can you improve the result in part (a) to some finite number of prime values that will guarantee $p(x)$ is irreducible?

14. Prove that there are infinitely many monic irreducible polynomials in $\mathbb{Z}_m[x]$. (*Hint.* Imitate the classical proof that there are infinitely many prime numbers.)

15. Show that every non-monic polynomial in $\mathbb{Z}_m[x]$ can be expressed as the product of a number in $\mathbb{Z}_m$ with a monic polynomial.

16. Determine the minimum modulus m given by the Landau-Mignotte bound that guarantees the mod m factorization of the following polynomials captures the factorization over the integers.

   (a) $x^4 - 14x^2 + 1$ (b) $x^4 + 5x^3 + 6x^2 + 5x + 1$

   (c) $x^4 - 2x^3 - x^2 - 2x + 1$ (d) $x^5 + x^4 + 1$

   (e) $x^6 - 9x^4 + 4x^2 - 36$ (f) $x^7 - x$

   (g) $x^8 - 16$ (h) $x^8 - 14x^6 + 5x^4 - 56x^2 + 4$

17. Suppose $p(x)$, $s(x)$, and $t(x)$ are polynomials in $\mathbb{Z}_m[x]$. Show that if $s(x)$ and $t(x)$ have no common factors, then

$$\gcd(p(x), s(x)\,t(x)) = \gcd(p(x), s(x)) \cdot \gcd(p(x), t(x)).$$

18. Show that the simple (non-repeated) roots of a polynomial are not roots of its derivative.

19. Factor the polynomials given below over $\mathbb{Z}_2[x]$ using the distinct degree factorization method. First find the distinct degree factors, and then split the factors of equal degree.

   (a) $x^6 + x^5 + x^2 + 1$

   (b) $x^7 + 1$

   (c) $x^{15} + 1$

   (d) $x^{21} + 1$

   (e) $x^{10} + x^9 + x^8 + x^3 + x^2 + x$

   (f) $x^{22} + x^{21} + x^{20} + x^{19} + x^{18} + x^{17} + x^{16} + x^7 + x^6 + x^5 + x^4 + x^3 + x^2 + x$

20. Use distinct degree factorization to perform the following factorizations modulo a prime. First find the distinct degree factors, then split the factors of equal degree.

    (a) $x^9 - x$ modulo 3. The result is the product of all monic irreducible polynomials of degrees 1 and 2 in $\mathbb{Z}_3[x]$.

    (b) $x^{27} - x$ modulo 3. This factorization gives the monic irreducible polynomials of degrees 1 and 3 in $\mathbb{Z}_3[x]$.

    (c) $x^{25} - x$ modulo 5. The result is the product of all monic irreducible polynomials of degrees 1 and 2 in $\mathbb{Z}_5[x]$.

21. Use the distinct degree method to factor the polynomials below in $\mathbb{Z}_5[x]$. Compare each result with the factorization over the integers, identifying the sources of the factors and coefficients in $\mathbb{Z}[x]$.

    (a) $x^4 + 4$ (b) $x^5 + x^4 + 1$ (c) $x^6 - 1$
    (d) $x^4 - 8x - 9$ (e) $x^5 + x + 1$ (f) $x^5 - x + 1$
    (g) $x^8 - 2x^6 + 2x - 1$ (h) $x^8 + 1$ (i) $x^8 - 1$

22. (a) Repeat Exercise 21 but do the factoring in $\mathbb{Z}_{11}[x]$.

    (b) Repeat again, but this time factor in $\mathbb{Z}_7[x]$. The polynomials in (b) and (e) have a repeated linear factor. With this information, can you detect these factors?

23. Use the Hensel method to linear lift the factorizations modulo a prime given below to produce the factorizations over the integers.

    (a) $x^6 - 65536 = (x^3 - 256)(x^3 + 256)$
    $\equiv x^6 - 2 = (x^3 - 3)(x^3 + 3) \pmod 7$

    (b) $x^3 - 37x^2 - 1366x + 9774 = (x + 27)(x^2 - 64x + 362)$
    $\equiv x^3 - x^2 - x = x(x^2 - x - 1) \pmod 3$

    (c) $x^4 - 81x^3 + 283x^2 - 2187x + 6912 = (x^2 + 27)(x^2 - 81x + 256)$
    $\equiv x^4 - x^3 - 2x^2 - 2x + 2 = (x^2 - x + 1)(x^2 + 2) \pmod 5$

    (d) $x^4 + 58x^3 - 3472x^2 + 1342208x - 92725248 = (x^2 - 40x + 9856)(x^2 + 98x - 9408)$
    $\equiv x^4 + 6x^3 - x^2 - 3x - 5 = (x^2 - x + 2)(x^2 - 6x + 4) \pmod{13}$

    (e) $x^6 - 256 = (x^3 - 16)(x^3 + 16)$
    $\equiv x^6 - 1 = (x - 1)(x + 1)(x^2 + x + 1)(x^2 - x + 1) \pmod 5$

    (f) $x^4 - 10x^2 + 1$ is irreducible over $\mathbb{Z}[x]$
    $\equiv x^4 + 1 = (x^2 - 2)(x^2 + 2) \pmod 5$

24. Factor the polynomials in Exercise 6 by performing the following steps:

    • For each polynomial $p(x)$, start by finding an "equivalent" monic polynomial q to use in factoring p. (This task is the subject of Exercise 6.)

    • Use distinct degree factorization to factor q modulo a prime. Work mod 3 for the polynomial in part (a); mod 13 for part (b); (c) mod 5; (d) mod 3; (e) mod 13; (f) mod 11.

    • Use the Hensel technique to lift the modular factorization of q found in the previous step to its factorization over the integers.

    • Deduce the factorization of the original polynomial p from the transformation in the first step and the result of the last step.

25. (Eisenstein's irreducibility criterion) Let m be prime and $p(x) = c_n x^n + c_{n-1}x^{n-1} + \ldots + c_1 x + c_0$ be a polynomial with integer coefficients having the following properties:

   1) c_n is not divisible by m,

   2) $c_{n-1}, \ldots, c_0$ are divisible by m,

   3) c_0 is not divisible by m^2.

   Show that p is irreducible over the integers.

Programming Projects

1. **(Schubert-Kronecker Factorization)** Write a procedure,

$$\text{SKFactor(<polynomial>,<variable>)},$$

   that factors a polynomial in one variable using the Schubert-Kronecker method. Test your code on the following polynomials:

$$
\begin{array}{ll}
x^4 + 4, & 25x^6 - 16, \\
x^5 + x + 1, & x^6 + 2x^4 + 4x^3 + x^2 + 4x + 4, \\
x^5 - x + 1, & x^8 - 2x^6 + 2x - 1, \\
y^6 - 1, & 2x^4 + 4x^3 + 2x + 8.
\end{array}
$$

   Try improving the efficiency of your procedure through judicious sampling of the points where the polynomial is evaluated.

2. **(Distinct Degree Factorization)** Write a pair of procedures that together achieve the modular factorization algorithm described in this chapter. The first,

$$\text{DDFactor(<polynomial>,<modulus>)},$$

   takes a polynomial mod m and a modulus m, and produces the distinct degree factorization described in Section 6.6.1. The second,

$$\text{EDFactor(<polynomial>,<degree>,<modulus>)},$$

   takes a polynomial modulo an odd modulus m that is the product of r factors of degree-d, and splits it into its r component factors as described in Section 6.6.2.

3. **(Hensel Lifting)** Write a procedure,

$$\text{Hensel(<polynomial>,<s\_mod\_m>,<t\_mod\_m>,<modulus>,<n\_iter>)},$$

   to accomplish the Hensel lifting technique described in Section 6.7. The procedure has three polynomial arguments: p in $\mathbb{Z}[x]$ and two factors, s and t, where $p = s\,t$ modulo a prime m. The fifth parameter gives the maximum number of iterations you should perform. Return the corresponding factors of p over the integers if the difference d goes to zero, otherwise return FAIL.

Chapter 7

Symbolic Integration

> *But just as much as it is easy to find the differential of a given quantity, so it is difficult to find the integral of a given differential. Moreover, sometimes we cannot say with certainty whether the integral of a given quantity can be found or not.*
>
> Johann Bernoulli (1667–1748)

> *Nature laughs at the difficulties of integration.*
>
> Pierre-Simon de Laplace (1749–1827)

One of the most impressive capabilities of computer algebra systems is their ability to compute integrals that would be difficult, if not impossible, to calculate by hand. Before CAS, complex tables of integrals were consulted to solve such problems. Interestingly, CAS have discovered many errors in these tables.

In this chapter, we explore how computer algebra systems perform *symbolic integration*. By the term "symbolic integration," we mean that we are looking for a "closed form" algebraic solution to an integration problem, rather than a numerical result. For example,

$$\int x \sin x \, dx = \sin(x) - x \cos(x)$$

is a symbolic indefinite integral and

$$\int_{\pi/3}^{\pi/2} x \sin x \, dx = 1 - \frac{1}{2}\sqrt{3} + \frac{1}{6}\pi$$

is a symbolic definite integral. We will not consider *numerical integration* problems like

$$\int_{\pi/3}^{\pi/2} \frac{\sin x}{x} \, dx = 0.3853033244,$$

although computer algebra systems have the capability to solve integrals numerically, as well. This last integral has no closed symbolic solution in terms of elementary functions, and we will be able to prove this fact using the mathematical tools developed here.

In the last paragraph, the term "elementary function" was used. What do we mean by this? An *elementary function* of a single variable is one that can be constructed from that variable and constants by any finite combination of the standard arithmetic operations (addition, subtraction, multiplication, and division), raising to powers and taking roots, taking logarithms and exponentials, forming trigonometric and hyperbolic functions and their inverses, and composing such operations.

The elementary functions are the ones encountered in a typical introductory calculus course. We could leave out the trigonometric and hyperbolic functions and their inverses since, as shown in Chapter 5, these can be expressed using (possibly complex) logarithms and exponentials.

7.1   Antidifferentiation

The symbolic indefinite integration problem can be posed as follows. We are given an elementary function $f(x)$ and asked to find another elementary function $g(x)$, if one exists, whose derivative is $f(x)$; that is, $g'(x) = f(x)$. Stated this way, we are determining the *antiderivative* g of f. Equivalently, we can phrase the problem in terms of integrals by finding the function g that satisfies

$$\int f(x)\,dx = g(x),$$

or more properly, $g(x) + C$, where C is an arbitrary constant. While calculus teachers usually insist on including the arbitrary constant C in the solution, we follow the tack of computer algebra systems and ignore the constant of integration.

We focus our attention on indefinite integration. In most (but not all) cases, we can easily solve a definite integral if we know the corresponding indefinite integral just by substituting the limits of integration.

There is a well-known set of rules for computing the symbolic derivatives of functions, which we list here:

1. $[f(x) \pm g(x)]' = f'(x) \pm g'(x)$ — sum/difference rule
2. $[f(x) g(x)]' = f'(x) g(x) + f(x) g'(x)$ — product rule
2'. $[c f(x)]' = c f'(x), \quad c$ constant — constant factors
3. $\left[\dfrac{f(x)}{g(x)}\right]' = \dfrac{f'(x) g(x) - f(x) g'(x)}{g(x)^2}$ — quotient rule
4. $[f(g(x))]' = f'(g(x)) g'(x)$ — chain rule
5. $[f(x)^{g(x)}]' = \left[g'(x) \ln f(x) + \dfrac{g(x) f'(x)}{f(x)}\right] f(x)^{g(x)}$ — exponentiation rule
5'. $[f(x)^n]' = n f'(x) f(x)^{n-1}, \quad n$ real — power rule
6. $[\ln f(x)]' = \dfrac{f'(x)}{f(x)}$ — logarithms
7. $[e^{f(x)}]' = f'(x) e^{f(x)}$ — exponentials

Of course, we could add other elementary functions, like the trigonometric and inverse trig functions, to the list but these can be written using logs and exponentials. If we want to differentiate an expression, we find its leading operator, invoke the appropriate rule and, if necessary, apply the process recursively to the result. Consequently, the derivative of an elementary function is an elementary function.

The first programs for differentiation predated the first successful effort at symbolic integration by almost a decade. In 1953, H. G. Kahrimanian (Temple University) and J. Nolan (MIT) independently wrote M.A. theses on the subject of automatic differentiation. Their work required substantial effort in view of the extremely limited computer memories available and the complete lack of high level programming languages at the time.

Unfortunately, no set of rules similar to those given for differentiation exists for integration. Moreover, we cannot run the rules for differentiation in reverse. Integration, instead, seems to be a random collection of techniques and special cases. Knowing the integrals of x^2 and e^x does not help us integrate their composition, e^{x^2} and, in fact, no closed form integral exists for this function.

The first successful attempt at a symbolic integration program was due to James R. Slagle (MIT) in 1961. For his Ph.D. dissertation, he wrote a program called SAINT (Symbolic Automatic INTegrator) that solved integration problems at the level of a freshman calculus student. Slagle's work predates the appearance of the first computer algebra system and is more properly classified in the realm of artificial intelligence rather than symbolic computation.

Several years later (1967) another MIT student, Joel Moses, wrote a successor called SIN (Symbolic INtegrator) for his Ph.D. thesis. While also relying largely on heuristics, this program took a far more algorithmic approach. SIN was much faster than SAINT and could solve a far wider range of integration problems. Moreover, while not a general purpose computer algebra system, SIN incorporated much of the infrastructure for manipulating and simplifying mathematical expressions that we normally associate with a CAS.

How do today's computer algebra systems go about the process of symbolic integration? Many believe that CAS solve all integration problems using a method attributed in name to Robert Risch (1969), and based on Liouville's theory of integration (c. 1830). In fact, CAS typically use a multistage approach to compute indefinite integrals. The early stages employ a mix of strategies similar to those taught in introductory calculus. These include table look-up, substitution, integration by parts, and several other heuristics. Perhaps surprisingly, a large number of integrals can be solved by these strategies. Rational functions that are not integrated in the early stages are solved via the methods presented in Sections 7.4–7.6. These techniques are an extension and generalization of the familiar partial fractions method but do not require a complete factorization of the denominator. The Risch algorithm is saved for last, when all else fails.

7.2   Heuristic Methods

Since computer algebra systems typically begin the process of symbolic integration by attempting to apply heuristics, it is appropriate that we examine some of these techniques. The SIN program, although outdated, is well-documented and provides a good case study for the way heuristics can be applied to solve symbolic integrals. SIN is divided into three stages:

1. The first stage incorporates the *derivative-divides* method, as well as two simple transformations. The derivative-divides technique is similar to the method of substitution taught in calculus. The two transformations involve 1) expressing the integral of a sum as a sum of integrals and 2) expanding an integrand consisting of an expression raised to a small, positive integer power.

2. At the second stage, the form of the integrand is examined to see whether one of eleven heuristic methods might be applied. Some of these techniques produce the integral directly, while others make a transformation that simplifies the form of the integrand. While the eleven heuristics include methods taught in calculus like trigonometric substitution, most are more advanced.

3. The third stage encompasses two techniques: integration by parts and the EDGE (EDucated GuEss) heuristic. The latter is based on Liouville's theory of integration and is a precursor to the Risch algorithm. The basic idea is to infer the form of the integral from the form of the integrand, and then apply the method of undetermined coefficients to obtain the specific solution.

According to Geddes et. al. (1992), who are authors of the Maple system, that program also consists of three phases:

1. The first phase includes integration of polynomials and table look-up for 35 simple forms (e.g., cos or exp with a linear argument).

2. Several heuristic methods are applied in the second phase. These include derivative-divides (substitution), integration by parts, partial fraction decomposition of rational functions when the degree of the denominator is < 7, and the integration of several special forms (e.g., $\frac{p(x)}{q(x)} f(ax + b)$).

3. The third phase consists of two general algorithmic methods: rational function integration and the Risch algorithm.

The authors of Maple state that the direct and heuristic methods in the first two phases are both fast and successful a large percentage of the time.

7.2.1 Derivative-Divides Method

The derivative-divides method is a variant of the method of integration by substitution taught in calculus. The idea is to determine whether derivatives of a subexpression of the integrand divide the rest of the integrand. If the integral is of the form

$$\int c \operatorname{op}(u(x)) \, u'(x) \, dx,$$

then its solution is simply $\int c \operatorname{op}(u(x)) \, u'(x) \, dx = c \int \operatorname{op}(y) \, dy \big|_{y=u}$. Here $\operatorname{op}(u)$ can be a trigonometric or arctrig function, an exponential, or a logarithm. Alternatively, $\operatorname{op}(u)$ can take the form $u(x), u^{-1}(x), u(x)^d$ where $d \neq -1$, or $d^{u(x)}$ where d is a constant. We present several applications of the technique.

Example 7.1. Use the derivative-divides method to solve the following integrals.

1. $\int 4 \cos(2x + 3) \, dx = 2 \sin(2x + 3)$

   We decompose the integrand: $\operatorname{op}(u) = \cos u, \quad u(x) = 2x + 3,$
   $\qquad\qquad\qquad\qquad\qquad\qquad\quad c = 2, \qquad\qquad u'(x) = 2.$

2. $\int \sin x \, \cos x \, dx = \frac{1}{2} \sin^2 x$

   This time we select: $\operatorname{op}(u) = u, \quad u(x) = \sin x,$
   $\qquad\qquad\qquad\qquad\quad c = 1, \qquad\quad u'(x) = \cos x.$

3. $\int x \, e^{x^2} \, dx = \frac{1}{2} e^{x^2}$

   Here we choose: $\operatorname{op}(u) = e^u, \quad u(x) = x^2,$
   $\qquad\qquad\qquad\qquad c = \frac{1}{2}, \qquad u'(x) = 2x.$

4. $\int \dfrac{e^x}{1+e^x}\, dx = \ln(1+e^x)$

This time we have: $\text{op}(u) = u^{-1}$, $u(x) = 1+e^x$,
                              $c = 1$, $u'(x) = e^x$.

5. $\int x\,(x^2+1)^{1/2}\, dx = \frac{1}{3}\,(x^2+1)^{3/2}$

Here we select: $\text{op}(u) = u^{1/2}$, $u(x) = x^2+1$,
                         $c = \frac{1}{2}$, $u'(x) = 2\,x$.

6. $\int \cos^2(e^x)\,\sin(e^x)\,e^x\, dx = -\frac{1}{3}\cos^3(e^x)$

We set: $\text{op}(u) = u^2$, $u(x) = \cos(e^x)$,
                 $c = -1$, $u'(x) = -e^x\,\sin(e^x)$.

$\square$

7.2.2   Integration by Parts

Integration by parts is useful when the integrand f can be regarded as the product of a function u, whose derivative is "simpler" than u, and another function v' that is easier to integrate than f itself. The integration by parts formula follows immediately from the product rule for differentiation and can be written

$$\int u(x)\,v'(x)\, dx = u(x)\,v(x) - \int v(x)\,u'(x)\, dx.$$

Unlike the derivative-divides method, which usually solves an integration problem, integration by parts merely transforms an integration problem into another, hopefully simpler, one.

The following example explores three paradigms that arise frequently when performing integration by parts. In the first problem, we transform the original integral into one that is basic and can be solved immediately. The second integral requires repeated integration by parts to solve. In the third, the original integration problem reoccurs when integration by parts is applied.

Example 7.2. Use integration by parts to solve the following integrals.

1. $\int x\,e^x\, dx$. Decomposing the integrand as

$$u(x) = x, \qquad\qquad\qquad v'(x) = e^x,$$
$$u'(x) = 1, \qquad\qquad\qquad v(x) = e^x,$$

leads to the simplification

$$\int x\,e^x\, dx = x\,e^x - \int e^x\, dx.$$

The exponential integral $\int e^x \, dx = e^x$ is elementary, so $\int x \, e^x \, dx = x \, e^x - e^x$. Had we selected $u = e^x$ instead,

$$u(x) = e^x, \qquad\qquad v'(x) = x,$$
$$u'(x) = e^x, \qquad\qquad v(x) = \tfrac{1}{2} x^2,$$

we would have transformed our integration problem into a more complex one,

$$\int x \, e^x \, dx = \tfrac{1}{2} x^2 \, e^x - \int \tfrac{1}{2} x^2 \, e^x \, dx.$$

2. $\int x^2 \, e^x \, dx$. By analogy with the last problem, we set

$$u(x) = x^2, \qquad\qquad v'(x) = e^x,$$
$$u'(x) = 2 \, x, \qquad\qquad v(x) = e^x,$$

yielding

$$\int x^2 \, e^x \, dx = x^2 \, e^x - \int 2 \, x \, e^x \, dx.$$

The power of x in the integrand on the right is one less than before, and we are left with a simpler integration problem. The new integral is essentially the one we just solved, and so repeated integration by parts gives

$$\int x^2 \, e^x \, dx = x^2 \, e^x - 2 \, x \, e^x + 2 \, e^x.$$

3. $\int \sin x \, \cos x \, dx$. Here we can't go wrong with either of the obvious choices: $u = \sin x$ or $u = \cos x$. Choosing

$$u(x) = \sin x, \qquad\qquad v'(x) = \cos x,$$
$$u'(x) = \cos x, \qquad\qquad v(x) = \sin x,$$

yields

$$\int \sin x \, \cos x \, dx = \sin^2 x - \int \sin x \, \cos x \, dx$$

We get back the original integral! This causes no problem whatsoever, since we can isolate the integral on the left side of the equation to obtain its solution,

$$\int \sin x \, \cos x \, dx = \tfrac{1}{2} \sin^2 x. \qquad\qquad \square$$

7.2.3 Elementary Transformations

Two heuristics we often apply when solving integration problems are: 1) to expand the integral of a sum as a sum of integrals, and 2) to expand products and quantities raised to small, positive integer powers in the integrand. An example of the first transformation is

$$\int (\sin x + \cos x)\, dx \rightarrow \int \sin x\, dx + \int \cos x\, dx,$$

and an example of the second, combined with the first, is

$$\int (x^2 + 1)^2\, dx \rightarrow \int x^4\, dx + 2 \int x^2\, dx + \int dx.$$

While these transformations are correct and generally useful, both can lead to trouble.

Consider the integral

$$\int \left(e^{x^2} + 2\, x^2\, e^{x^2} \right) dx = x\, e^{x^2},$$

which can be readily verified by differentiating the right side. Neither $\int e^{x^2}\, dx$ nor $\int x^2\, e^{x^2}\, dx$, however, have elementary closed forms! In this case, expansion hides the integrability of the function.

Another example where we are better off without expansion is

$$\int x\, (x^2 + 1)^6\, dx$$

While this problem can be solved with expansion, it is much easier to apply the derivative-divides method.

7.3   Rational Function Integration

The method taught in introductory calculus to integrate a rational function consists of the following steps:

1. factor the denominator into linear and irreducible quadratic factors,

2. find the partial fraction decomposition, and

3. integrate each partial fraction.

The second step involves solving a system of linear equations, where the number of equations and unknowns is equal to the degree of the denominator. When performing the third step, each linear factor produces a logarithmic term in the answer. Integrating each quadratic factor requires a trigonometric substitution or a reduction formula. We illustrate the method with an example.

Example 7.3. Find

$$\int \frac{x^6 + 7x^5 + 15x^4 + 32x^3 + 23x^2 + 25x - 3}{x^6 + 4x^5 + 10x^4 + 16x^3 + 17x^2 + 12x + 4} \, dx$$

using the "traditional" method.

Solution. The denominator factors as

$$x^6 + 4x^5 + 10x^4 + 16x^3 + 17x^2 + 12x + 4 = (x+1)^2 \, (x^2 + x + 2)^2.$$

The partial fraction decomposition is given by

$$1 - \frac{7}{(x+1)^2} + \frac{4}{x+1} - \frac{2x+1}{(x^2+x+2)^2} - \frac{x-3}{x^2+x+2}.$$

Note that when the integrand is not a "proper" rational function,[†] as is the case here, the decomposition includes a polynomial part. Integrating each term of the sum,

$$\int dx - \int \frac{7}{(x+1)^2} \, dx + \int \frac{4}{x+1} \, dx - \int \frac{2x+1}{(x^2+x+2)^2} \, dx - \int \frac{x-3}{x^2+x+2} \, dx,$$

we obtain the integral,

$$x + \frac{7}{x+1} + 4 \ln(x+1) + \frac{1}{x^2+x+2}$$
$$- \frac{1}{2} \ln(x^2 + x + 2) + \sqrt{7} \arctan\left(\tfrac{\sqrt{7}}{7}(2x+1)\right).$$

The fifth term of the partial fraction decomposition contributes the last two terms of the result. ☐

Observe that the polynomial part of the answer, x, comes from the polynomial term in the partial fraction expansion. The rational part of the answer, $\frac{7}{x+1} + \frac{1}{x^2+x+2}$, comes from the terms in the expansion whose denominator is raised to the power two (in general, two or more). The transcendental part of the answer comes from the terms in the partial fraction expansion whose denominator is raised to the power one. This transcendental part is expressed using logarithms and arctangents, although logarithms alone suffice if we are willing to use complex numbers since

$$\arctan u = \tfrac{1}{2} i \left(\ln(1 - iu) - \ln(1 + iu) \right).$$

The form of the solution is not coincidental. In 1812, Pierre-Simon de Laplace showed that the integral of a rational function can always be expressed as the sum of a rational function and a finite number of multiples of logarithms of rational functions. The proof essentially mirrors the calculus heuristic.

[†] A *proper rational function* is one in which the degree of the polynomial in the numerator is strictly less than the degree of the polynomial in the denominator.

Theorem 7.1 (Laplace's Theorem). *The integral of a rational function* $f(x)$ *with rational coefficients can be expressed as*

$$\int f(x)\,dx = R_0(x) + \sum_i c_i \ln R_i(x),$$

where R_0 *is a rational function with rational coefficients, the* c_i *are algebraic numbers (possibly complex), and* $R_1, R_2, \dots$ *are rational functions with coefficients that are (possibly complex) algebraic numbers.*

Unfortunately, the method for integrating rational functions taught in introductory calculus is fraught with difficulty. It is often impossible to produce the desired factorization of the denominator since polynomials of degree greater than four are not, in general, solvable by radicals. The procedure is also unnecessarily complex and may involve a great deal of needless algebraic manipulation. Moreover, the procedure can fail to produce solutions where elementary integrals exist.

Consider the following integrals:

(1) $\displaystyle\int \frac{8\,x^5 - 10\,x^4 + 5}{(2\,x^5 - 10\,x + 5)^2}\,dx = \frac{1 - x}{2\,x^5 - 10\,x + 5}.$

(2) $\displaystyle\int \frac{4\,x^5 - 1}{(x^5 + x + 1)^2}\,dx = -\frac{x}{x^5 + x + 1}.$

(3) $\displaystyle\int \frac{4\,x^4 + 4\,x^3 + 16\,x^2 + 12\,x + 8}{x^6 + 2\,x^5 + 3\,x^4 + 4\,x^3 + 3\,x^2 + 2\,x + 1}\,dx$

$\qquad\qquad = -\dfrac{x^2 - x + 4}{x^3 + x^2 + x + 1} + 3\arctan x.$

(4) $\displaystyle\int \frac{x}{x^4 + 1}\,dx = \tfrac{1}{2}\arctan x^2.$

In the first, the denominator is not solvable in terms of radicals. We cannot even get started except by using numerical approximations to the roots. In the second and third examples, the denominator factors over the integers but the factorization is not obvious, nor is it used to express the answer. In the fourth problem, the denominator does not factor over the integers but its roots are easily seen to be the fourth roots of -1: $\frac{\sqrt{2}}{2}(\pm 1 \pm i)$. The partial fraction expansions of integrals (2), (3), and (4) are quite involved. Even after factorization, a great deal of algebra and integration remains. Moreover, the integrals of the terms in the partial fraction expansions do not correspond to the terms in the simple solutions shown here.

In the next three sections, we examine the techniques used by computer algebra systems to integrate rational functions. We describe two methods to determine the rational part of the result, one due to Charles Hermite (1872) and the other to Ellis Horowitz (1971). We also present a technique developed independently by Michael Rothstein (1976) and Barry Trager (1976) to express the transcendental (logarithmic) part.

7.4 Hermite's Method

A rational function can be expressed in the form $P/Q + R$ where (1) P, Q, and R are polynomials, (2) P and Q have no common factors, and (3) the degree of P is less than the degree of Q. The polynomial R is trivially integrated. From Laplace's Theorem, the integral of the proper rational function P/Q can be written as a rational function plus a sum of logarithms.

Hermite showed how the rational part of the integral can be determined without factoring the denominator Q or finding its roots, as we are taught in calculus! Instead the technique begins with the *square-free decomposition* of the denominator (see Section 6.4.2). It then performs a partial fraction expansion based on this representation. Finally, the extended Euclidean algorithm and repeated integration by parts are used to determine both (1) the rational part of the integral and (2) the derivative of the logarithmic part.

Suppose we want to determine $\int \frac{P(x)}{Q(x)}\, dx$, where degree($P$) < degree($Q$) and $\gcd(P, Q) = 1$. We begin by performing a square-free decomposition of the denominator,

$$Q(x) = Q_1(x)\, Q_2^2(x)\, Q_3^3(x) \cdots Q_n^n(x),$$

where each Q_i has only simple (non-repeated) roots and no pair, Q_i, Q_j, has a common factor (i.e., $\gcd(Q_i, Q_j) = 1$).

Using the extended Euclidean algorithm, we next find polynomials A, B such that

$$\gcd(Q_1, Q_2^2 Q_3^3 \cdots Q_n^n) = 1 = B\, Q_1 + A\, Q_2^2 Q_3^3 \cdots Q_n^n.$$

Multiplying both sides by the numerator P and dividing by the denominator, Q, we have

$$\frac{P}{Q} = \frac{A\, P}{Q_1} + \frac{B\, P}{Q_2^2 Q_3^3 \cdots Q_n^n}.$$

When performing this step, it may be the case that the terms on the right are improper rational functions. If this occurs, we may safely perform long division and discard the quotients since they will be canceled by other quotients that arise later in the decomposition. Since both terms on the right have denominators whose degree is less than that of Q, our integration problem is simplified. We repeat the procedure for each square-free factor to arrive eventually at

$$\frac{P}{Q} = \frac{A_1}{Q_1} + \frac{A_2}{Q_2^2} + \frac{A_3}{Q_3^3} + \ldots + \frac{A_n}{Q_n^n},$$

where each term in the sum is a proper rational function.

Example 7.4. Apply the square-free partial fraction decomposition procedure to the rational function

$$\int \frac{x^4 + 6x^3 + 7x^2 + 6x + 4}{x^6 + 2x^5 + 4x^4 + 4x^3 + 4x^2 + 2x + 1} \, dx.$$

Solution. Here we have:

$$P = x^4 + 6x^3 + 7x^2 + 6x + 4, \quad Q = x^6 + 2x^5 + 4x^4 + 4x^3 + 4x^2 + 2x + 1,$$

The square-free factorization of the denominator Q is

$$Q = (x^2 + 1)(x^2 + x + 1)^2 = Q_1 Q_2^2.$$

Using the extended Euclidean algorithm, we find

$$\gcd(Q_1, Q_2^2) = 1 = (x^2 + 2x + 2) Q_1 - 1 \cdot Q_2^2$$

Multiplying both sides by P and dividing by $Q = Q_1 Q_2^2$, we have

$$\frac{P}{Q} = -\frac{P}{Q_1} + \frac{(x^2 + 2x + 2) P}{Q_2^2}$$

$$= -\frac{x^4 + 6x^3 + 7x^2 + 6x + 4}{x^2 + 1}$$

$$+ \frac{x^6 + 8x^5 + 21x^4 + 32x^3 + 30x^2 + 20x + 8}{(x^2 + x + 1)^2}.$$

Both terms on the right are improper rational functions, so we apply long division,

$$\frac{P}{Q} = -(x^2 + 6x + 6) + \frac{2}{x^2 + 1} + (x^2 + 6x + 6) - \frac{x^2 - 2x - 2}{(x^2 + x + 1)^2}$$

$$= \frac{2}{x^2 + 1} - \frac{x^2 - 2x - 2}{(x^2 + x + 1)^2}.$$

Magically, the polynomial parts of the two improper rational functions cancel! We are left with the desired square-free partial fraction expansion. □

After completing the square-free partial fraction expansion, we reduce each integral, $\int \frac{A_m}{Q_m} \, dx$ for $m > 1$, as follows. Since Q_m has no repeated factors, Q_m and its derivative Q_m' have no common factors. Therefore we can apply the extended Euclidean algorithm to find polynomials U and V such that

$$\gcd(Q_m, Q_m') = 1 = U Q_m + V Q_m'.$$

As a result, we can express A_m as

$$A_m = C Q_m + D Q_m',$$

where $C = A_m U$ and $D = A_m V$ are polynomials. Dividing by Q_m^m, we have

$$\int \frac{A_m}{Q_m^m} \, dx = \int \frac{C Q_m + D Q_m'}{Q_m^m} \, dx$$

$$= \int \frac{C}{Q_m^{m-1}} \, dx + \int \frac{D Q_m'}{Q_m^m} \, dx.$$

Using integration by parts with

$$u = D, \qquad\qquad v' = Q_m'/Q_m^m,$$

$$u' = D', \qquad\qquad v = -\frac{1}{(m-1) Q_m^{m-1}},$$

the second integral can be expressed as

$$\int \frac{D Q_m'}{Q_m^m} \, dx = -\frac{D}{(m-1) Q_m^{m-1}} + \int \frac{D'/(m-1)}{Q_m^{m-1}} \, dx,$$

and so we find

$$\int \frac{A_m}{Q_m^m} \, dx = -\frac{D}{(m-1) Q_m^{m-1}} + \int \frac{C + D'/(m-1)}{Q_m^{m-1}} \, dx. \qquad (7.1)$$

We have reduced the degree of the denominator by one and arrived at a similar, simpler integral. As with the partial fraction decomposition, the rational functions on the right side may be improper. Whenever this occurs, we perform long division and, as before, discard the quotient.

We continue this way, reducing the degree of the denominator by one at each step, until we arrive eventually at

$$\int \frac{A_m}{Q_m^m} \, dx = R_m + \int \frac{S_m}{Q_m} \, dx,$$

where R_m is a rational function and S_m is a polynomial. The integral on the right has no rational part since all the roots of Q_m are simple (non-repeated).

Repeating the procedure for each term $m > 1$ in the square-free partial fraction expansion, we have

$$\int \frac{P}{Q} \, dx = \int \frac{A_1}{Q_1} \, dx + R_2 + \ldots + R_m + \int \frac{S_2}{Q_2} \, dx + \ldots + \int \frac{S_m}{Q_m} \, dx.$$

The rational part of the solution is $R_2 + \ldots + R_m$. The remaining integrals give rise to the logarithmic part.

Algorithm 7.1 presents Hermite's method in its entirety.

Example 7.5. Complete the application of Hermite's algorithm to the integral,

$$\int \frac{x^4 + 6 x^3 + 7 x^2 + 6 x + 4}{x^6 + 2 x^5 + 4 x^4 + 4 x^3 + 4 x^2 + 2 x + 1} \, dx.$$

procedure $Hermite(P, Q)$
{Input: polynomials P, Q where $\deg(P) < \deg(Q)$, $\gcd(P, Q) = 1$}
{Output: rational function $RationalPart$, integral(s) $LogPart$
      where $RationalPart$ is the rational part of $\int \frac{P}{Q}$,
         $LogPart$ is the logarithmic part of $\int \frac{P}{Q}$ }

  $(Q_1, Q_2, \ldots, Q_n) := SquareFree(Q)$
  $(A_1, A_2, \ldots, A_n) := PartialFraction(P, Q_1, Q_2, \ldots, Q_n)$
    {Note: $\frac{P}{Q} = \frac{A_1}{Q_1} + \frac{A_2}{Q_2^2} + \ldots + \frac{A_n}{Q_n^n}$}
  $RationalPart := 0$
  $LogPart := \int \frac{A_1}{Q_1}$
  for $m := 2$ **to** n **do**
    $(R_m, S_m) := HermiteReduce(A_m, Q_m, m)$
    $RationalPart := RationalPart + R_m$
    $LogPart := LogPart + \int \frac{S_m}{Q_m}$
  end do
  return$(RationalPart, LogPart)$
end procedure

procedure $HermiteReduce(A, Q, m)$
{Input: polynomials A, Q and integer m where $m > 1$}
{Output: rational function R and polynomial S
      where $\int \frac{A}{Q^m} = R + \int \frac{S}{Q}$ }
  $S := A;$ $R := 0$
  $(U, V) := ExtendedEuclid(Q, Q')$ {$\gcd(Q, Q') = 1 = UQ + VQ'$}
  for $n := m$ **downto** 2 **do**
    $C := S \cdot U;$ $D := S \cdot V;$ $E := C + \dfrac{D'}{n-1}$
    $S := rem(E, Q^{n-1})$
    $R := -\dfrac{rem(D, Q^{n-1})}{(n-1) Q^{n-1}} + R$
  end do
  return(R, S)
end procedure

ALGORITHM 7.1: Hermite's method

Solution. In Exercise 7.4 we found

$$\frac{P}{Q} = \frac{A_1}{Q_1} + \frac{A_2}{Q_2^2} = \frac{2}{x^2+1} - \frac{x^2-2x-2}{(x^2+x+1)^2}.$$

The integral of the first term is completely transcendental (logarithmic). The second term produces the rational part of the result.

To reduce the integral $\int \frac{A_2}{Q_2^2}\,dx$, we first calculate $Q_2' = 2x+1$ and apply the extended Euclidean algorithm to find

$$\gcd(Q_2, Q_2') = 1 = \tfrac{4}{3}Q_2 - \tfrac{1}{3}(2x+1)Q_2'.$$

So $C = -\tfrac{4}{3}A_2 = \tfrac{4}{3}(x^2-2x-2)$, $D = -\tfrac{1}{3}(2x+1)A_2 = -\tfrac{1}{3}(2x^3-3x^2-6x-2)$, and $D' = -2(x^2-x-1)$.

We obtain the actual reduction by substituting $m = 2$ into (7.1), the formula derived using integration by parts,

$$\int \frac{A_2}{Q_2^2}\,dx = -\frac{D}{Q_2} + \int \frac{C+D'}{Q_2}\,dx$$

$$= -\frac{1}{3}\frac{2x^3-3x^2-6x-2}{x^2+x+1} + \int \frac{2}{3}\frac{x^2+x+1}{x^2+x+1}\,dx. \qquad (7.2)$$

Applying long division to the improper fractions, we find

$$\int \frac{A_2}{Q_2^2}\,dx = -\frac{1}{3}(2x-5) + \frac{x-1}{x^2+x+1} + \frac{2}{3}\int dx$$

$$= \frac{x-1}{x^2+x+1} + c,$$

where c is an arbitrary constant of integration. Alternatively, we could divide the two improper fractions in (7.2) and discard the quotients, which gives the desired result,

$$\int \frac{A_2}{Q_2^2}\,dx = \frac{x-1}{x^2+x+1},$$

immediately. Notice that in this example, $\int \frac{A_2}{Q_2^2}\,dx$ is completely rational and does not contribute a logarithmic term to the solution.

Putting together the results of the square-free partial fraction decomposition and the previous step, we have

$$\int \frac{P}{Q}\,dx = \frac{x-1}{x^2+x+1} + \int \frac{2}{x^2+1}\,dx. \qquad \Box$$

7.5   Horowitz' Method

Hermite's method is best suited to small problems solved by manual calculation. The disadvantage is that it needs several sub-algorithms (square-free factorization, partial fraction decomposition, the extended Euclidean algorithm for polynomial GCDs) that require some fairly complicated programming. Horowitz proposed an alternative way to produce the same result by solving an appropriate system of simultaneous linear equations. This method was actually discovered by the Russian mathematician Mikhail Ostrogradsky in the mid 1800's. The procedure runs more quickly than Hermite's algorithm and, as a result, is the method of choice for implementers of computer algebra systems.

As with Hermite's algorithm, we are given two polynomials, P and Q, where degree(P) < degree(Q) and gcd$(P, Q) = 1$. From Laplace's Theorem, we can write

$$\int \frac{P(x)}{Q(x)}\, dx = \frac{P_1(x)}{Q_1(x)} + \int \frac{P_2(x)}{Q_2(x)}\, dx, \tag{7.3}$$

where P_1/Q_1 is the rational part of the integral and P_2/Q_2 is the integrand of the logarithmic part. Moreover if $Q = \prod_{i=1}^{r} q_i^i$ is the square-free factorization of Q (see Section 6.4.2), the repeated factors of Q enter into Q_1 as $Q_1 = \prod_{i=2}^{r} q_i^{i-1}$, and $Q_2 = \prod_{i=2}^{r} q_i$ is the product of the square-free factors without multiplicities. (Why?) We can identify the repeated and square-free factors by calculating

$$Q_1 = \gcd(Q, Q') \qquad \text{and} \qquad Q_2 = Q/Q_1.$$

Thus we know how to find the denominator polynomials Q_1 and Q_2 in (7.3).

How do we calculate the numerator polynomials, P_1 and P_2? Differentiating both sides of (7.3) gives

$$\frac{P}{Q} = \frac{P_1'}{Q_1} - \frac{P_1\, Q_1'}{Q_1^2} + \frac{P_2}{Q_2},$$

and multiplying by $Q = Q_1\, Q_2$ to clear denominators, we get

$$P = P_1'\, Q_2 - P_1 \frac{Q_1'\, Q_2}{Q_1} + P_2\, Q_1.$$

Since Q_1 divides $Q_1'\, Q_2$ (why?), we can define the polynomial $S = Q_1'\, Q_2/Q_1$ and we have

$$P = P_1'\, Q_2 - P_1\, S + P_2\, Q_1. \tag{7.4}$$

In Equation (7.4), P is given and we have shown how to calculate Q_1, Q_2, and S. Furthermore, P_1/Q_1 and P_2/Q_2 are proper rational functions.

procedure $Horowitz(P, Q)$
{Input: polynomials P, Q where $\deg(P) < \deg(Q), \gcd(P, Q) = 1$}
{Output: rational function $RationalPart$, integral $LogPart$}
    {P_1/Q_1 is the rational part of $\int \frac{P}{Q}$
    P_2/Q_2 is the integrand of the logarithmic part}
    $Q_1 = \gcd(Q, Q')$; $Q_2 := Q/Q_1$
    $S := Q_1' Q_2/Q_1$
    $m := degree(Q_1)$; $n := degree(Q_2)$
    $P_1 := a_{m-1}x^{m-1} + \ldots + a_0$; $P_2 := b_{n-1}x^{n-1} + \ldots + b_0$
    $Solve(P = P_1' Q_2 - P_1 S + P_2 Q_1, \{a_{m-1}, \ldots, a_0, b_{n-1}, \ldots, b_0\})$
    $(P_1, P_2) := Substitute(\{a_{m-1}, \ldots, a_0, b_{n-1}, \ldots, b_0\}, (P_1, P_2))$
    $RationalPart := P_1/Q_1$
    $LogPart := \int P_2/Q_2$
    return$(RationalPart, LogPart)$
end procedure

ALGORITHM 7.2: Horowitz' method

So if $degree(Q_1) = m$ and $degree(Q_2) = n$, we can express the unknown polynomials P_1 and P_2 as

$$P_1 = a_{m-1} x^{m-1} + a_{m-2} x^{m-2} + \ldots + a_0,$$
$$P_2 = b_{n-1} x^{n-1} + b_{n-2} x^{n-2} + \ldots + b_0,$$

and solve Equation (7.4) by the method of undetermined coefficients. Note that $degree(Q) = degree(Q_1) + degree(Q_2) = m + n$, and we have a system of $m + n$ simultaneous linear equations in $m + n$ unknowns.

The entire procedure is outlined in Algorithm 7.2.

Example 7.6. Apply Horowitz' algorithm to the integral in Examples 7.4 and 7.5,

$$\int \frac{x^4 + 6x^3 + 7x^2 + 6x + 4}{x^6 + 2x^5 + 4x^4 + 4x^3 + 4x^2 + 2x + 1} \, dx.$$

Solution. Again we have

$$P = x^4 + 6x^3 + 7x^2 + 6x + 4,$$
$$Q = x^6 + 2x^5 + 4x^4 + 4x^3 + 4x^2 + 2x + 1.$$

We begin by calculating

$$Q' = 6x^5 + 10x^4 + 16x^3 + 12x^2 + 8x + 2.$$

Next, we set

$$Q_1 = \gcd(Q, Q') = x^2 + x + 1,$$
$$Q_2 = Q/Q_1 = x^4 + x^3 + 2x^2 + x + 1,$$

and form

$$S = Q_1' Q_2/Q_1 = 2x^3 + x^2 + 2x + 1.$$

Since $m = \text{degree}(Q_1) = 2$ and $n = \text{degree}(Q_2) = 4$, $\text{degree}(P_1) = 1$ and $\text{degree}(P_2) = 3$ so

$$P_1 = a_1 x + a_0,$$
$$P_2 = b_3 x^3 + b_2 x^2 + b_1 x + b_0.$$

Now we form the polynomial $P_1' Q_2 - P_1 S + P_2 Q_1$,

$$a_1 (x^4 + x^3 + 2x^2 + x + 1) - (a_1 x + a_0)(2x^3 + x^2 + 2x + 1)$$
$$+ (b_3 x^3 + b_2 x^2 + b_1 x + b_0)(x^2 + x + 1),$$

which, collecting the coefficients by degree, is equal to

$$b_3 x^5 + (b_3 + b_2 - a_1) x^4 + (b_3 + b_2 + b_1 - 2a_0) x^3$$
$$+ (b_2 + b_1 + b_0 - a_0) x^2 + (b_1 + b_0 - 2a_0) x + (b_0 + a_1 - a_0).$$

Equating the coefficients with those of P, we obtain the system of equations

$$b_3 = 0, \qquad b_3 + b_2 - a_1 = 1, \qquad b_3 + b_2 + b_1 - 2a_0 = 6,$$
$$b_2 + b_1 + b_0 - a_0 = 7, \qquad b_1 + b_0 - 2a_0 = 6, \qquad b_0 + a_1 - a_0 = 4,$$

whose solution is

$$a_1 = 1, \quad a_0 = -1, \quad b_3 = 0, \quad b_2 = 2, \quad b_1 = 2, \quad b_0 = 2.$$

Therefore, we have $P_1 = x - 1$ and $P_2 = 2x^2 + 2x + 2$. Observing that

$$\frac{P_1}{Q_1} = \frac{x-1}{x^2 + x + 1} \quad \text{and} \quad \frac{P_2}{Q_2} = \frac{2x^2 + 2x + 2}{x^4 + x^3 + 2x^2 + x + 1} = \frac{2}{x^2 + 1},$$

our integral can be expressed as

$$\int \frac{x^4 + 6x^3 + 7x^2 + 6x + 4}{x^6 + 2x^5 + 4x^4 + 4x^3 + 4x^2 + 2x + 1} \, dx = \frac{x-1}{x^2 + x + 1} + \int \frac{2}{x^2 + 1} \, dx,$$

where the remaining integral is entirely logarithmic. $\qquad\qquad\square$

7.6 Rothstein-Trager Method

Having found the rational part of the integral using either Hermite's method or Horowitz' method, we are left with the problem of determining the transcendental, or logarithmic, part,

$$\int \frac{P(x)}{Q(x)}\, dx = \sum_i c_i \ln R_i(x), \qquad (7.5)$$

where Q has no repeated factors, degree(P) < degree(Q), and gcd(P,Q) = 1. In general, the c_i are complex constants and the R_i are rational functions. Without loss of generality, we assume Q is monic and degree(Q) = r.

If we know the (possibly complex) roots a_i of Q, then we can factor Q,

$$Q(x) = (x - a_1)(x - a_2) \cdots (x - a_r),$$

use partial fractions to express the integrand,

$$\frac{P(x)}{Q(x)} = \frac{c_1}{x - a_1} + \frac{c_2}{x - a_2} + \ldots + \frac{c_r}{x - a_r},$$

and integrate each term in the expansion,

$$\int \frac{P(x)}{Q(x)}\, dx = \sum_{i=1}^{r} c_i \ln(x - a_i).$$

Terms with the same coefficient c_i can then be combined using the addition law for logarithms, $\ln a \pm \ln b = \ln(a\, b^{\pm 1})$, to yield (7.5).

Unfortunately, this approach presents several practical difficulties. Abel's Theorem asserts that we cannot always find closed form formulas for the roots of Q when its degree is more than four. Furthermore, when the degree of the denominator is either three or four, the closed forms may be unwieldy. Even if we know the roots of Q and their sizes are reasonable, the "simplest" form of the solution may not use these roots.

The problem of choosing an appropriate form to express the logarithmic part of a rational integral is explored in the following example.

Example 7.7. Consider the ways the following strictly logarithmic integrals can be expressed.

1. $\int \dfrac{1}{x^3 + x}\, dx$. The roots of $x^3 + x$ are $0, +i, -i$ and the partial fraction expansion of the integrand is

$$\frac{1}{x^3 + x} = \frac{1}{x} - \frac{1}{2}\frac{1}{x - i} - \frac{1}{2}\frac{1}{x + i}$$

$$= \frac{1}{x} - \frac{x}{x^2 + 1}.$$

The integral can be expressed several ways,

$$\int \frac{1}{x^3 + x}\, dx = \ln x - \tfrac{1}{2}\ln(x - i) - \tfrac{1}{2}\ln(x + i)$$

$$= \ln x - \tfrac{1}{2}\ln(x^2 + 1)$$

$$= \ln \frac{x}{\sqrt{x^2 + 1}}.$$

The first solution uses the roots of the denominator explicitly but requires three logarithms. Just two logarithms and no complex numbers are used in the second solution. The third also avoids complex numbers and eliminates another logarithm at the expense of introducing an algebraic quantity.

2. $\int \dfrac{1}{x^3 - 7x}\, dx$. The roots of the denominator are $0, +\sqrt{7}, -\sqrt{7}$, but $\sqrt{7}$ is not required to express the solution.

$$\int \frac{1}{x^3 - 7x}\, dx = -\tfrac{1}{7}\ln x + \tfrac{1}{14}\ln(x - \sqrt{7}) + \tfrac{1}{14}\ln(x + \sqrt{7})$$

$$= -\tfrac{1}{7}\ln x + \tfrac{1}{14}\ln(x^2 - 7).$$

3. $\int \dfrac{1}{x^2 - 2}\, dx$. The roots of the denominator are $\pm\sqrt{2}$. Although similar to the previous integral, $\sqrt{2}$ is needed to express the answer.

$$\int \frac{1}{x^2 - 2}\, dx = \tfrac{\sqrt{2}}{4}\ln(x - \sqrt{2}) - \tfrac{\sqrt{2}}{4}\ln(x + \sqrt{2})$$

$$= \frac{\sqrt{2}}{4}\ln\left(\frac{x - \sqrt{2}}{x + \sqrt{2}}\right).$$

□

How can we determine the c_i and R_i in (7.5) *without* explicitly finding the roots of Q? This dilemma was resolved by Rothstein (1976) and Trager (1976), who independently developed a technique to determine the logarithmic part of a rational integral. Moreover, the solution given by their method is "simplest" in the sense that it requires the fewest number of logarithms and does not introduce any unnecessary algebraic quantities. The following theorem forms the basis of the algorithm.

Theorem 7.2 (Rothstein-Trager Theorem). *Suppose*

$$\int \frac{P(x)}{Q(x)}\, dx = \sum_{i=1}^{m} c_i \ln R_i, \tag{7.6}$$

where $\mathrm{degree}(P) < \mathrm{degree}(Q)$, $\gcd(P, Q) = 1$, *and* Q *is square-free. Then the* c_i *are the (possibly complex) numbers* c *such that* $P - c\,Q'$ *and* Q *have a common factor, and the corresponding* $R_i = \gcd(P - c_i\,Q', Q)$.

Proof. Differentiating both sides of (7.6), we find that $P/Q = \sum_{i=1}^{m} c_i \, R_i'/R_i$ and

$$P \, R_1 \, R_2 \, \cdots \, R_m = Q \sum_{i=1}^{m} c_i \, R_1 \, \cdots \, R_{i-1} \, R_i' \, R_{i+1} \, \cdots \, R_m.$$

P and Q have no common factors since $\gcd(P, Q) = 1$. Also $\gcd(R_i, R_i') = 1$, otherwise R_i would be a repeated factor of Q. Similarly, $\gcd(R_i, R_j) = 1$ for $i \neq j$, otherwise their gcd would be a repeated factor of Q. All of this implies

$$Q = R_1 \, R_2 \, \cdots \, R_m$$

and

$$P = \sum_{i=1}^{m} c_i \, R_1 \, \cdots \, R_{i-1} \, R_i' \, R_{i+1} \, \cdots \, R_m.$$

Differentiating the expression for Q,

$$Q' = \sum_{i=1}^{m} R_1 \, \cdots \, R_{i-1} \, R_i' \, R_{i+1} \, \cdots \, R_m.$$

For any of the c_i,

$$P - c_i \, Q' = \sum_{i \neq j} (c_j - c_i) \, R_i \, R_1 \, \cdots \, R_{j-1} \, R_j' \, R_{j+1} \, \cdots \, R_m.$$

Observe that R_i is the only R in every term of the sum. Therefore, we have

$$\gcd(P - c \, Q', Q) = \begin{cases} R_i & \text{if } c = c_i, \\ 1 & \text{otherwise.} \end{cases} \qquad \square$$

So the problem of determining the logarithmic part of the integral of a rational function reduces to that of finding the common factors of two polynomials. This is usually solved by finding the *resultant* of the polynomials.

It is, however, possible to avoid computing the resultant by observing that $\gcd(P - c \, Q', Q)$ is a polynomial in c that must be 0. Applying the extended Euclidean algorithm gives

$$\gcd(P - c \, Q', Q) = 1 = \frac{S_N(x, c)}{S_D(c)} (P - c \, Q') + \frac{T_N(x, c)}{T_D(c)} Q, \qquad (7.7)$$

where the coefficients S and T are polynomials in x but rational functions of c. The numerators, in general, depend on both x and c, but the denominators depend only on c. Multiplying both sides of (7.7) by the least common multiple of the denominators, we see that $\text{lcm}(S_D(c), T_D(c))$ is the polynomial in c whose roots are the c_i. For each c_i, we obtain the polynomial $P - c_i Q'$ by substitution. Another polynomial gcd calculation gives the corresponding $R_i = \gcd(P - c_i \, Q', Q)$.

The Rothstein-Trager procedure is given in Algorithm 7.3.

procedure *RothsteinTrager*(P, Q)
{Input: polynomials P, Q where $\deg(P) < \deg(Q)$, $\gcd(Q, Q') = 1$ }
{Output: *LogPart*, where $LogPart = \int \frac{P}{Q} = \sum_i c_i \ln R_i$ }
　　$(S, T) := ExtendedEuclid(P - c\,Q', Q)$
　　　{ $GCD(P - c\,Q', Q) = 1 = \frac{S_N(c)}{S_D(c)}(P - c\,Q') + \frac{T_N(c)}{T_D(c)}Q'$ }
　　$cEqn := LCM(Denom(S), Denom(T))$
　　$cSoln := Solve(cEqn, c)$　　　{ $cSoln = \{c_1, \dots, c_n\}$ }
　　$LogPart := 0$
　　for $i := 1$ **to** $|cSoln|$ **do**
　　　$R_i := GCD(P - c_i\,Q', Q)$
　　　$LogPart := c_i \ln R_i + LogPart$
　　end do
　　return$(LogPart)$
end procedure

ALGORITHM 7.3: Rothstein-Trager method

Example 7.8. Compute the rational integrals in Example 7.7 using the Rothstein-Trager method.

1. $\displaystyle \int \frac{1}{x^3 + x}\, dx$. Here we have $P = 1$, $Q = x^3 + x$, and $Q' = 3x^2 + 1$. The extended Euclidean computation gives

$$\gcd(P - c\,Q', Q) = 1 = -\frac{3\,c\,x^2 + 2\,c + 1}{2\,c^2 - c - 1}\left(1 - c\,(3\,x^2 + 1)\right)$$
$$- 9\,\frac{c^2\,x}{2\,c^2 - c - 1}\,(x^3 + x),$$

so $\mathrm{lcm}(S_D(c), T_D(c)) = 2c^2 - c - 1$. The roots of this polynomial are $c_1 = 1, c_2 = -\frac{1}{2}$. With $c_1 = 1$ we have

$$R_1 = \gcd(P - c_1\,Q', Q) = \gcd(-3\,x^2, x^3 + x) = x,$$

and with $c_2 = -\frac{1}{2}$ we have

$$R_2 = \gcd(P - c_2\,Q', Q) = \gcd(-\tfrac{3}{2}\,x^2 + \tfrac{3}{2}, x^3 + x) = x^2 + 1.$$

Therefore, the integral is

$$\int \frac{1}{x^3 + x}\, dx = \sum_i c_i \ln R_i = \ln x - \tfrac{1}{2}\ln(x^2 + 1).$$

2. $\int \dfrac{1}{x^3 - 7\,x}\,dx$. This time $P = 1$, $Q = x^3 - 7x$, and $Q' = 3x^2 - 7$. The extended Euclidean computation reveals

$$\gcd(P - c\,Q', Q) = 1 = -\frac{3\,c\,x^2 - 14\,c + 1}{98\,c^2 - 7\,c - 1}\left(1 - c\,(3\,x^2 - 7)\right)$$
$$-9\,\frac{c^2\,x}{98\,c^2 - 7\,c - 1}\,(x^3 - 7\,x),$$

so $\operatorname{lcm}(S_D(c), T_D(c)) = 98c^2 - 7c - 1$, whose roots are $c_1 = -\frac{1}{7}, c_2 = -\frac{1}{14}$. With $c_1 = -\frac{1}{7}$ we find

$$R_1 = \gcd(P - c_1\,Q', Q) = \gcd(\tfrac{3}{7}\,x^2, x^3 - 7\,x) = x,$$

and with $c_2 = \frac{1}{14}$ we find

$$R_2 = \gcd(P - c_2\,Q', Q) = \gcd(-\tfrac{3}{14}\,x^2 + \tfrac{3}{2}, x^3 - 7\,x) = x^2 - 7,$$

So the integral is

$$\int \frac{1}{x^3 - 7\,x}\,dx = \sum_i c_i \ln R_i = -\tfrac{1}{7} \ln x + \tfrac{1}{14} \ln(x^2 - 7).$$

3. $\int \dfrac{1}{x^2 - 2}\,dx$. Here $P = 1$, $Q = x^2 - 2$, $Q' = 2x$, and the result of the extended Euclidean computation is

$$\gcd(P - c\,Q', Q) = 1 = -\frac{2\,c\,x + 1}{8\,c^2 - 1}\,(1 - 2\,c\,x) - 4\,\frac{b^2}{8\,c^2 - 1}\,(x^2 - 2).$$

The polynomial to be solved is $8c^2 - 1$, whose roots are $c_1 = \frac{\sqrt{2}}{4}$ and $c_2 = -\frac{\sqrt{2}}{4}$. $c_1 = \frac{\sqrt{2}}{4}$ gives

$$R_1 = \gcd(P - c_1\,Q', Q) = \gcd(-\tfrac{1}{2}\sqrt{2}\,x + 1, x^2 - 2) = x - \sqrt{2},$$

and $c_2 = -\frac{\sqrt{2}}{4}$ gives

$$R_2 = \gcd(P - c_2\,Q', Q) = \gcd(\tfrac{1}{2}\sqrt{2}\,x + 1, x^2 - 2) = x + \sqrt{2}.$$

(Note that $\gcd(-\frac{1}{2}\sqrt{2}\,x + 1, x^2 - 2)$ can be calculated by working with the monic polynomial $x - \sqrt{2}$ instead of $-\frac{1}{2}\sqrt{2}\,x + 1$. Letting $\alpha = \sqrt{2}$, we find $\gcd(x - \alpha, x^2 - \alpha^2) = x - \alpha$. This is a multivariate gcd computation.) Thus the integral is

$$\int \frac{1}{x^2 - 2}\,dx = \sum_i c_i \ln R_i = \tfrac{\sqrt{2}}{4} \ln(x - \sqrt{2}) - \tfrac{\sqrt{2}}{4} \ln(x + \sqrt{2}). \qquad \square$$

The second integral in the preceding example demonstrates that the Rothstein-Trager algorithm may express the solution using fewer algebraic quantities than would be required by working with the roots of the denominator. This has two important computational advantages. It avoids the cost of factoring the denominator, and it also avoids the overhead of manipulating algebraic numbers.

Example 7.9. Find $\int \dfrac{x}{x^4 + 1}\, dx$ using the Rothstein-Trager method.

Solution. Here we have $P = x$, $Q = x^4 + 1$, and $Q' = 4x^3$. The extended Euclidean algorithm gives

$$\gcd(P - cQ', Q) = 1 = -\frac{x^3 - 4cx}{16c^2 + 1}(1 - 4cx^3) - \frac{4c^3 - 16c^2 - 1}{16c^2 + 1}(x^4 + 1).$$

The solutions to $16c^2 + 1$ are $c_1 = -i/4$, $c_2 = i/4$, and the corresponding R_i are

$$R_1 = \gcd(P - c_1 Q', Q) = \gcd(x + i x^3, x^4 + 1) = x^2 - i,$$
$$R_2 = \gcd(P - c_2 Q', Q) = \gcd(x - i x^3, x^4 + 1) = x^2 + i.$$

Therefore, the integral can be expressed as

$$\int \frac{x}{x^4 + 1}\, dx = \sum_i c_i \ln R_i = -\frac{i}{4} \ln(x^2 - i) + \frac{i}{4}\ln(x^2 + i)$$

$$= -\frac{i}{4} \ln\left(\frac{x^2 - i}{x^2 + i}\right) \qquad \text{since } \ln a - \ln b = \ln \tfrac{a}{b}$$

$$= \tfrac{1}{2} \arctan x^2 + C \qquad \text{since } \arctan u = \tfrac{1}{2i} \ln\left(\tfrac{u-i}{u+i}\right) - \tfrac{\pi}{2}.$$

A calculus student would arrive at the solution expressed with the arctangent by the substitution $u = x^2$. If the roots of the denominator, $\frac{\sqrt{2}}{2}(\pm 1 \pm i)$, are used as the starting point, the same solution can be found.

$$\int \frac{x}{x^4 + 1}\, dx = \tfrac{1}{2}\left(\arctan(\sqrt{2}\,x - 1) - \arctan(\sqrt{2}\,x + 1)\right).$$

$$= \tfrac{1}{2} \arctan(-\tfrac{1}{x^2})$$

$$\text{since } \arctan(a) \pm \arctan(b) = \arctan \tfrac{a\pm b}{1 \mp ab}$$

$$= -\tfrac{1}{2}(\tfrac{\pi}{2} - \arctan x^2)$$

$$= \tfrac{1}{2} \arctan x^2 + C \qquad \qquad \square$$

We close our discussion of rational function integration with a paradoxical observation. In calculus, we are taught that the key to solving such problems is finding the roots of the denominator. The Rothstein-Trager method shows that, indeed, the integration of the logarithmic part depends on the solvability of a polynomial—but of one different from the denominator!

7.7 Liouville's Principle

The problem of determining whether a given elementary function has an integral that can be expressed in "closed form" and, if so, finding one has a long and interesting history. The two principal inventors of the calculus, Isaac Newton and Gottfried von Liebniz, took different approaches to this question. Newton rejected the use of transcendental quantities like the logarithm. For him, the result had to be expressed solely in algebraic terms. His *De Quandratura Curvarum* (1704) includes infinite series solutions to integrals for functions like $1/(ax + b)$ and $\sqrt{ax^2 + bx + c}$. Liebniz, on the other hand, did not share Newton's prejudice against non-algebraic quantities. His objection was, instead, to nonfinite representations and he embraced the use of transcendental functions to express solutions in "finite terms."

The most important contributor to the problem of integration in finite terms is undoubtedly Joseph Liouville (1809–1882). In a series of papers published between 1833 and 1841, he considered the integrability of many important classes of functions and, in so doing, he developed a theory of integration that has served as the basis for all subsequent work. Liouville was able to resolve a long standing conjecture that a particular class of elliptic integrals has no closed form solution. He was also able to show that both $\int e^x/x\, dx$ and $\int \sin(ax)/(1 + x^2)\, dx$ are not elementary.

Liouville's main result, which is reminiscent of Laplace's Theorem, states that if an algebraic function $f(x)$ is integrable in finite terms, its antiderivative is the sum of an algebraic function and a finite number of logarithms of algebraic functions. Here an *algebraic* quantity is taken in the usual sense of a solution to some polynomial equation.

Theorem 7.3 (Liouville's Theorem). *(1) If f is an algebraic function of $x, \theta_1, \ldots, \theta_m$, where $\theta_1, \ldots, \theta_m$ are functions whose derivatives are each algebraic functions of $x, \theta_1, \ldots, \theta_m$, then $\int f(x, \theta_1, \ldots, \theta_m)\, dx$ is elementary if and only if*

$$\int f(x, \theta_1, \ldots, \theta_m)\, dx = V_0(x, \theta_1, \ldots, \theta_m) + \sum_i c_i \ln V_i(x, \theta_1, \ldots, \theta_m),$$

where the c_i are constants and the V_i are algebraic functions.

(2) If $f(x, \theta_1, \ldots, \theta_m)$ and $\theta'_1, \ldots, \theta'_m$ are rational functions of $x, \theta_1, \ldots, \theta_m$, then the V_i are also rational functions.

The proof is based on the fact that when we differentiate an exponential, we get back an exponential with the same argument. When we differentiate a logarithm of degree higher than one, we get back the same logarithm. As a result, no new exponential terms and no new logarithmic terms except in linear form, can be part of the integral of an algebraic function. Trigonometric functions pose no difficulty since they can be expressed as complex exponentials.

The following example illustrates the fundamental concept underlying Liouville's Theorem.

Example 7.10. Suppose

$$f(x, \theta_1, \ldots, \theta_7) = f(x, e^x, \ln x, e^{e^x}, \ln(\ln(x)), \sin(x), \cos(x), \cos e^x)$$

is an algebraic function of its arguments. Then the first part of Liouville's theorem applies to the integral, $\int f(x, \theta_1, \ldots, \theta_7) \, dx$, since

$$\theta_1' = e^x = \theta_1, \quad \theta_2' = \frac{1}{x}, \quad \theta_3' = e^x e^{e^x} = \theta_1 \theta_3, \quad \theta_4' = \frac{1}{x \ln x} = \frac{1}{x \theta_2},$$

$$\theta_5' = \cos x = \theta_6, \quad \theta_6' = -\sin x = -\theta_5, \quad \theta_7' = -e^x \sin e^x = -\theta_1 \sqrt{1 - \theta_7^2},$$

are algebraic functions of $x, \theta_1, \ldots, \theta_7$. The second part of Liouville's Theorem does not apply, however, since θ_7' is not a rational function of $x, \theta_1, \ldots, \theta_7$.
   If

$$g(x, \theta_1, \ldots, \theta_6) = g(x, e^x, \ln x, e^{e^x}, \ln(\ln(x)), \sin(x), \cos(x))$$

is a rational function of its arguments, then the second part of Liouville's Theorem does apply. Here $\theta_1', \ldots, \theta_6'$ are all rational functions of $x, \theta_1, \ldots, \theta_6$, and the integral of g is elementary if and only if

$$\int g(x, \theta_1, \ldots, \theta_6) \, dx = V_0(x, \theta_1, \ldots, \theta_6) + \sum_i c_i \ln V_i(x, \theta_1, \ldots, \theta_6),$$

where the c_i are constants and the V_i are rational functions of $x, \theta_1, \ldots, \theta_6$. $\quad\square$

Over a century elapsed before the next major contribution to theory of integration in finite terms. Alexander Ostrowski (1946) used the idea of an algebraic field extension to generalize Liouville's work and to widen the class of functions to which it applies. Soon thereafter, Joseph Ritt (1948) published a monograph that has become the classical account of the subject. Ritt's work promoted and extended the algebraic approach taken by Ostrowski. In addition, it brought the subject up to date by summarizing the important developments and simplifying the presentation. Maxwell Rosenlicht (1968) published the first entirely algebraic presentation of Liouville's theory. Building on this work, Robert Risch in his Ph.D. dissertation at Berkeley (1968) took an algorithmic approach by showing how the general problem of integration in finite terms can be reduced to a decidable question in the theory of algebraic functions. Since then, a number of researchers have made important contributions by closing some unresolved cases and by implementing the algorithm.

The Risch algorithm is quite involved and an entire book by Manuel Bronstein (1997, 2006) is devoted to a scholarly exposition of the topic. At the risk of oversimplification, we present an introduction to the method in the remainder of this chapter. Our goals are to impart some of the flavor of the procedure and to use it to prove that some very simple elementary integrals have no closed form solution. We first consider the integration of logarithms and then the integration of exponentials. A complete treatment would also require a discussion of the integration of algebraic functions and of mixed integrals combining logarithms, exponentials, and algebraic quantities.

7.8   Risch Algorithm: Logarithmic Case

Suppose our integrand f is a rational function of both the integration variable x and a logarithmic quantity in x, $\theta = \ln u$. As with a rational function of a single variable, the integrand can be split into a polynomial in θ whose coefficients depend only on x, plus a proper rational function. The square-free factorization of the denominator is obtained and used as the basis for a partial fraction decomposition. This process allows us to express the integrand as

$$f(x,\theta) = A_m(x)\,\theta^m + A_{m-1}(x)\,\theta^{m-1} + \ldots + A_0(x) + \sum_{i=1}^{n} \frac{P_i(x,\theta)}{Q_i^i(x,\theta)},$$

where the A_i do not contain θ, P_i and Q_i are polynomials in θ whose coefficients are also free of θ, $\text{degree}(P_i) < \text{degree}(Q_i)$, and the Q_i have only simple (non-repeated) roots. The forms that the integrals of the polynomial and rational parts must take if the integral has a closed form solution are given by Liouville's Theorem. In the remainder of this section, we show how to obtain these integrals.

7.8.1   Polynomial Part

Suppose we want to integrate a polynomial in some logarithmic quantity $\theta = \ln u$,
$$A_m\,\theta^m + A_{m-1}\,\theta^{m-1} + \ldots + A_0,$$
where the coefficients A_i may depend on integration variable x but not on θ. Liouville's principle tells us that the integral, if it exists, is another polynomial in θ of degree at most $m+1$,

$$I(x,\theta) = \int A_m\,\theta^m + A_{m-1}\,\theta^{m-1} + \ldots + A_0\, dx$$
$$= B_{m+1}\,\theta^{m+1} + B_m\,\theta^m + \ldots + B_0,$$

where again the coefficients B_i do not contain θ, and only B_0 may introduce a new logarithmic quantity. Differentiating the expression for the integral and observing that $\theta' = u'/u$ since θ is logarithmic, we have

$$I'(x, \theta) = B'_{m+1} \theta^{m+1} + \left((m+1) B_{m+1} \frac{u'}{u} + B'_m \right) \theta^m$$

$$+ \left(m B_m \frac{u'}{u} + B'_{m-1} \right) \theta^{m-1} + \ldots + \left(2 B_2 \frac{u'}{u} + B'_1 \right) \theta + B_1 \frac{u'}{u} + B'_0.$$

We determine the B_i by comparing coefficients of powers of θ, starting with θ^{m+1}.

Since the integrand has no θ^{m+1} term, B'_{m+1} must be a constant, say b_{m+1}. Therefore,

$$A_m = (m+1) b_{m+1} \frac{u'}{u} + B'_m$$

and, integrating both sides of this equation, we have

$$\int A_m \, dx = (m+1) b_{m+1} \theta + B_m.$$

This integral is easier than the original one since it does not involve θ and can be evaluated by applying the algorithm recursively. If the integral does not exist, then neither does the original integral. If it does exist, then the only logarithmic term it can contain is $\theta = \ln u$ since the B_i do not involve θ for $i > 0$. Otherwise, the integral can have no closed form solution.

Suppose

$$\int A_m \, dx = c\theta + d(x)$$

for some constant c. Then, by comparing coefficients, $b_{m+1} = c/(m+1)$ and $B_m = d(x) + b_m$, where b_m is a constant of integration. At this and each subsequent iteration, we discover the constant term of the higher coefficient and the current coefficient up to a constant. We continue by substituting $d(x) + b_m$ for B_m to obtain an integral for B_{m-1}. If all the B_i can be found in this fashion without violating the restrictions on the form of the B_i, then the integral exists and we obtain its solution.

The following simple example demonstrates much of the machinery of the algorithm.

Example 7.11. Find $\int \ln x \, dx = \int \theta \, dx$, where $\theta = \ln x$ and $\theta' = 1/x$.

Solution. The integrand may be expressed as $A_1 \theta + A_0$, where $A_1 = 1$ and $A_0 = 0$. From Liouville's Theorem the integral, if it exists, is of the form

$$I(x, \theta) = B_2 \theta^2 + B_1 \theta + B_0$$

and

$$I'(x, \theta) = B'_2 \theta^2 + \left(\frac{2 B_2}{x} + B'_1 \right) \theta + \frac{B_1}{x} + B'_0.$$

There is no θ^2 term in the integrand, so $B_2' = 0$ and B_2 is a constant, b_2. Comparing the coefficients of the θ terms, we see that

$$A_1 = \frac{2\,B_2}{x} + B_1',$$

where $A_1 = 1$ and $B_2 = b_2$. Integrating this equation, we have

$$\int dx = x + \text{const} = 2\,b_2\,\theta + B_1.$$

Therefore it must be that $b_2 = 0$ and $B_1 = x + b_1$, where b_1 is a constant. Finally, comparing A_0 to the constant term in $I'(x, \theta)$, we have

$$A_0 = \frac{B_1}{x} + B_0',$$

$$0 = \frac{x + b_1}{x} + B_0' = 1 + \frac{b_1}{x} + B_0'.$$

Rearranging this equation and integrating, we find

$$\int - dx = -x + \text{const} = b_1\,\theta + B_0,$$

so $b_1 = 0$ and $B_0 = -x + b_0$, where b_0 is a constant. Since $B_2 = 0, B_1 = x, B_0 = -x + b_0$, we arrive at the solution

$$\int \ln x \, dx = I(x, \theta) = x \ln x - x + b_0. \qquad \square$$

The next example shows what happens when the integral has no closed form.

Example 7.12. Show $\displaystyle\int \frac{\ln(x-1)}{x} \, dx$ has no elementary solution.

Solution. Letting $\theta = \ln(x-1)$, we have $\theta' = 1/(x-1)$ and the integrand can be expressed as $A_1\theta + A_0$, where $A_1 = 1/x$ and $A_0 = 0$. Liouville's principle tells us that the integral, if it exists, must be of the form

$$I(x, \theta) = B_2\,\theta^2 + B_1\,\theta + B_0$$

and so

$$I'(x, \theta) = B_2'\,\theta^2 + \left(\frac{2\,B_2}{x-1} + B_1'\right)\theta + \frac{B_1}{x-1} + B_0'.$$

Since there is no θ^2 term in the integrand, $B_2' = 0$ and $B_2 = b_2$, a constant. Comparing the coefficients of the θ terms, we have

$$A_1 = \frac{2\,B_2}{x-1} + B_1',$$

where $A_1 = 1/x$ and $B_2 = b_2$. Integrating this equation, we have

$$\int \frac{1}{x}\, dx = \ln x + \text{const} = 2\, b_2 \ln(x-1) + B_1.$$

So $b_2 = 0$ since $\ln(x-1)$ does not appear on the left. But this is impossible since $B_1 = \ln x + b_1$ would introduce a new logarithmic quantity into the solution. Therefore the integral has no elementary closed form. ☐

One final example illustrates how multiple logarithmic extensions are handled by the algorithm.

Example 7.13. Find $\displaystyle\int \left(\ln(x-1) + \frac{1}{x-1} \right) \ln x\, dx$.

Solution. Let $\theta_1 = \ln(x-1)$ and $\theta_2 = \ln x$. We regard the integrand as a polynomial in one of these logarithms with coefficients that are functions of x and the other logarithm. Choosing to express the integrand as a polynomial in θ_2, we have

$$A_1(x, \theta_1)\, \theta_2 + A_0(x, \theta_1),$$

where $A_1 = \theta_1 + \frac{1}{x-1}$ and $A_0 = 0$. With this representation, we seek an integral of the form

$$I(x, \theta_1, \theta_2) = B_2(x, \theta_1)\, \theta_2^2 + B_1(x, \theta_1)\, \theta_2 + B_0(x, \theta_1).$$

Since $\theta_2' = 1/x$, the form of I' is

$$I'(x, \theta_1, \theta_2) = B_2'\, \theta_2^2 + \left(\frac{2\, B_2}{x} + B_1' \right) \theta_2 + \frac{B_1}{x} + B_0'.$$

As in the previous two examples, $B_2' = 0$ and so $B_2 = b_2$ is a constant. Comparing the coefficients of the θ_2 terms, we have

$$\ln(x-1) + \frac{1}{x-1} = \frac{2\, b_2}{x} + B_1'.$$

The logarithm on the left side is simpler to integrate than the original integral. It can be solved by applying the algorithm recursively and, doing so, we find

$$x \ln(x-1) - x + \text{const} = 2\, b_2\, \theta_2 + B_1,$$

giving $b_2 = 0$ and $B_1 = x\, \theta_1 - x + b_1$, where b_1 is a constant.

Comparing $A_0 = 0$ to the "constant" term in I', we see that

$$0 = \frac{x \ln(x-1) - x + b_1}{x} + B_0'.$$

Rearranging the terms and integrating gives

$$-\int (\ln(x-1) - 1)\, dx = \int \left(\frac{b_1}{x} + B_0' \right) dx \tag{7.8}$$

$$-(x-1)\ln(x-1) + (x-1) + x + \text{const} = b_1\, \theta_2 + B_0,$$

so $b_1 = 0$ and $B_0 = -(x-1)\,\theta_1 + 2\,x + C$. Again, we apply the integration procedure recursively when evaluating the left side of (7.8).

With $B_2 = 0$, $B_1 = x\ln(x-1) - x$, and $B_0 = -(x-1)\ln(x-1) + 2x + C$, we at last have

$$\int \left(\ln(x-1) + \frac{1}{x-1}\right)\ln x\, dx =$$
$$(x\ln(x-1) - x)\ln x - (x-1)\ln(x-1) + 2\,x + C. \qquad \square$$

This example illustrates an important feature of the Risch algorithm. When integrating a function that involves several transcendental quantities, $\theta_1, \ldots, \theta_m$, we place an ordering on them. If θ_m is last in the ordering, we regard f as function of θ_m with coefficients that involve x and the other transcendental quantities. In the course of solving an integral, the Risch algorithm may be called recursively to evaluate other integrals, but these are always simpler in the sense of involving one fewer transcendental quantity than the current (sub)integral being evaluated.

Moreover, the transcendental quantities must be *linearly independent* of one another in the sense of the Structure Theorem presented in Chapter 5. In Example 7.13, the functions $\theta_1 = \ln(x-1)$ and $\theta_2 = \ln x$ are independent. However, $\theta_3 = \ln x^2$ is not independent of these other logarithms since $\ln x^2 = 2\ln x = 2\theta_2$.

7.8.2 Logarithmic Part

We now turn our attention to the problem of integrating a proper rational function, $\frac{P(x,\theta)}{Q(x,\theta)}$, involving a logarithmic quantity $\theta = \ln u$. $P(x,\theta)$ and $Q(x,\theta)$ can be regarded as polynomials in θ with coefficients that may be rational functions of the integration variable x. We first consider the case where Q is square-free (i.e., Q has no repeated roots), and later turn our attention to the case where the denominator is of the form $Q^m(x,\theta)$.

Observe that if the denominator can be factored into a product of linear factors involving θ, the integral is easily determined. In this case, Q can be written in the form

$$Q(x,\theta) = (\theta - A_1(x))\,(\theta - A_2(x)) \cdots (\theta - A_r(x)),$$

where the A_i are rational functions of x, and we perform a partial fraction decomposition to express the integrand as

$$\frac{P(x,\theta)}{Q(x,\theta)} = \frac{R_1(x)}{\theta - A_1(x)} + \frac{R_2(x)}{\theta - A_2(x)} + \ldots + \frac{R_r(x)}{\theta - A_r(x)},$$

where the R_i are again rational functions of x. The integral of each term exists only when

$$R_i = c_i\,(\theta - A_i)', \qquad c_i \text{ constant},$$

which can be checked easily by differentiation. If a constant c_i is found, then the integral of the term is given by a new logarithmic quantity, $c_i \ln(\theta - A_i)$.

As we saw earlier with the heuristic approach to rational function integration, such a procedure is inadequate to deal with most cases. The Rothstein-Trager method, however, can be modified to deal with the situation. The following is the analogue of Theorem 7.2.

Theorem 7.4. *Suppose $P(x,\theta)/Q(x,\theta)$ is a rational function with $\theta = \ln u$ such that $\mathrm{degree}(P) < \mathrm{degree}(Q)$, $\gcd(P,Q) = 1$, and Q is square-free. Then the integral of P/Q has a closed form solution given by*

$$\int \frac{P(x,\theta)}{Q(x,\theta)}\, dx = \sum_i c_i \ln R_i$$

if and only if all the roots c_i of $\gcd(P - cQ', Q)$ are constants. The sum is taken over all of the roots and $R_i = \gcd(P - c_iQ', Q)$.

We present two simple examples to illustrate the approach. The first integral exists and is found by the method, while the second cannot be expressed in elementary terms.

Example 7.14. Determine $\displaystyle\int \frac{1}{x\ln x}\, dx$.

Solution. Letting $\theta = \ln x$, we have $P = 1/x$ and $Q = \theta, Q' = 1/x$. The extended Euclidean gcd calculation is

$$\gcd(y - cy, \theta) = 1 = \frac{S_N(x,\theta,c)}{S_D(x,c)}(P - cQ') + \frac{T_N(x,\theta,c)}{T_D(x,c)}Q$$

$$= \frac{1}{y - cy}(y - cy) + 0\cdot\theta,$$

where $y = 1/x$ and the division is performed with respect to θ. Following the rational case, $\mathrm{lcm}(S_D(x,c), T_D(x,c)) = y - cy$ and solving for c we find $c = 1$, a constant. The corresponding value of R is $\gcd(0,\theta) = \theta$ and so

$$\int \frac{1}{x\ln x}\, dx = c\ln R = \ln\theta = \ln\ln x.$$

The naïve heuristic can also be applied to this integral since the denominator Q is trivially in linearly factored form with respect to θ. Again $P = 1/x$ and $Q = \theta$, and $A = 0$. The integral exists since $R = c(\theta - A)' = c/x$ and $P/Q = R/(\theta - A)$ implies $c = 1$, a constant. Therefore,

$$\int \frac{1}{x\ln x}\, dx = c\ln\theta = \ln\ln x. \qquad \square$$

Example 7.15. Show that $\int \dfrac{1}{\ln x}\, dx$ has no elementary closed form.

Solution. With $\theta = \ln x$, we have $P = 1$ and $Q = \theta, Q' = 1/x$. Applying the extended Euclidean algorithm,

$$\gcd(1 - cy, \theta) = 1 = \frac{S_N(x, \theta, c)}{S_D(x, c)}(P - cQ') + \frac{T_N(x, \theta, c)}{T_D(x, c)}Q$$

$$= \frac{1}{1 - cy}(1 - cy) + 0 \cdot \theta,$$

where $y = 1/x$ and the division is performed with respect to θ. $\mathrm{lcm}(S_D, T_D) = 1 - cy$, and solving for c we find $c = 1/y = x$, which is not a constant. Therefore, the integral is not elementary.

Again, the naïve heuristic can be applied to this simple problem. $R = c\theta' = c/x$, and $P/Q = R/\theta$ implies $c = x$. Since c is not a constant, the integral has no elementary closed form. $\qquad\square$

7.8.3 Rational Part

The Risch algorithm can be thought of as a generalization of the techniques of Hermite and Rothstein-Trager to the problem of integrating transcendental functions. The following example shows how these procedures are brought to bear in obtaining the integral of a more complicated rational function of a logarithm.

Example 7.16. Find $\int \dfrac{2\theta^3 - 2\theta^2 + (-2x^2 + \frac{2}{x})\theta + 2x^2 - 2x}{(\theta^2 - x^2)^2}\, dx$, where $\theta = \ln x$.

Solution. The denominator is already in square-free form and the partial fraction expansion of the integrand with respect to θ is

$$\frac{\frac{2\theta}{x} - 2x}{(\theta^2 - x^2)^2} + \frac{2\theta - 2}{\theta^2 - x^2}. \tag{7.9}$$

In the notation of Hermite's algorithm, the first term is $\frac{A_2(x,\theta)}{Q_2^2(x,\theta)}$, where $A_2 = \frac{2\theta}{x} - 2x$ and $Q_2 = \theta^2 - x^2$. Continuing as in the case of a simple rational function, we calculate $Q_2' = \frac{2\theta}{x} - 2x$ and apply the extended Euclidean algorithm to express A_2 as linear combination of Q_2 and Q_2',

$$A_2 = 0 \cdot Q_2 + 1 \cdot Q_2'.$$

So we see that $\int \frac{A_2}{Q_2^2}\, dx = \int \frac{Q_2'}{Q_2^2}\, dx$, resulting immediately in

$$\int \frac{\frac{2\theta}{x} - 2x}{(\theta^2 - x^2)^2}\, dx = -\frac{1}{\theta^2 - x^2} = -\frac{1}{\ln^2 x - x^2}.$$

The integral of the second term in (7.9) is transcendental and must be expressed completely in terms of new logarithms. Setting $P_2 = 2\theta - 2$ and performing the extended Euclidean algorithm to calculate $\gcd(P_2 - cQ_2', Q_2)$, we have

$$\frac{(c-x)\theta + cx^2 - x}{2x(1 + c^2 x^2 - c^2 - x^2)}(P_2 - cQ_2') + \frac{(x-c)^2}{x^2(1 + c^2 x^2 - c^2 - x^2)}Q_2 = 1.$$

The roots of $\mathrm{lcm}(S_D, T_D) = 2x^2(1 + c^2 x^2 - c^2 - x^2)$ are $c_1 = 1, c_2 = -1$, and the corresponding values of R are

$$R_1 = \gcd(P_2 - Q_2', Q_2) = \gcd\left((2 - \tfrac{2}{x})\theta + 2x - 2, \theta^2 - x^2\right) = x + \theta,$$
$$R_2 = \gcd(P_2 + Q_2', Q_2) = \gcd\left((2 + \tfrac{2}{x})\theta - 2x - 2, \theta^2 - x^2\right) = x - \theta,$$

where the divisions in the gcd calculations are performed with respect to θ. Therefore,

$$\int \frac{2\theta - 2}{\theta^2 - x^2}\, dx = \sum_i c_i \ln R_i = \ln(x + \ln x) - \ln(x - \ln x). \qquad \square$$

7.9 Risch Algorithm: Exponential Case

The integration of exponential quantities is more involved than the integration of logarithms because the derivative of an exponential is an exponential of the same degree. As with our study of the Risch algorithm in the logarithmic case, we divide our discussion into sections. The first focuses on the integration of polynomials of exponentials and the second on the integration of rational functions of exponentials.

7.9.1 Polynomials of Exponentials

Suppose θ is an exponential quantity, $\theta = e^u$, where u is rational function of the integration variable x. Note that each term in the polynomial $\sum_k A_k(x)\theta^k$ may be treated separately since $\theta^k = (e^u)^k = e^{ku}$ can be considered a new exponential quantity, $\theta_1 = e^v$, where $v = ku$ is another rational function of x. With this observation, the following special case of Liouville's Theorem gives the form the integral $\int f(x)\, e^u\, dx$ takes when it is elementary.

Theorem 7.5 (Liouville's Theorem, Exponential Case). *Suppose $f(x)$ and $u(x)$ are rational functions of x where $u(x)$ is not a constant. Then $\int f(x)\, e^{u(x)}\, dx$ exists if and only if there is a rational function $R(x)$ such that*

$$\int f(x)\, e^{u(x)}\, dx = R(x)\, e^{u(x)} + C, \tag{7.10}$$

where C is an arbitrary constant of integration.

Differentiating both sides of (7.10), we uncover an equivalent condition for the integrability in closed elementary form of the function $f(x)\,e^{u(x)}$—that there exists a rational function $R(x)$ satisfying

$$f(x) = R'(x) + R(x)\,u'(x). \tag{7.11}$$

If $R(x) = P(x)/Q(x)$ then, differentiating R with the quotient rule and grouping all the terms on the right (except possibly the denominator of u') over a single common denominator, (7.11) becomes

$$f(x) = \frac{P'(x)\,Q(x) - P(x)\,Q'(x) + u'(x)\,P(x)\,Q(x)}{Q^2(x)}.$$

Formula (7.11) is known as the *Risch differential equation*. The integration of a polynomial in an exponential quantity boils down to solving an instance of this differential equation. It may seem that solving a differential equation is a harder problem computationally than solving the original integral. The restrictions on the form of both the integrand and the solution, however, simplify the task considerably.

We present several examples to show how the Risch algorithm goes about integrating the product of a rational function and an exponential.

Example 7.17. Use Theorem 7.5 to find $\int x\,e^{x^2}\,dx$.

Solution. This integral gives the Risch differential equation

$$x = R' + 2\,x\,R.$$

Suppose that the denominator of $R = P/Q$ is not a constant. Then we have

$$x = \frac{P'\,Q - P\,Q' + 2\,x\,P\,Q}{Q^2}.$$

This is impossible since the right side has a non-trivial denominator but the left side has no denominator. So Q is a constant. We can assume, without loss of generality, that $Q = 1$ since we can absorb a constant value of Q into the denominators of the coefficients of P. This leaves us with

$$x = P' + 2\,x\,P.$$

Examining the degrees of both sides, we see that P must be a constant and, clearly, $P = 1/2$. Therefore,

$$\int x\,e^{x^2}\,dx = P\,e^{x^2} = \tfrac{1}{2}\,e^{x^2}. \qquad \square$$

The next examples demonstrate that two simple exponential integrals have no closed form solutions.

Example 7.18. Show that $\int e^{x^2}\, dx$ has no elementary closed form.

Solution. The Risch differential equation associated with this integral is

$$1 = R' + 2\,x\,R.$$

As in the previous example, if we assume the denominator of R is not a constant, we have

$$1 = \frac{P'\,Q - P\,Q' + 2\,x\,P\,Q}{Q^2},$$

which is impossible. Without loss of generality, we let $Q = 1$ and arrive at

$$1 = P' + 2\,x\,P,$$

which is impossible for any polynomial P. So the integral has no elementary closed form. $\qquad\square$

Example 7.19. Show that $\int \dfrac{e^x}{x}\, dx$ is not elementary.

Solution. Here the Risch differential equation is

$$\frac{1}{x} = R' + R$$

and, with $R = P/Q$, we have

$$\frac{1}{x} = \frac{P'\,Q - P\,Q' + P\,Q}{Q^2}.$$

But then $Q = \sqrt{x}$, which is not a polynomial, so the integral has no elementary solution. $\qquad\square$

Our final example brings into play the method of undetermined coefficients.

Example 7.20. Find $\int \dfrac{2\,x^3 - 2\,x^2 - 1}{(x-1)^2}\, e^{x^2}\, dx$.

Solution. The Risch differential equation associated with this integral is

$$\frac{2\,x^3 - 2\,x^2 - 1}{(x-1)^2} = R' + 2\,x\,R$$

which, with the substitution $R = P/Q$ has the form

$$\frac{2\,x^3 - 2\,x^2 - 1}{(x-1)^2} = \frac{P'\,Q - P\,Q' + 2\,x\,P\,Q}{Q^2}.$$

Equating the denominators of both sides, we have $Q = x - 1, Q' = 1$ giving the following polynomial relationship for the numerator,

$$2\,x^3 - 2\,x^2 - 1 = P'\,(x-1) - P + 2\,x\,P\,(x-1).$$

Examining the degrees of both sides, P must be linear: $P = a_1 x + a_0$, $P' = a_1$. Substituting these values, expanding products, and collecting like terms, we have

$$2\,x^3 - 2\,x^2 - 1 = 2\,a_1\,x^3 + 2\,(a_0 - a_1)x^2 - 2\,a_0\,x - (a_0 + a_1).$$

Equating the coefficients of the terms with like powers, we get the system of equations

$$2\,a_1 = 2, \quad 2\,(a_0 - a_1) = -2, \quad -2\,a_0 = 0, \quad -(a_0 + a_1) = -1,$$

whose solution is $a_1 = 1$, $a_0 = 0$, and so $P = x$. Therefore,

$$\int \frac{2\,x^3 - 2\,x^2 - 1}{(x-1)^2}\, e^{x^2}\, dx = \frac{P}{Q}\, e^{x^2} = \frac{x}{x-1}\, e^{x^2}. \qquad \square$$

7.9.2 Rational Functions of Exponentials

Having dealt with the integration of polynomials of exponentials, we turn our attention to the integration of a rational function, $\frac{P(x,\theta)}{Q(x,\theta)}$, of an exponential quantity, $\theta = e^u$. As in the case of integrating logarithms, we consider both P and Q to be polynomials in θ with coefficients that may be rational functions of the integration variable x. We focus on the case where P/Q is proper and Q is square-free.

There are complications that arise in this problem that do not occur in the integration of purely rational functions or rational functions of logarithms. First, observe that a square-free polynomial θ and its derivative are not relatively prime. For example, if $\theta = e^x$ then θ' is also e^x, and $\gcd(\theta, \theta') = e^x \neq 1$. Another difficulty is that if Q is a polynomial in an exponential quantity, Q'/Q need not be proper rational function, as was the case in the other situations we considered. Again with $\theta = e^x$, suppose $Q = \theta^2 + 1$. Then $Q' = 2\theta\theta' = 2\theta^2$ and $\text{degree}(Q) = \text{degree}(Q')$, so Q'/Q is not proper.

As a result, the form of the integral is not exactly the same as it is in both the purely rational and logarithmic cases, $\sum_i c_i \ln R_i$, and an additional summation must be added to compensate for these difficulties. The following theorem gives the form of the integral.

Theorem 7.6. *Suppose $P(x,\theta)/Q(x,\theta)$ is a rational function with $\theta = e^u$ such that $\text{degree}(P) < \text{degree}(Q)$, $\gcd(P,Q) = 1$, and Q is a square-free polynomial of degree n. Then the integral of P/Q has a closed form solution given by*

$$\int \frac{P(x,\theta)}{Q(x,\theta)}\, dx = \sum_i c_i \ln R_i - n\,u \sum_i c_i$$

if and only if all the roots c_i of $\gcd(P - c(Q' - nu'Q), Q)$ are constants. The sums are taken over all of the roots and $R_i = \gcd(P - c_i(Q' - nu'Q), Q)$.

At this juncture, a simple example should suffice to illustrate how the Rothstein-Trager method is adapted to integrate rational functions of exponentials in light of the preceding theorem.

Example 7.21. Find $\displaystyle\int \frac{1}{e^x + 1}\, dx$.

Solution. Letting $\theta = e^x$ so $u = x$, $u' = 1$, we have $P = 1$, $Q = \theta + 1$, $Q' = \theta$, and $n = $ degree$(Q) = 1$. As a result of applying the extended Euclidean algorithm, we find

$$\gcd(P - c\,(Q' - n\,u'\,Q), Q) = \gcd(1 + c, \theta) = 1 = \frac{1}{1+c}\,(1+c) + 0 \cdot \theta.$$

The least common multiple of the denominators of the two coefficients is $1 + c$, whose root is a constant, $c = -1$, so the integral is elementary. The corresponding value of R is

$$R = \gcd(P - c\,(Q' - n\,u'\,Q), Q) = \gcd(0, \theta + 1) = \theta + 1.$$

Therefore,

$$\int \frac{1}{e^x + 1}\, dx = c \ln R - c\,n\,u = -\ln(\theta + 1) + x$$

$$= x - \ln(e^x + 1). \qquad \square$$

7.10 Exercises

1. Compute the derivative $f'(x)$ of each of the following functions f.

    (a) x^x (b) $\ln(\ln x)$ (c) $x^x \ln(\ln x)$ (d) $\dfrac{x^x}{\ln(\ln x)}$

    (e) $x^{\ln(\ln x)}$ (f) $\ln(\ln x^x)$ (g) $x^{(x^x)}$ (h) $\ln(\ln(\ln x))$

2. Calculate the following integrals with the derivative-divides method.

    (a) $\displaystyle\int e^x\, e^{e^x}\, dx$ (b) $\displaystyle\int \frac{e^{\sqrt{x}}}{\sqrt{x}}\, dx$

    (c) $\displaystyle\int \frac{\ln x}{x}\, dx$ (d) $\displaystyle\int \frac{\ln(\ln x)}{x \ln x}\, dx$

    (e) $\displaystyle\int e^x \cos(e^x)\, dx$ (f) $\displaystyle\int e^x \sin(e^x) \cos(e^x)\, dx$

    (g) $\displaystyle\int \frac{x + 1}{(3x^2 + 6x + 5)^{1/3}}\, dx$ (h) $\displaystyle\int \ln(\cos x) \tan(x)\, dx$

3. Calculate the following integrals using integration by parts.

(a) $\int x \cos x \, dx$
(b) $\int \ln x \, dx$
(c) $\int \sqrt{x} \, \ln x \, dx$

(d) $\int \dfrac{\ln(\ln x)}{x} \, dx$
(e) $\int (\ln x)^3 \, dx$
(f) $\int x^3 \, e^{x^2} \, dx$

(g) $\int \cos(\ln x) \, dx$
(h) $\int x \cos x \, e^x \, dx$
(i) $\int \sec^3 x \, dx$

4. The integral

$$\mathrm{Ei}(x) \equiv \int \frac{e^x}{x} \, dx$$

has no elementary closed form. None of the following related integrals are elementary as well, but each can be expressed with the Ei function.

(a) $\int \dfrac{e^x}{x^2} \, dx$
(b) $\int e^x \ln x \, dx$
(c) $\int x \, e^x \ln x \, dx$

(d) $\int \mathrm{Ei}(x) \, dx$
(e) $\int \dfrac{e^x}{x} \mathrm{Ei}(x) \, dx$

Use integration by parts to obtain solutions to these integrals in terms of $\mathrm{Ei}(x)$.

5. Show that the following integrals have an elementary closed form solution although neither term in each integrand does.

(a) $\int (x^x + x^x \ln x) \, dx$
(b) $\int \left(\dfrac{e^x}{x} - \dfrac{e^x}{x^2} \right) \, dx$

(c) $\int \left(\dfrac{\cos x}{x} - \dfrac{\sin x}{x^2} \right) \, dx$
(d) $\int \left(\dfrac{1}{\ln x} - \dfrac{1}{\ln^2 x} \right) \, dx$

6. Show that the arguments to each of the following functions f meet the criteria for the first part of Liouville's Theorem. Then determine whether the arguments meet the criteria for the second part of the theorem.

(a) $f(x, e^x)$
(b) $f(x, \sin x)$
(c) $f(x, \sin x, \cos x)$

(d) $f(x, e^x, \sin x)$
(e) $f(x, e^x, \ln x)$
(f) $f(x, e^x, \sin x, \cos x, \ln x)$

7. Use Hermite's method to determine the rational part and the integrand of the logarithmic part of the following integrals.

(a) $\int \dfrac{2x^2 + 3x - 1}{x^3 + x^2 - x - 1} \, dx$
(b) $\int \dfrac{2x + 1}{x^3 - 3x^2 + 3x - 1} \, dx$

(c) $\int \dfrac{2x^2 + x + 1}{(x + 3)(x - 1)^2} \, dx$
(d) $\int \dfrac{x^3 + 7x^2 - 5x + 5}{(x - 1)^2 (x + 1)^3} \, dx$

(e) $\int \dfrac{x^3 + x + 2}{x^4 + 2x^2 + 1} \, dx$
(f) $\int \dfrac{3x}{(x^2 + x + 1)^3} \, dx$

(g) $\int \dfrac{8x^5 - 10x^4 + 5}{(2x^5 - 10x + 5)^2} \, dx$
(h) $\int \dfrac{4x^5 - 1}{(x^5 + x + 1)^2} \, dx$

8. Repeat Exercise 7 using Horowitz' method.

9. Use the Rothstein-Trager method to evaluate the following integrals.

(a) $\displaystyle\int \frac{1}{x^3 - x}\, dx$

(b) $\displaystyle\int \frac{1}{x^3 + 1}\, dx$

(c) $\displaystyle\int \frac{1}{x^2 - 8}\, dx$

(d) $\displaystyle\int \frac{x+1}{x^2 + 1}\, dx$

(e) $\displaystyle\int \frac{1}{x^2 + x + 1}\, dx$

(f) $\displaystyle\int \frac{x}{x^2 + x + 1}\, dx$

(g) $\displaystyle\int \frac{x}{x^4 - 4}\, dx$

(h) $\displaystyle\int \frac{x^3}{x^4 + 1}\, dx$

(i) $\displaystyle\int \frac{1}{x^4 + 1}\, dx$

10. Defend or attack the following position:

All real polynomials in a single variable can be factored "numerically" into (at worst) quadratic factors over the reals, and into (at worst) linear factors over the complex numbers. This trivializes the problem of rational function integration.

11. Decide whether each of the following integrals of a polynomial in a logarithm is elementary. If so, evaluate the integral using the Risch algorithm.

(a) $\displaystyle\int \ln^2(x-1)\, dx$

(b) $\displaystyle\int \ln(x-1)^2\, dx$

(c) $\displaystyle\int \ln^3 x\, dx$

(d) $\displaystyle\int x \ln x\, dx$

(e) $\displaystyle\int x^2 \ln x\, dx$

(f) $\displaystyle\int x \ln^2 x\, dx$

(g) $\displaystyle\int \frac{\ln x}{x}\, dx$

(h) $\displaystyle\int \frac{\ln^2 x}{x}\, dx$

(i) $\displaystyle\int \frac{\ln(x+a)}{x+b}\, dx$

12. Prove the Liouville-Hardy theorem (1905):

Theorem. *If $f(x)$ is a rational function, then $\int f(x) \ln x\, dx$ is elementary if and only if there exists a rational function $g(x)$ and a constant C such that $f(x) = C/x + g'(x)$.*

13. Decide whether each of the following rational functions of a logarithm has an elementary integral. If so, evaluate the integral using the Risch algorithm.

(a) $\displaystyle\int \frac{1}{x \ln^2 x}\, dx$

(b) $\displaystyle\int \frac{1}{x \ln^3 x}\, dx$

(c) $\displaystyle\int \frac{\ln x - 1}{\ln^2 x}\, dx$

(d) $\displaystyle\int \frac{1}{x^2 \ln x}\, dx$

(e) $\displaystyle\int \frac{2 \ln x}{x\left(\ln^2 x + 1\right)^2}\, dx$

(f) $\displaystyle\int \frac{1}{x \ln x \ln \ln x}\, dx$

14. Consider the integral

$$\int \frac{2x\left(1 - \ln x\right)}{\left(\ln^2 x - x^2\right)^2}\, dx,$$

whose integrand is a rational function of $\theta = \ln x$.

(a) Find the part of the integral that is rational in θ and the integrand of the logarithmic part of the integral.

(b) Show that the logarithmic part of the integral does not have an elementary closed form.

15. Decide whether each of the following integrals of a rational function times an exponential is elementary. If so, evaluate the integral using the Risch algorithm.

(a) $\displaystyle\int x^2\, e^{x^2}\, dx$

(b) $\displaystyle\int x^3\, e^{x^2}\, dx$

(c) $\displaystyle\int x^2\, e^{x^3}\, dx$

(d) $\displaystyle\int \frac{(x-1)\, e^x}{x}\, dx$

(e) $\displaystyle\int \frac{e^x}{x^2}\, dx$

(f) $\displaystyle\int \frac{(x-2)\, e^x}{x^3}\, dx$

(g) $\int \dfrac{(x^2 - x - 1)\, e^x}{(x-1)^2}\, dx$ (h) $\int \dfrac{(4x^4 + 4x - 1)\,(e^{x^2} - 1)\,(e^{x^2} + 1)}{(x+1)^2}\, dx$

16. For which values of the non-negative integer n is $\displaystyle\int x^n\, e^{x^2}\, dx$ elementary? What is the general form of the solution when it exists?

17. Show that
$$\int \frac{x^2 + a\,x + b}{(x-1)^2}\, e^x\, dx = \frac{x + a + 1}{x - 1}\, e^x$$
has the elementary solution given if and only if $b = -2a - 3$.

18. Show that none of the following integrals have an elementary solution by relating each to an integral of the form $\int e^{t^2}\, dt$ or $\int t^2\, e^{t^2}\, dt$.

(a) $\displaystyle\int \sqrt{\ln x}\, dx$ (b) $\displaystyle\int \frac{1}{\sqrt{\ln x}}\, dx$ (c) $\displaystyle\int \frac{e^x}{\sqrt{x}}\, dx$

19. Show that none of the following integrals have an elementary solution by relating each to an integral of the form $\int \frac{e^t}{t}\, dt$.

(a) $\displaystyle\int e^{e^x}\, dx$ (b) $\displaystyle\int \frac{1}{\ln x}\, dx$ (c) $\displaystyle\int \ln(\ln x)\, dx$ (d) $\displaystyle\int \frac{\sin x}{x}\, dx$

Hint. Use integration by parts and (b) for (c).

20. Decide whether each of the following integrals of a rational function in an exponential is elementary. If so, evaluate the integral using the Risch algorithm.

(a) $\displaystyle\int \frac{1}{(e^x)^2 + 3}\, dx$ (b) $\displaystyle\int \frac{1}{e^{2x} - 1}\, dx$ (c) $\displaystyle\int \frac{x}{e^{2x} - 1}\, dx$

(d) $\displaystyle\int \frac{e^x}{e^{2x} - 1}\, dx$ (e) $\displaystyle\int \frac{1}{e^{x^2} - 1}\, dx$ (f) $\displaystyle\int \frac{x}{e^{x^2} - 1}\, dx$

21. Compute $\displaystyle\int \frac{1}{e^{3x} - 1}\, dx$ first using $\theta_1 = e^{3x}$, and then using $\theta_2 = e^x$. Show that the two results are equivalent.

22. The following integrals involve two transcendental quantities. Decide whether each is elementary and, if so, determine its solution using the Risch algorithm.

(a) $\displaystyle\int \ln(\ln x)\, dx$ (b) $\displaystyle\int x^x\, dx$ (c) $\displaystyle\int (1 + \ln(x))\, x^x\, dx$

Note. $x^x = e^{x \ln x}$.

23. Prove the following theorem relating the integral of a function and its inverse.

Theorem. *If f and f^{-1} are inverse functions on some closed interval, then*
$$\int f(x)\, dx = x\, f(x) - g(f(x)),$$
where $g(x) = \int f^{-1}\, dx$. Moreover, if f and f^{-1} are both elementary functions, then $\int f(x)\, dx$ is elementary if and only if $\int f(x)\, dx$ is elementary.

24. Use the theorem of Exercise 23 to show the following integrals are not elementary.

(a) $\displaystyle\int \sqrt{\ln x}\, dx$ (b) $\displaystyle\int \frac{1}{\ln x}\, dx$

25. Calculate the trigonometric integrals

(a) $\int \cos x \, dx$ (b) $\int \sin x \, dx$ (c) $\int \tan x \, dx$

by performing the following steps:

1) Convert the integrand to a complex exponential.

2) Transform the integral to a rational one with the substitution $y = e^{ix}$, $dy = ie^{ix} \, dx$.

3) Integrate the resulting rational function of y.

4) Reverse the substitutions to express the result in terms of trig functions.

Programming Projects

1. **(Differentiation)** Write a procedure,

$$\text{Derivative(<expression>,<variable>)},$$

that calculates the derivative of a symbolic expression with respect to the specified variable.

Test your code on the examples in Exercise 1.

2. **(Hermite's Method)** Write a set of procedures that together implement Hermite's algorithm for rational function integration. The first,

$$\text{PartialFraction(<numer\_poly>,<denom\_poly>,<variable>)},$$

performs the square-free partial fraction decomposition with respect to the variable specified. The second,

$$\text{Hermite(<numer\_poly>,<denom\_poly>,<variable>)},$$

and third,

$$\text{HermiteReduce(<numer\_poly>,<denom\_poly>,<variable>)},$$

correspond to the procedures with these names in Algorithm 7.1. **Hermite** should return two separate components: the rational part of the result, and an integral (or sum of integrals) that when evaluated gives the logarithmic part of the solution. Be sure to check that the polynomials sent to each procedure meet the conditions required by the algorithms.

3. **(Horowitz Method)** Write a procedure to implement Horowitz' algorithm for rational function integration,

$$\text{Horowitz(<numer\_poly>,<denom\_poly>,<variable>)},$$

Your procedure should return two items: a rational function, and an integral that gives the logarithmic part of the solution. Be sure to check that the input polynomials meet the conditions required by the algorithm.

4. **(Rothstein-Trager Method)** Write a procedure,

    RothsteinTrager(<numer_poly>,<denom_poly>,<variable>),

    that computes the logarithmic part of the integral of a rational function. Be sure that degree(numer_poly) < degree(denom_poly) and that denom_poly is square-free.

5. **(Rational Function Integration)** Write a driver procedure,

    RationalIntegrate(<expression>,<variable>),

    to accomplish rational function integration. Split expression into a polynomial plus a proper rational function, then integrate both parts. Incorporate your procedures from Programming Projects 2, 3, and 4. If expression is not a rational function of variable, return FAIL.

Chapter 8

Gröbner Bases

The art of doing mathematics consists in finding that special case which contains all the germs of generality.

David Hilbert (1862–1943)
Quoted in Constance Reid, *Hilbert*, 1970

Simplicity, simplicity, simplicity! I say, let your affairs be as two or three, and not a hundred or a thousand; instead of a million count half a dozen, and keep your accounts on your thumb-nail.

Henry David Thoreau (1817–1862)
Walden, 1854

Many fundamental problems in mathematics, the natural sciences and engineering can be formulated in terms of systems of nonlinear multivariate polynomials. Gröbner bases provide a uniform, powerful approach for solving such problems. Examples include the solution of algebraic systems of equations, symbolic integration and summation of expressions involving special functions, and solution of linear boundary value problems (differential equations). The technique has been successfully applied to other areas as well including geometric theorem proving, graph coloring, and linear integer optimization. Perhaps surprisingly, Gröbner bases have also been used to solve problems in areas as far reaching as the construction and analysis of nonlinear cryptosystems, robotics, software engineering, finding genetic relationships between species, and the solution of Sudoku puzzles.

The Gröbner basis method is a very powerful tool in computer algebra. All general purpose computer algebra systems, such as Maple and Mathematica, provide an implementation of Gröbner bases, and several special purpose mathematical systems also offer support for Gröbner basis calculations.

The goal of this chapter is to provide a "broad brush" introduction to the subject of Gröbner bases including the mathematics behind them. To focus on key ideas instead of computational details, Maple is used as a calculator throughout the presentation. The questions that will be addressed here include: (1) What is a Gröbner basis? (2) Why are Gröbner bases useful? (3) How are Gröbner bases calculated?

8.1   Solution of Polynomial Equations

The notion of Gröbner bases was introduced by Bruno Buchberger in his 1965 Ph.D. thesis. Buchberger named them for his thesis supervisor, Wolfgang Gröbner. (Both Gröbner and its transliterated spelling, Groebner, are used in textbooks and the literature.) Although some of the ideas underpinning this work predate his thesis, Buchberger was the first to present a systematic treatment of the subject including a computational algorithm. In 1997, he received an award from the Association for Computing Machinery (ACM) recognizing the importance of this contribution to mathematics and computer science.

Gröbner bases provide a way to transform a set F of multivariate polynomials into a second set G that has certain "nice" properties that F does not possess. For example, problems that are difficult to solve in terms of F become relatively easy to solve with G. Moreover, the algorithm for computing G from F is fairly easy to understand and reasonably simple to implement.

As an example, suppose we want to solve the following system of four nonlinear quadratic equations in three unknowns:

$$x\,y = 0, \qquad x^2 + y^2 = 1, \qquad z^2 - x^2 = 1, \qquad y^2 + z^2 = 2.$$

Gröbner bases work with polynomials instead of equations so the system is written as a list of polynomials that are implicitly equated to zero. Here's how to compute the Gröbner basis in Maple.

```
> F := [ x*y,  x^2 + y^2 - 1,  z^2 - x^2 - 1,  y^2 + z^2 - 2 ];
```

$$F := [x\,y, \, x^2 + y^2 - 1, \, z^2 - x^2 - 1, \, y^2 + z^2 - 2]$$

```
> G := Groebner[Basis](F, plex(z,y,x));
```

$$G := [x^4 + 1, \, x^2 + y^2 - 1, \, -x\,y + z]$$

The `plex` parameter is required to compute the Gröbner basis with respect to a particular lexicographical ordering of terms. Here's the basis for the same set F with respect to a different ordering.

```
> Groebner[Basis](F, plex(x,y,z));
```

$$[z^4 - 2\,z^2 + 2, \, y^2 + z^2 - 2, \, -y\,z^3 + 2\,x]$$

What is the relationship between a Gröbner basis G and the input basis F? The main property is summarized by the following theorem.

Theorem 8.1. (Gröbner Basis). *Let $F = \{f_1, f_2, \ldots f_s\}$ be the input basis (a set of polynomials with real or complex coefficients) and let $G = \{g_1, g_2, \ldots g_t\}$ be the output, or Gröbner, basis (also a set of polynomials). Then the real and complex solutions of the two systems F and G are identical, even up to the multiplicity of the solutions.*

So a Gröbner basis simplifies a set of polynomials, but it doesn't actually solve the system. Observe, however, that both of the bases calculated above "triangularize" the system so it is much easier to solve. For example, the first polynomial in G

```
> G[1];
```

$$x^4 + 1$$

involves only x, so it can be solved directly

```
> solve(G[1]=0, x);
```

$$\frac{\sqrt{2}}{2} + \frac{I\sqrt{2}}{2}, \quad -\frac{\sqrt{2}}{2} + \frac{I\sqrt{2}}{2}, \quad -\frac{\sqrt{2}}{2} - \frac{I\sqrt{2}}{2}, \quad \frac{\sqrt{2}}{2} - \frac{I\sqrt{2}}{2}$$

The second polynomial involves both x and y, but not z,

```
> G[2];
```

$$x^2 + y^2 - 1$$

It is quadratic in y, so each of the four solutions for x yields two solutions for y—a total of eight solutions. Since the last polynomial

```
> G[3];
```

$$-xy + z$$

is linear in z, it is easily rearranged to give a single z for each (x, y) pair.

What if our system of equations has no solutions? A very simple example is the pair of linear equations

$$2x + 8y = 5, \qquad x + 4y = 2.$$

Multiplying the second equation by 2 and subtracting the result from the first gives $0 = 1$. This means that the system of equations is equivalent to the system $1 = 0$, which is inconsistent and has no solutions.

```
> F := [2*x + 8*y - 5, x + 4*y - 2]:
> G := Groebner[Basis](F, plex(x,y));
```

$$G := [1]$$

In general, a system F has a (reduced) Gröbner basis $G = [1]$ if and only the corresponding set of equations $\{f_1 = 0, f_2 = 0, \ldots, f_s = 0\}$ has no solution.

8.2 Mathematical Applications

Over the years, Gröbner basis theory has gained much attention outside the mathematical community due to the wide variety of applications that have been found in many areas of science and engineering. In all of these applications, the general methodology is to reformulate a particular problem as a question about sets of multivariate polynomials. After performing one or more Gröbner basis computations, the properties of Gröbner bases are used to answer the question. This section provides just a taste of the kinds of mathematical problems that can be solved with Gröbner bases.

8.2.1 Linear Algebra

The simplest types of polynomial equations are, of course, linear equations. In the case where all of the polynomials in the input basis have degree one, computing a Gröbner basis is equivalent to solving the system by Gaussian elimination.

Consider a system of three simultaneous linear equations in three unknowns.

$$3x + 2y - z = 1$$
$$2x - 2y + 4z = -2$$
$$-x + y - z = 0$$

To solve the system, all we need to do is to convert the equations to a list of polynomials

```
> F := [ 3*x + 2*y - z - 1, 2*x - 2*y + 4*z + 2, -x + y - z ];
```

$$F := [3x + 2y - z - 1, 2x - 2y + 4z + 2, -x + y - z];$$

and compute their Gröbner basis.

```
> G := Groebner[Basis](F, plex(x,y,z)); solve(F);
```

$$G := [z + 1, 5y + 3, 5x - 2]$$
$$\{x = \frac{2}{5}, y = -\frac{3}{5}, z = -1\}$$

The solution can be read off from the Gröbner basis!

Now consider a system of five linear equations in four unknowns. Let's examine how this system is solved using standard techniques of linear algebra.

$$
\begin{aligned}
w + x \quad\quad\quad &= 0 \\
w \quad\quad - z &= 1 \\
y + z &= 1 \\
x + y + 2z &= 0 \\
w + x + y + z &= 1
\end{aligned}
$$

The coefficient matrix A and right hand side vector b of the system are

```
> A,b := Matrix([[1,1,0,0],[1,0,0,-1],[0,0,1,1],[0,1,1,2],
        [1,1,1,1]]), Vector([0,1,1,0,1]);
```

$$
A, b := \begin{bmatrix} 1 & 1 & 0 & 0 \\ 1 & 0 & 0 & -1 \\ 0 & 0 & 1 & 1 \\ 0 & 1 & 1 & 2 \\ 1 & 1 & 1 & 1 \end{bmatrix}, \begin{bmatrix} 0 \\ 1 \\ 1 \\ 0 \\ 1 \end{bmatrix}
$$

The solution of the system involves transforming the augmented matrix $[A \mid b]$ to reduced row echelon form through a series of *elementary row operations*.[†]

```
> with(LinearAlgebra):
  ReducedRowEchelonForm( < A | b > );
```

$$
\begin{bmatrix} 1 & 0 & 0 & -1 & 1 \\ 0 & 1 & 0 & 1 & -1 \\ 0 & 0 & 1 & 1 & 1 \\ 0 & 0 & 0 & 0 & 0 \\ 0 & 0 & 0 & 0 & 0 \end{bmatrix}
$$

Converting back to equations in w, x, y, z, this corresponds to the reduced system

$$
w - z = 1, \quad\quad x + z = -1, \quad\quad y + z = 1.
$$

The system has infinitely many solutions that have been parameterized in terms of z. These solutions are obtained by assigning values to z.

Now let's enter the equations into Maple as a list of polynomials

```
> F := [w+x, w-z-1, y+z-1, x+y+2*z, w+x+y+z-1];
```

$$
F := [w + x, w - z - 1, y + z - 1, x + y + 2z, w + x + y + z - 1]
$$

and calculate the Gröbner basis.

[†]The *elementary row operations* are: (1) interchanging two rows, (2) multiplying one row by a non-zero number, and (3) adding a multiple of one row to a different row. A matrix is in *reduced row echelon form* if: (1) all rows consisting entirely of zeros are at the bottom, (2) the first non-zero entry from the left in each row is a 1 (called the leading 1), (3) the leading 1 in each row is to the right of all the leading 1's in the rows above it, and (4) each leading 1 is the only non-zero entry in that column.

```
> G := Groebner[Basis](F,plex(w,x,y,z));
```

$$G := [y + z - 1,\ x + z + 1,\ w - z - 1]$$

The Gröbner basis gives the same parameterized solutions as the (unique) reduced row echelon form! In general, the reduced row echelon form for a linear system is a (reduced) Gröbner basis, and the Gröbner is in reduced row echelon form. One way to understand Gröbner bases is that they are a natural generalization of reduced row echelon form to non-linear systems.

8.2.2 Greatest Common Divisors

Gröbner bases can be used to compute the greatest common divisor of polynomials in one or several variables. The method is not, however, as computationally efficient as applying Euclid's algorithm directly.

Suppose we want to determine the greatest common divisor of two single variable polynomials, $p(x)$ and $q(x)$. It follows from Theorem 8.1 that if the input basis F consists of these two polynomials, the output will be just a single polynomial in x whose roots (real and complex) are exactly the same as those of p and q. This polynomial is precisely $\gcd(p, q)$.

```
> p := x^5 - 20*x^3 + 30*x^2 + 19*x - 30;
  q := x^6 - 6*x^5 - 5*x^4 + 90*x^3 - 176*x^2 + 96*x;
```

$$p := x^5 - 20\,x^3 + 30\,x^2 + 19\,x - 30$$

$$q := x^6 - 6\,x^5 - 5\,x^4 + 90\,x^3 - 176\,x^2 + 96\,x$$

```
> G := Groebner[Basis]([p,q], plex(x)); gcd(p,q);
```

$$G := [x^3 - 6\,x^2 + 11\,x - 6]$$

$$x^3 - 6\,x^2 + 11\,x - 6$$

The situation is somewhat different with regard to multivariate polynomials. Here, we need to introduce a new variable, say t, that does not appear in either polynomial and which will be eliminated in the Gröbner basis computation. Also, the procedure for finding $\gcd(p, q)$ involves first finding the least common multiple, $\text{lcm}(p, q)$, and then applying the well known formula relating the gcd to the lcm, $\gcd(p, q) = pq\,/\,\text{lcm}(p, q)$. To obtain the lcm, compute the Gröbner basis of $t \cdot p$ and $(1 - t) \cdot q$ with respect to lexicographical order. The lcm will be the first element of the list.

```
> p := (x-1)^3 * (x+y)^4 * (x-y);
  q := (x+1)^3 * (x+y)^3 * (x-y);
  F := [t*p,(1-t)*q]:
```

$$p := (x - 1)^3\,(x + y)^4\,(x - y)$$

$$q := (x + 1)^3\,(x + y)^3\,(x - y)$$

```
> G := Groebner[Basis](F, plex(t,x,y)):
  LCM := factor(G[1]);
```

$$LCM := (x-1)^3 \, (x+1)^3 \, (x-y) \, (x+y)^4$$

```
> GCD := p*q/LCM; gcd(p,q);
```

$$GCD := (x+y)^3 \, (x-y)$$
$$(x+y)^3 \, (x-y)$$

8.2.3 Implicitization of Parametric Equations

The pair of equations $x = f(t)$ and $y = g(t)$ are said to be in *parametric form*. Whenever a real value is assigned to t, the equations determine a point in the xy-plane. It is sometimes desirable to eliminate the parameter t to obtain an implicit equation describing the relationship between x and y. Gröbner bases provide a mechanism for doing this when $f(t)$ and $g(t)$ can be expressed as rational functions of t.

For example, the parametric form for a unit circle is given by the pair of equations

$$x(t) = \frac{1-t^2}{1+t^2} \quad \text{and} \quad y(t) = \frac{2t}{1+t^2}$$

for $-\infty < t < +\infty$. Clearing denominators gives $(1+t^2)\,x = 1 - t^2$ and $(1+t^2)\,y = 2t$. Converting these equations to polynomials,

```
> F := [ (1+t^2)*x-(1-t^2), (1+t^2)*y-2*t ];
```

$$F := [\,(t^2+1)\,x + t^2 - 1, \, (t^2+1)\,y - 2t\,]$$

F consists of two equations in three unknowns: x, y, and t. Computing a Gröbner basis under the order $t > x > y$ (or $t > y > x$) forces the elimination of t.

```
> Groebner[Basis](F, plex(t,x,y));
```

$$G := [\,x^2 + y^2 - 1, \, ty + x - 1, \, tx + t - y\,]$$

Voilà! The first polynomial in the list gives the implicit equation for the unit circle, $x^2 + y^2 = 1$.

In general when a system of parametric equations involves the variables $x_1, x_2, \ldots t_1, t_2, \ldots$, use plex($t_1, t_2, \ldots x_1, x_2, \ldots$) to eliminate $t_1, t_2, \ldots$.

8.2.4   Integer Programming

Integer programming is a method to achieve the best outcome (e.g., maximum profit, lowest cost) in a mathematical model represented by linear integer relationships. The solution of such systems with Gröbner bases represents a generalization of the simplex method that is typically used to solve these problems.

Consider the problem of finding the minimum number of coins (pennies, nickels, dimes, and quarters) to make change for some amount of money, say $1.17. Stated as an optimization problem, the goal is to minimize $P+N+D+Q$ subject to the constraints that $P+5\,N+10\,D+25\,Q = 117$ and that P, N, D, Q are integers.

The key idea in solving this problem with Gröbner bases is use polynomials that encode the number of coins as exponents of the variables. The input basis gives the values of N, D, Q in terms of pennies P, the basic unit of money.

```
> F := [ P^5 - N, P^10 - D, P^25 - Q ];
```

$$F := [\,P^5 - N,\ P^{10} - D,\ P^{25} - Q\,]$$

In order to minimize the number of coins, which appear as exponents, the Gröbner basis is computed with respect to total degree order rather than lexicographical order. More on the difference in the next section.

```
> G := Groebner[Basis](F, grlex(P,N,D,Q));
```

$$G := [\,N^2 - D,\ D^3 - N\,Q,\ D^2\,N - Q,\ P^5 - N\,]$$

This list gives the conversions used to reduce the number of coins: a dime for two nickels, a nickel and a quarter for three dimes, and so forth. The total number of coins always decreases.

Because **grlex** is specified, the reduction of a polynomial with respect to G yields a polynomial whose total degree is smallest among all polynomials equivalent to the original one. As a result, the optimal way of exchanging $1.17 is found by reducing some arbitrary way to pay this amount.

```
> Groebner[NormalForm](P^117, G, grlex(P,N,D,Q));
```

$$D\,N\,P^2\,Q^4$$

```
> Groebner[NormalForm](P^7*N^2*D^5*Q^2, G, grlex(P,N,D,Q));
```

$$D\,N\,P^2\,Q^4$$

The optimal solution is $(P, N, D, Q) = (2, 1, 1, 4)$.

8.3 General Polynomial Division

8.3.1 Ordering of Monomials

When Maple is asked to perform a Gröbner basis computation, a special parameter is required. The two types we have encountered are `plex` and `grlex`, and each of these takes an argument that is an ordered list of variables. In this section, we consider the meaning of this parameter and why it is important when we divide polynomials.

Consider the following (randomly ordered) polynomial of two variables in expanded form,

$$x^2 + y^2 + 4\,x\,y^2 + 5\,x + 2 + 7\,x\,y.$$

Each of the six terms is called a *monomial*, and each monomial is expressed as a constant coefficient times a product of variables raised to powers. (The powers of the constant term itself are all zero.)

One ordering of the terms is to sort them in descending order by degree of one of the variables, say x,

$$x^2 > (4\,x\,y^2,\, 5\,x,\, 7\,x\,y) > (y^2,\, 2).$$

There is one term of degree 2 in x, three of degree 1, and two of degree 0. Now for each set of terms of the same degree in x, we sort them by descending degree in y,

$$x^2 > 4\,x\,y^2 > 7\,x\,y > 5\,x > y^2 > 2$$

to achieve a total ordering of all the terms. This is called *pure lexicographical order*, and is specified in Maple by `plex(x, y)`.

```
> sort(x^2 + y^2 + 4*x*y^2 + 5*x + 2 + 7*x*y,order=plex(x,y));
```

$$x^2 + 4\,x\,y^2 + 7\,x\,y + 5\,x + y^2 + 2$$

If there were a third variable z, we would sort all the terms in each $x^i\,y^j$ by descending order of their degree in z.

Another way to order terms is to sort them first in descending order by total degree,

$$4\,x\,y^2 > (x^2,\, y^2,\, 7\,x\,y) > 5\,x > 2.$$

There is one term of degree 3, three of degree 2, one of degree 1, and one of degree 0. For each set of terms of the same degree, we break ties sorting by lexicographical order,

$$4\,x\,y^2 > x^2 > 7\,x\,y > y^2 > 5\,x > 2.$$

This is called *graded lexicographical order*, and is specified in Maple by `grlex(x, y)`.

```
> sort(x^2 + y^2 + 4*x*y^2 + 5*x + 2 + 7*x*y,order=grlex(x,y));
```

$$x^2 + 4\,x\,y^2 + 7\,x\,y + 5\,x + y^2 + 2$$

By the way, this is the order that Maple prefers to use if not directed otherwise.

Our interest in Gröbner bases is motivated by the fact that they have "nicer" properties than other systems of polynomials. Depending on what properties are of interest in a particular context, the Gröbner basis of a given system is computed with respect to a specific ordering of monomials. For example, lexicographical order (**plex**) is used most frequently because it produces a system that is triangularized, and therefore easier to solve by a method similar to Gaussian elimination. Graded lexicographical ordering (**grlex**) is used in applications like the integer programming example of Section 8.2.4, where ordering the total degree of a monomial is more important than the sequence of exponents in that monomial.

8.3.2   Issues in Polynomial Division

We are now ready to tackle some of the basic issues involved when performing polynomial division.

The procedure taught in school for doing long division of two univariate polynomials involves arranging the terms of both the dividend and the divisor in decreasing powers of the variable. Have you ever wondered what would happen if the division were performed with the terms arranged by increasing degree? First, try dividing $f(x) = x^3 + 1$ by $g(x) = x + 1$ with the terms arranged in the usual order. You should get a quotient $q(x) = x^2 - x + 1$ and a remainder $r = 0$. Then reverse the terms in f and g, and divide $f(x) = 1 + x^3$ by $g(x) = 1 + x$. After three iterations of the procedure, we get the same quotient and remainder as before.

Now change the divisor to $g(x) = x - 1$ and divide f by g with the terms in the usual order. You should get a quotient $q(x) = x^2 + x + 1$ and a remainder $r = 2$. Finally, reverse the terms of f and g, and divide $f(x) = 1 + x^3$ by $g(x) = -1 + x$. The quotient looks like $q(x) = -1 - x - x^2 - 2x^3 - 2x^4 - 2x^5 - \ldots$, but the iterations go on forever!

Whenever two polynomials in a single variable are divided with their terms arranged in increasing order of powers, the usual division procedure works correctly if the remainder $r = 0$ but doesn't terminate if $r \neq 0$. So for polynomials in one variable, the only possible term ordering for both the divided and the divisor is

$$1 < x < x^2 < x^3 < \ldots .$$

The situation is different, however, for polynomials in two or more variables where many term orderings are possible.

When computing Gröbner bases, we will need to divide one polynomial by a set of two or more polynomials. This process is called *general division*. (Sometimes, this is referred to as *multiple* or *multivariable division*.) Suppose,

for example, we divide a polynomial f by two polynomials $[g_1, g_2]$. This produces a quotient q_1 with respect to g_1, another quotient q_2 with respect to g_2, and a final remainder r satisfying

$$f = q_1 \cdot g_1 + q_2 \cdot g_2 + r.$$

The division process depends on both the order of the divisors and a monomial ordering with respect to which the division will take place.

Let's step through an example, illustrated in Figure 8.1. We will employ a scheme similar to that for long division of polynomials in one variable. The two divisors are listed to the left of the radical symbol used for general division, the two quotients are listed on the top, and the dividend goes in its usual place.

Example 8.1. Divide $f = x^2 y + x y^2 + y^2$ by $[g_1, g_2] = [x y - 1, y^2 - 1]$, ordering terms lexicographically with $x > y$.

Solution. The terms of f, g_1, g_2 are already sorted with respect to $\texttt{plex}(x, y)$ order. The division algorithm always works with leading terms and we have: $\mathrm{LT}(f) = x^2 y$, $\mathrm{LT}(g_1) = x y$, $\mathrm{LT}(g_2) = y^2$. Since $\mathrm{LT}(g_1)$ divides $\mathrm{LT}(f)$, this gives the first quotient term, $q_1 = x$. As with normal long division, we multiply x by g_1 and subtract from f to obtain a new dividend, $f - x g_1$.

The leading term of this new dividend, $x y^2$, is also divisible by $\mathrm{LT}(g_1)$, giving a quotient term y. We add this to q_1, multiply by g_1 and subtract to get our next dividend, $f - x g_1 - y g_1 = x + y^2 + y$. While neither $\mathrm{LT}(g_1)$ nor $\mathrm{LT}(g_2)$ divides $\mathrm{LT}(x + y^2 + y) = x$, this is not the final remainder since $\mathrm{LT}(g_2)$ divides y^2. If we move x to the remainder, we can continue dividing.

Dividing $\mathrm{LT}(g_2)$ into y^2 gives a quotient term of 1 for q_2. Multiplying this by g_2 and subtracting from the current dividend leaves $y + 1$. Since neither $\mathrm{LT}(g_1)$ nor $\mathrm{LT}(g_2)$ divides y, we add $y + 1$ to the remainder and we are through dividing. The final remainder is $r = x + y + 1$ and the final quotients are $q_1 = x + y$ and $q_2 = 1$. We have expressed the original dividend as

$$x^2 y + x y^2 + y^2 = (x + y) \cdot (x y - 1) + 1 \cdot (y^2 - 1) + (x + y + 1). \qquad \square$$

The general division procedure is presented formally as Algorithm 8.1. All steps are performed with respect to some fixed ordering of monomials. Each time through the **while**-loop, one of two things happens. If the leading term of some divisor f_i divides the leading term of the remaining dividend h, the algorithm proceeds as in normal long division. If the leading term of no f_i divides the leading term of h, then the algorithm adds the leading term of h to the remainder and deletes it from the remaining dividend. The algorithm then continues to work with this new dividend.

The remainder and the quotients produced are sensitive to the order in which the divisors are presented to the algorithm. Let's do another example, this time with the help of Maple.

$$q_1 = x + y$$
$$q_2 = 1$$

$$
\begin{array}{ll}
g_1 = x\,y - 1 & \sqrt{x^2 y + x\,y^2 + y^2} \quad = f \\
g_2 = y^2 - 1 & \underline{x^2 y - x} \\
& x\,y^2 + x + y^2 \qquad = f - x\,g_1 \\
& \underline{x\,y^2 - y} \\
& x + y^2 + y \qquad\quad = f - (x+y)\,g_1 \\
& \underline{y^2 + y} \qquad x \to r \\
& y^2 - 1 \\
& \underline{y + 1} \qquad\qquad = f - (x+y)\,g_1 - g_2 \\
& 0 \qquad y + 1 \to r \quad r = x + y + 1
\end{array}
$$

FIGURE 8.1: Division of $f = x^2\,y + x\,y^2 + y^2$ by $[g_1, g_2] = [x\,y - 1, y^2 - 1]$.

Example 8.2. Use Maple to divide $x\,y^2 + y^2$ by $[g_1, g_2] = [x\,y + y, x\,y + x]$, ordering terms lexicographically with $x > y$.

Solution. The following Maple commands from the Groebner package do the job.

```
> f := x*y^2 + y^2: g := [x*y + y, x*y + x]:
  r := Groebner[NormalForm](f, g, plex(x,y), 'q'); q;
```

$$r = 0$$
$$[y, 0]$$

The remainder is 0, the first quotient is y and the second is 0. Let's check by expanding the result to see if we get f back.

```
> expand(q[1]*g[1] + q[2]*g[2] + r);
```

$$x\,y^2 + x\,y$$

Now let's see what happens if the order of the divisors is reversed.

```
> f := x*y^2 + y^2: g := [x*y + x, x*y + y]:
  r := Groebner[NormalForm](f, g, plex(x,y), 'q'); q;
```

$$r := y^2 + x$$
$$[y - 1, 0]$$

This time the remainder is $y^2 + x$, the first quotient is $y - 1$ and the second is 0. Since this is a different remainder, we'd better check.

```
> expand(q[1]*g[1] + q[2]*g[2] + r);
```

$$x\,y^2 + y^2$$

Everything checks. What is going on?   □

procedure *GeneralDivide*(f, G)
{Input: Dividend polynomial f, divisor polynomials $G = [g_1, \ldots, g_s]$}
{Output: Remainder r, quotients $[q_1, \ldots, q_s]$ }
    $h := f$ {initialize remaining dividend}
    $q_1 := 0; \ldots; q_s := 0$ {initialize quotients to zero}
    while $h \neq 0$ **do**
        divoccurred := *false*
        for $i := 1$ **to** s **while** *divoccurred* = *false* **do**
            if $\text{LT}(g_i)$ divides $\text{LT}(h)$ **then**
                $q_i := q_i + \text{LT}(h)/\text{LT}(g_i)$ {add term to quotient q_i}
                $h := h - \text{LT}(h)/\text{LT}(g_i) \cdot g_i$ {subtract product from h}
                divoccurred := *true*
            end if
            if *divoccurred* = *false* **then**   {$\text{LT}(h)$ has no divisor}
                $r := r + \text{LT}(h)$ {add $\text{LT}(h)$ to remainder}
                $h := h - \text{LT}(h)$ {remove $\text{LT}(h)$ from dividend}
            end if
        end do
    return$(r, [q_1, \ldots, q_s])$
end procedure

ALGORITHM 8.1: General polynomial division

It is disturbing that the remainder depends on the order that the divisions are performed. The problem, however, is not caused by the division algorithm but with the basis containing the divisors. It is not a "good basis;" it is not a Gröbner basis.

When an equivalent Gröbner basis is used, we get a remainder of zero regardless of the order the divisions are performed. Let's find the new basis,

```
> f := x*y^2 + y^2:  g := [x*y + x, x*y + y]:
  G := Groebner[Basis](g, plex(x,y));
```

$$G := [y^2 + y, \, x - y]$$

and perform the divisions in both orders.

```
> r := Groebner[NormalForm](f,[G[1],G[2]],plex(x,y),'q'); q:
  q[1]*G[1]+q[2]*G[2]+r = expand(q[1]*G[1] + q[2]*G[2] + r);
```

$$r := 0$$

$$[x, \, -y]$$

$$(y^2 + y)\,x - (x - y)\,y = x\,y^2 + y^2$$

```
> r := Groebner[NormalForm](f,[G[2],G[1]],plex(x,y),'q'); q:
  q[1]*G[2]+q[2]*G[1]+r = expand( q[1]*G[2] + q[2]*G[1] + r );
```

$$r := 0$$

$$[y^2, y]$$

$$(y^2 + y)\, y + (x - y)\, y^2 = x\, y^2 + y^2$$

This time the remainders are the same! The quotients, however, are sensitive to the order in which the divisors are presented.

The following theorem summarizes a key point of this discussion. We rely heavily on it in the next section where we show how to construct Gröbner bases.

Theorem 8.2. (Linear Combinations). *If $G = \{g_1, g_2, \ldots g_s\}$ is a Gröbner basis and f is a polynomial, then the remainder when f is divided by the elements of G is zero, remainder$(f, G) = 0$, if and only if f can be expressed (not uniquely) as a linear combination of the basis elements,*

$$f = h_1 \cdot g_1 + h_2 \cdot g_2 + \ldots + h_s \cdot g_s,$$

for some polynomials h_i.

Suppose we have an arbitrary set of polynomials F and want to determine whether some polynomial f can be expressed as a linear combination (with polynomial coefficients) of elements from F. This appears to be a hopeless task since there are infinite numbers of possible combinations. In fact, it was this problem that provided the motivation for Buchberger's thesis. His solution was to develop the notion of a Gröbner basis G equivalent to F in the sense that the polynomials expressible as linear combinations of elements from F and from G are precisely the same.[‡] Armed with a Gröbner basis equivalent to F, Theorem 8.2 provides a straightforward way to test any polynomial f.

8.4   Construction of Gröbner Bases

8.4.1   S-polynomials

The question addressed in this section is how, starting with a set of polynomials F, we can construct, in a systematic (algorithmic) way, a Gröbner basis G equivalent to F. Buchberger solved this problem by defining a particular linear combination of two polynomials called an S-polynomial. Starting

---

[‡]In the terminology of abstract algebra, the set of all polynomials that can be expressed as a linear combination of elements from F is called the *ideal* generated by F.

with $G = F$, his procedure searches G for pairs of elements (g_1, g_2) whose S-polynomial, spoly(g_1, g_2), does not reduce to zero over G; i.e., whose remainder found through generalized division is not zero. If no pair exists, we are done and G is the Gröbner basis we seek. If we find a pair with an S-polynomial whose remainder $r \neq 0$, we add r to G and continue iterating.

Theorem 8.3. (Buchberger's Theorem). *A set of polynomials G is a Gröbner basis if and only if the S-polynomial of any two elements of G reduces to zero; i.e., for all pairs (g_i, g_j) in G,*

$$\text{remainder}(\text{spoly}(g_1, g_2), G) = 0.$$

What is an S-polynomial? The *S-polynomial* of p and q is defined as

$$\text{spoly}(p, q) = \text{lcm}(\text{LT}(p), \text{LT}(q)) \left(\frac{p}{\text{LT}(p)} - \frac{q}{\text{LT}(q)} \right),$$

where LT denotes the leading term of a polynomial. The inclusion of the lcm factor forces the cancellation of the LT terms in the denominators.[§] For example, if

$$p = 2\, x^2\, y + x\, y^4 \quad \text{and} \quad q = x^2 + y + 1,$$

then with respect to lexicographical order with $x > y$ (i.e., `plex(x, y)`), their S-polynomial is

$$\text{spoly}(p, q) = \text{lcm}(2\, x^2\, y, x^2) \left(\frac{2\, x^2\, y + x\, y^4}{2\, x^2\, y} - \frac{x^2 + y + 1}{x^2} \right)$$
$$= x\, y^4 - 2\, y^2 - 2\, y.$$

Ordering the same polynomials by total degree with $x > y$ (i.e., `grlex(x, y)`), the S-polynomial is

$$\text{spoly}(p, q) = \text{lcm}(x\, y^4, x^2) \left(\frac{2\, x^2\, y + x\, y^4}{x\, y^4} - \frac{x^2 + y + 1}{x^2} \right)$$
$$= -y^5 + 2\, x^3\, y - y^4.$$

Since the S-polynomial of p and q is a linear combination of the two polynomials, its remainder upon division by a Gröbner basis G containing p and q must be zero by Theorem 8.2.

[§]The lcm factor is often expressed in the literature as lcm(LP(p), LP(q)), where LP denotes the leading power product of a polynomial; i.e., the leading term with its coefficient removed. The two lcms, and consequently the corresponding S-polynomials, differ by at most a constant factor. The advantage of the definition used here is that if p and q are both polynomials with integer coefficients, their S-polynomial also has integer coefficents. Either definition produces the same reduced Gröbner basis. Maple adopts the convention presented here.

8.4.2   An Example

Before presenting Buchberger's algorithm formally, we illustrate how it works with a simple example. Suppose the input basis consists of two polynomials: $x\,y - 2\,y$ and $x^2 - 2\,y^2$.

```
> F := [x*y - 2*y, x^2 - 2*y^2];
```

$$F := [\,x\,y - 2\,y,\ x^2 - 2\,y^2\,]$$

Since there are only two polynomials in F, we begin by checking this pair to see if it already constitutes a Gröbner basis.

```
> s12 := Groebner[SPolynomial](F[1],F[2],plex(x,y));
  r := Groebner[NormalForm](s12,F,plex(x,y));
```

$$s12 := 2\,y^3 - 2\,x\,y$$
$$r := 2\,y^3 - 4\,y$$

Since the remainder r is not zero, we add it F. (Actually we divide r by its leading coefficient and add this monic polynomial to F. More on this point shortly.)

```
> F := [F[1], F[2], r/2];
```

$$F := [\,x\,y - 2\,y,\ x^2 - 2\,y^2,\ y^3 - 2\,y\,]$$

Now there are three polynomials in F and three pairs to check: (f_1, f_2), (f_1, f_3), (f_2, f_3). Although only two pairs are new, we'll check all three for completeness.

```
> s12 := Groebner[SPolynomial](F[1],F[2],plex(x,y)):
  Groebner[NormalForm](s12,F,plex(x,y));
```

$$0$$

This time the same S-polynomial reduces to zero since we've extended the basis F. Now let's check the two new pairs.

```
> s13 := Groebner[SPolynomial](F[1],F[3],plex(x,y));
  Groebner[NormalForm](s13,F,plex(x,y));
```

$$s13 := -2\,y^3 + 2\,x\,y$$

$$0$$

```
> s23 := Groebner[SPolynomial](F[2],F[3],plex(x,y));
  Groebner[NormalForm](s23,F,plex(x,y));
```

$$s23 := -2\,y^5 + 2\,x^2\,y$$

$$0$$

Since all three remainders are zero, the extended basis is the Gröbner basis we seek.

procedure *GröbnerBasis*(F)
{Input: Set of polynomials F}
{Output: Gröbner basis G equivalent to F}
    $G := F$ {initialize G to F}
    $C := G \times G$ {form set of (unordered) critical pairs}
    while $C \neq \emptyset$ **do**   {any critical pairs left?}
      Choose a pair (f, g) from C
      $C := C \setminus \{(f, g)\}$ {remove the pair from C}
      $h := \text{remainder}(\text{spoly}(f, g), G)$ {reduce the S-polynomial}
      if $h \neq 0$ **then**   {spoly(f, g) does not reduce to 0}
        $C := C \cup (G \times \{h\})$ {add new pairs to C}
        $G := G \cup \{h\}$ {place h in G}
      end if
    end do
    return(G)
end procedure

ALGORITHM 8.2: Buchberger's algorithm

```
> F; Groebner[Basis](F,plex(x,y));
```

$$[\, x\,y - 2\,y,\ x^2 - 2\,y^2,\ y^3 - 2\,y\,]$$
$$G := [\, y^3 - 2\,y,\ x\,y - 2\,y,\ x^2 - 2\,y^2\,]$$

The polynomials in the basis we constructed are the same as those in the one given by Maple. Had we worried about sorting by lexicographical order when we inserted the new element, we would have achieved the same result.

8.4.3   Buchberger's Algorithm

We are now ready to describe Buchberger's algorithm. The procedure takes as input a system of polynomials F and produces its Gröbner basis G with respect to some ordering of monomials. Pseudo-code for the procedure appears as Algorithm 8.2.

As a first step, G is initialized with F and the set of all (unordered) pairs of polynomials is generated. These "critical pairs" are assigned to a variable C. Then we enter a loop that is iterated as long as there are critical pairs left in C. At each iteration, one pair is removed from C, and its S-polynomial is computed and reduced with respect to the elements of G. If the remainder h is zero, we go on and process the next critical pair. If it is not zero, we add new critical pairs (g_i, h) to C for every element g_i currently in G, and we add h to G.

```
procedure ReducedBasis(F)
{Input: Set of polynomials F constituting a Gröbner basis}
{Output: An equivalent reduced Gröbner basis G}
    G := F   {initialize G to F}
    for g in G do    {reduce each g in G}
        g := g / LC(g)   {make monic; divide by coeff of leading term}
        h := remainder(g, G \ {g})   {reduce g w.r.t. to rest of G}
        if h ≠ g then    {does g reduce to itself?}
            G := G \ {g}   {no, remove it from G}
        end if
    end do
    return(G)
end procedure
```

ALGORITHM 8.3: Reduced Gröbner basis

8.4.4   Reduced Gröbner Bases

The simple version of Buchberger's algorithm just presented has several flaws. One is that it never deletes elements from the Gröbner basis under construction; it just adds new elements to the initial system F. Another defect is that it does not produce a unique, or canonical, representation for a Gröbner basis. Suppose, for example, we are given two equivalent but structurally different systems of polynomials, F and F'. If we compute their Gröbner bases, G and G' respectively, using the algorithm, there is no way to tell, in general, whether the two systems are equivalent. We would like a procedure that produces the same Gröbner basis for every equivalent system.

This leads to the notion of a *reduced Gröbner basis*. If the polynomial systems F and F' are equivalent, then their reduced Gröbner bases are identical, i.e., $G = G'$. Given a Gröbner basis G that meets the criterion of Buchberger's Theorem, G is a reduced Gröbner basis if for every element g in G,

$$\text{remainder}(g, G \setminus \{g\}) = g \quad \text{and} \quad g \text{ is monic.}$$

Fortunately, it is very easy to alter the algorithm presented in the previous section to produce a reduced system. As new elements are added to G, we divide the polynomials by their leading coefficients to make them monic. At the end of the procedure, we reduce each element of the basis constructed with respect to all the other elements. Algorithm 8.3 presents pseudo-code for the reduction procedure.

In Section 8.3.2, we considered the system $F = [x\,y + y, x\,y + x]$, whose Gröbner basis, as computed by Maple, is $G = [y^2 + y, x - y]$. If we ran the algorithm on F without reduction, the basis returned would be the union of the two sets,

```
> F := [x*y + y, x*y + x, y^2 + y, x - y];
```

$$F := [\, x\,y + y,\ x\,y + x,\ y^2 + y,\ x - y \,]$$

Let's reduce each of the polynomials with respect to the other elements in the set.

```
> Groebner[NormalForm](F[1],[F[2],F[3],F[4]],plex(x,y));
```

$$0$$

```
> Groebner[NormalForm](F[2],[F[1],F[3],F[4]],plex(x,y));
```

$$0$$

```
> Groebner[NormalForm](F[3],[F[1],F[2],F[4]],plex(x,y));
```

$$y^2 + y$$

```
> Groebner[NormalForm](F[4],[F[1],F[2],F[3]],plex(x,y));
```

$$x - y$$

So the first two polynomials, the ones in the initial basis F, can be eliminated, leaving the two elements computed through reduction of S-polynomials! This would not have been possible with the version of the algorithm presented in the previous subsection.

8.4.5 Efficiency Considerations

Although Buchberger's algorithm is very simple, its termination is not trivial to prove. The algorithm discovers new critical pairs as it proceeds, and some of these may lead to even further non-zero reductions. In his thesis, Buchberger was able to prove that only finitely many extensions are possible to the basis under construction. The leading term of some new polynomial h discovered along the way cannot be a multiple of the leading term of another polynomial already in G. This is a consequence of the fact that h is reduced (remaindered) with respect to the elements already in G and leads to the termination of the algorithm.

What about the computational complexity of constructing Gröbner bases? When the elements of the input basis F contain n variables and the set of polynomials has total degree not exceeding d, the degrees of the polynomials in the Gröbner basis G have been shown to be at most $2\left(\frac{1}{2}d^2 + d\right)^{2^{n-1}}$. This bound is doubly exponential in n but only polynomial in d.

The bound is stated in terms of the complexity of the problem rather than any particular algorithm. It has direct implications on the memory requirements and running time not only of Buchberger's algorithm, but any procedure for computing Gröbner bases.

The following example should give a sense of the limitations of Gröbner base computation. A system of n quadratic equations (implying all monomials are of degree two or less) in n unknowns has 2^n solutions in general. Trying to solve such a system using lexicographical monomial order can result in a triangular system with one polynomial of degree 2^n. This is too large to compute for even small values of n. Here is a system of four quadratics in four unknowns.

```
> f1 := w^2 + x^2 + y^2 + z^2 - 1:
  f2 := w^2 + 2*x^2 - w*y + 2*x*z + 2:
  f3 := y^2 + 2*z^2 + w*z - 2*x*y + 2:
  f4 := w*x + w*y + w*z - 1:
  F := [f1,f2,f3,f4];
```

$$F := [\, w^2 + x^2 + y^2 + z^2 - 1,\; w^2 - w\,y + 2\,x^2 + 2\,x\,z + 2,$$
$$w\,z - 2\,x\,y + y^2 + 2\,z^2 + 2,\; w\,x + w\,y + w\,z - 1\,]$$

```
> G := Groebner[Basis](F, plex(w,x,y,z)); G[1];
```

The first element of the Gröbner basis is a polynomial in z of degree $2^4 = 16$.

$$189440\,z^{16} + 418816\,z^{14} - 1000000\,z^{12} + 6544032\,z^{10} + 7190712\,z^8$$
$$- 53268084\,z^6 + 155558195\,z^4 - 164616517\,z^2 + 95746225$$

We can never hope to compute the basis for a general quadratic system with $n = 100$ variables, let alone store a polynomial of degree 2^{100}. On the other hand, a linear system of 100 equations in 100 unknowns is easy.

In practice, the ordering used for the terms can have a significant effect on efficiency. Degree orders (e.g., `grlex`) typically exhibit better performance than elimination orders (e.g., `plex`). To take advantage of this, techniques have been developed that allow a basis computed initially with respect to a cheaper order to be transformed to a more expensive one.

Gröbner bases can be used to solve problems that are generally considered computationally hard. Therefore, it should not be surprising that the worst case algorithmic complexity of Buchberger's algorithm is very high. Despite this unavoidable worst case behavior, in cases of practical importance it is usually possible to obtain Gröbner bases in a reasonable time. Many non-trivial, interesting polynomial systems encountered in practice have a lot of structure that makes their Gröbner bases relatively easy to compute. Consequently, the method has been used successfully in a wide variety of applications across many fields of mathematics, science, and engineering.

This chapter has presented an introduction to the concept of Gröbner bases. There is much more to the subject—more properties, more applications. The approach taken here has been to provide a gentle introduction, avoiding concepts of abstract algebra such as ideals and varieties. For more information, the textbook by Cox, Little and O'Shea (2015) is the authoritative source on the subject.

8.5 Exercises

1. Use a computer algebra system to find a Gröbner basis for each of the following polynomial systems. Compute each basis with respect to `plex` order, then use the elimination method to solve the resulting systems.

   (a) $x^2 + yz + x = 0$
   $y^2 + xz + y = 0$
   $z^2 + xy + z = 0$

   (b) $x^2 + y + z = 1$
   $x + y^2 + z = 1$
   $x + y + z^2 = 1$

   (c) $x^2 + y^2 + z^2 = 1$
   $x^2 + z^2 = y$
   $x = z$

   (d) $x^2 + 2y^2 = 3$
   $x^2 + xy = 3$

2. Solve the following systems of linear equations by hand using Gaussian elimination. Perform a series of elementary row operations to transform each system to reduced row echelon form.

   (a) $x + y + 2z = 9$
   $2x + 4y - 3z = 1$
   $3x + 6y - 5z = 0$

   (b) $3x + 2y - z = -15$
   $5x + 3y + 2z = 0$
   $3x + y + 3z = 11$

   (c) $3x + 5y - 4z = 7$
   $3x + 2y - 4z = 1$
   $6x + y - 8z = -4$

   (d) $2x + 3y - z = 9$
   $x - y = 1$
   $3x + 7y - 2z = 17$

3. For which values of a does the following linear system have no solutions? Exactly one solution? Infinitely many solutions?

$$x + 2y - 3z = 4$$
$$3x - y + 5z = 2$$
$$4x + y + (a^2 - 14)z = a + 2$$

4. Perform a Gröbner basis computation on a computer algebra system to determine the greatest common divisor of each of the following collections of univariate polynomials.

   (a) $\gcd(\, x^4 + 2\,x^3 - 13\,x^2 - 14\,x + 24,\; x^4 - 2\,x^3 - 13\,x^2 + 14\,x + 24\,)$

   (b) $\gcd(\, x^5 + x^4 - 2\,x^3 - 2\,x^2 + x + 1,\; x^5 - 3\,x^4 + 2\,x^3 + 2\,x^2 - 3\,x + 1,$
        $x^5 - x^4 - 2\,x^3 + 2\,x^2 + x - 1\,)$

   (c) $\gcd(\, x^8 - 1,\; x^{16} - 1,\; x^{32} - 1,\; x^{64} - 1,\; x^{128} - 1\,)$

   (d) $\gcd(\, x^8 - 1,\; x^{10} - 1,\; x^{12} - 1,\; x^{14} - 1,\; x^{16} - 1\,)$

   (e) $\gcd(\, x^3 + 1,\; x^9 + 1,\; x^{27} + 1,\; x^{81} + 1,\; x^{243} + 1\,)$

   (f) $\gcd(\, x^{22} + x^{11} + 1,\; x^{26} + x^{13} + 1,\; x^{34} + x^{17} + 1,\; x^{38} + x^{19} + 1,\; x^{46} + x^{23} + 1\,)$

5. Perform a Gröbner basis computation on a computer algebra system to determine the greatest common divisor of each of the following pairs of multivariate
   polynomials.

   (a) $\gcd(\, 5\,x\,y - 5\,y^2 - 7\,x + 7\,y,\; 2\,x^2 - x\,y - y^2\,)$

   (b) $\gcd(\, x^7 + x^4\,y^3 - x^3\,z^4 - y^3\,z^4,\; x^7 - x^4\,y^3 + x^3\,z^4 - y^3\,z^4\,)$

   (c) $\gcd(\, x^4 + 3\,x^3\,z + x\,y^3 + 3\,y^3\,z,\; x^3\,y - 2\,x^3\,z + y^4 - 2\,y^3\,z\,)$

   (d) $\gcd(\, (z^4 + (x^4 + y^4)\,z^3 + 1)\,(x^4 + y^4 + z^4 - 2\,z + 1),$
        $(z^4 + (x^4 + y^4)\,z^3 + 1)\,(x^4 - y^4 + z^4 + 2\,y + 1)\,)$

   (e) $\gcd(\, 3\,x^8 + 5\,x^4\,y^4 - 3\,x^4\,z^4 - 5\,y^4\,z^4,\; 7\,x^8 - 2\,x^4\,y^4 - 7\,x^4\,z^4 + 2\,y^4\,z^4\,)$

   (f) $\gcd(\, 3\,x^7 - 12\,x^5\,z^2 + 5\,x^4\,y^3 + 12\,x^3\,z^4 - 20\,x^2\,y^3\,z^2 + 20\,y^3\,z^4,$
        $5\,x^7 - 20\,x^5\,z^2 - 3\,x^4\,y^3 + 20\,x^3\,z^4 + 12\,x^2\,y^3\,z^2 - 12\,y^3\,z^4\,)$

6. This exercise illustrates by example how Buchberger's algorithm computes the
   greatest common divisor of two polynomials in one variable.

   (a) Apply Algorithm 8.2 to the find the gcd of

   $$f = 3\,x^4 + 6\,x^3 - 6\,x - 3 \quad \text{and} \quad g = 5\,x^4 - 10\,x^3 + 10\,x - 5.$$

   Use a computer algebra system to find the S-polynomial and perform the
   reduction at each step.

   (b) Reduce the Gröbner basis found in part (a). There should be just one
   element—the greatest common divisor.

   (c) Apply Euclid's algorithm to find $\gcd(f, g)$ and compare the sequence of
   remainders to the computations performed in part (a).

7. Repeat the steps in Exercise 6 to find the greatest common divisor of

   $$f = x^{14} + x^7 + 1 \quad \text{and} \quad g = x^{10} + x^5 + 1.$$

8. The goal of this exercise is to prove that the reduced Gröbner basis for two polynomials in a single variable gives their greatest common divisor, up to a constant factor.

(a) Suppose Buchberger's algorithm is applied to compute the greatest common divisor of

$$f(x) = a_m x^m + a_{m-1} x^{m-1} + \ldots + a_0, \quad g(x) = b_n x^n + b_{n-1} x^{m-1} + \ldots + n_0,$$

where $m \geq n$. Show that the first S-polynomial is given by

$$s_1(f, g) = b_n \left(f - \frac{\mathrm{LT}(f)}{\mathrm{LT}(g)} g \right),$$

and that the first remainder computed by Euclid's algorithm is

$$r_1 = f - \frac{\mathrm{LT}(f)}{\mathrm{LT}(g)} g.$$

These are equivalent up to a constant factor.

(b) Buchberger's algorithm starts with $G = [f, g]$ and appends s_1 to G, giving $G = [f, g, s_1]$. If $s_1 = 0$, we are done. Otherwise we reduce s_1 with respect to G, and continue iterating until $s_t = 0$. Show that we need only reduce s_1 with respect to g and, by induction, s_k with respect to s_{k-1} at each iteration.

(c) Buchberger's algorithm produces a Gröbner basis $G = [f, g, s_1, \ldots, s_{t-1}]$. Show that the equivalent reduced basis is $G = [s'_{t-1}]$, where s'_{t-1} is s_{t-1} divided by its leading coefficient. Also show that s'_{t-1} differs from $r_{t-1} = \gcd(f, g)$, the final non-zero remainder in Euclid's algorithm, by at most a constant factor. (s'_{t-1} is the gcd produced by the monic remainder sequence in Chapter 4.)

9. The trigonometric parameterization of the unit circle is given by

$$x(\theta) = \cos \theta \quad \text{and} \quad y(\theta) = \sin \theta$$

for $-\pi < \theta < \pi$. Show how this trigonometric parameterization can be transformed into the rational parameterization given in Section 8.2.3,

$$x(t) = \frac{1 - t^2}{1 + t^2} \quad \text{and} \quad y(t) = \frac{2t}{1 + t^2},$$

through the change of parameters $\theta = 2 \arctan t$ and its inverse $t = \tan \frac{\theta}{2}$, where $-\infty < t < \infty$.

10. A sphere of radius a can be parameterized by

$$x(u, v) = a \frac{2u}{u^2 + v^2 + 1}, \quad y(u, v) = a \frac{2v}{u^2 + v^2 + 1}, \quad z(u, v) = a \frac{u^2 + v^2 - 1}{u^2 + v^2 + 1}.$$

Perform a Gröbner basis calculation to find the implicit equation for the surface of the sphere.

11. (a) The equation of a four-leaf rose in polar coordinates is $r = \sin(2\theta)$. (See the plot on the left in Figure 8.2.) Using an identity for $\sin(2\theta)$, derive a trigonometric parameterization for the rose in Cartesian coordinates by observing that $x(\theta) = r \cos \theta$ and $y(\theta) = r \sin \theta$.

(b) Transform the trigonometric parameterization of the rose in part (a) to a rational parameterization using the identities

$$\cos \theta = \frac{1 - t^2}{1 + t^2} \quad \text{and} \quad \sin \theta = \frac{2t}{1 + t^2}$$

derived in Exercise 9.

(c) Perform a Gröbner basis calculation to find the implicit equation of the rose in Cartesian coordinates. You should obtain $(x^2 + y^2)^3 = 4x^2 y^2$.

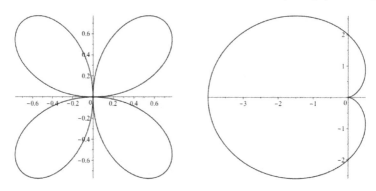

FIGURE 8.2: Four-leaf rose of Exercise 11 and cardioid of Exercise 12.

12. (a) This problem is similar to Exercise 11 except you will be working with a cardioid, whose equation in polar coordinates is $r = 2(1 - \cos \theta)$. (See the plot on the right in Figure 8.2.) Derive a representation for the cardioid in Cartesian coordinates.

(b) Transform the trigonometric parameterization of the cardioid in part (a) to a rational parameterization.

(c) Perform a Gröbner basis calculation to find the implicit equation of the cardioid in Cartesian coordinates. You should obtain $(x^2 + y^2)^2 = 4y^2 - 4x(x^2 + y^2)$.

13. (a) The smallest unit of currency in Bitlogna is the bit. Coins are issued in powers of two bits. The coins currently in circulation have values of 1, 2, 4, 8, 16, 32, and 64 bits. Find the minimum number of coins necessary to purchase items costing 100, 117, and 199 bits.

(b) The 64-bit coin is called a "quarter" because it is worth one-fourth the smallest unit of paper money, the byte. What is the smallest number of pieces of currency (bits and bytes) required to purchase items costing 1.75, 2.375, and 2.546875 bytes?

(c) *Extra credit.* Are bitcoins an official currency anywhere other than Bitlogna? What is the price of a shave and a haircut in Bitlogna?

14. A certain country issues postage stamps in denominations of 1, 3, 9, 27, and 81 cents. Find the minimum number of stamps necessary to pay 100, 117, 199, and 500 cents postage.

15. Try solving the change making problem in Section 8.2.4 using the simple lexicographical order plex(P,N,D,Q) instead of grlex to form the Gröbner basis. Explain the resulting set of coin exchanges and discuss why this monomial order does not produce the desired result.

16. The real numbers a, b, c satisfy the relations

$$a + b + c = 3, \qquad a^2 + b^2 + c^2 = 9, \qquad a^3 + b^3 + c^3 = 24.$$

Find the value of $a^4 + b^4 + c^4$.

17. Express the following polynomials in the lexicographical order plex(x, y, z).

(a) $f(x, y, z) = 2x + 2y + z + x^2 - z^2 + y^2$

(b) $f(x, y, z) = 2x^2 y^8 - 3x^5 y z^4 + xy z^3 - xy^4$

(c) $f(x, y, z) = -zyx + zx^3 + y^3 x + z^3 y - y^2 x^3 - z^3 x^2 - z^2 y^3 + z^2 y^2 x^2$

(d) $f(x, y, z) = x^3 y z^2 - y z^5 - x^5 z^2 + x^2 z^5 - x^3 y^5 + y^5 z^3 + x^5 y^4 - x^2 y^4 z^3$

(e) $f(x, y, z) = -z^4 + y^4 + 2y z^2 + 6x z^2 + 6x y^2 - y^2 - 6xy$

(f) $f(x, y, z) = z^2 + 2y z + 2x z + y^2 + 2xy + x^2 - z^3 + 3y z^2 - 3y^2 z + y^3$

18. Express the polynomials in Exercise 17 using the graded lexicographical order grlex(x, y, z).

19. Perform by hand the division in Example 8.1 but with the order of the divisors reversed; i.e., divide $f = x^2 y + xy^2 + y^2$ by $[g_2, g_1] = [y^2 - 1, xy - 1]$. Check your answer by expanding the products of the quotients and the corresponding divisors plus the remainder.

20. Apply the generalized division procedure by hand to divide f by two polynomials, g_1 and g_2, using plex(x, y) order. First perform the divisions using the order $[g_1, g_2]$, then the order $[g_2, g_1]$. In each case, find the quotients q_1, q_2 and express f as

$$f = q_1 g_1 + q_2 g_2 + r.$$

Note that the final remainder may depend on the order the divisions are performed.

(a) $f = xy^2 + 1$ by $[g_1, g_2] = [xy + 1, y + 1]$

(b) $f = x^2 y + xy^2 + y^2$ by $[g_1, g_2] = [xy - 1, y^2 - 1]$

(c) $f = x^3 + 2x^2 y - 5x + y^3 - 3y$ by $[g_1, g_2] = [xy - 1, x^2 + y^2 - 4]$

(d) $f = 2x^3 + x^2 + xy$ by $[g_1, g_2] = [x^3 - xy, x + y^2]$

21. In this exercise, we examine the general division algorithm when used with a Gröbner basis. Divide xy by $[y - z, x + z]$, then reverse the order of the basis elements and divide xy by $[x + z, y - z]$. Perform both divisions by hand. You should get the same remainder but the quotients are different for the two divisions. This illustrates that uniqueness of the remainder is the best one can expect.

22. Determine whether each of the polynomials listed below can be written as a linear combination of elements from the Gröbner basis

$$G = [\, x\,z - z, \; y - z, \; 2\,z^2 - z \,].$$

(That is, determine whether each of the polynomials is a member of the "ideal" generated by G.) Either express each polynomial as a linear combination in the sense of Theorem 8.2 or show why it is impossible to do so. A computer algebra system should be used to perform the necessary calculations.

(a) $f = x\,y - y$ (b) $f = x^2 - 2\,y^2$ (c) $f = 2\,y^2 - y$

(d) $f = 2\,x^3 - 4\,y^3$ (e) $f = 4\,y^3 - z$ (f) $f = x\,z + x + 2\,z^2$

(g) $f = 4\,x^2\,y^2\,z - z$ (h) $f = 8\,x^2\,y^2\,z^2 - z$ (i) $f = x\,z - 2\,y + 6\,z^2 - 2\,z$

23. Apply Buchberger's algorithm to find a Gröbner basis for each of the following sets of polynomials with respect to $\texttt{plex}(x,y)$ order. In addition, reduce the basis by removing any redundant polynomials. Use a computer algebra system to compute S-polynomials and perform general division.

(a) $F = [\, x\,y + y, \; x\,y + x \,]$ (b) $F = [\, x\,y - 1, \; x^2 + y^2 - 4 \,]$

(c) $F = [\, x^2\,y - 1, \; x + y^2 - 4 \,]$ (d) $F = [\, x^2 + x\,y + y^2, \; x\,y \,]$

(e) $F = [\, x^3\,y + x^2 - 1, \; x + y^2 + 1 \,]$ (f) $F = [\, x^2\,y - 1, \; x\,y^2 - x \,]$

(g) $F = [\, x^2 - 2, \; x^2 + 2\,x\,y + 2, \; y^2 - 2 \,]$ (h) $F = [\, x^2 + y, \; x^4 + 2\,x^2\,y + y^2 + 3 \,]$

24. Use a computer algebra system to compute the Gröbner basis for the system

$$[\, x^5 + y^4 + z^3 - 1, \; x^3 + y^3 + z^2 - 1 \,]$$

with respect to the lexicographical order $\texttt{plex}(x,y,z)$ and the graded lexicographical order $\texttt{grlex}(z,y,x)$. Compare the bases constructed with regard to their complexity (e.g., the number, size, and degree of elements).

Programming Projects

1. **(General Polynomial Division)** Write a procedure,

   $\qquad$ `GeneralDivide(<polynomial>,[<p-list>],<order>)` ,

   that implements the general polynomial division algorithm described in Section 8.3.2. `<polynomial>` is the dividend, `<p-list>` is a list of polynomial divisors separated by commas, and `<order>` is a total ordering defined on monomials. The parameter `<order>` should be in the format used by your computer algebra system so it can be passed directly to any support routines invoked.

   The procedure should return a list,

   $\qquad$ `[<remainder>,[<q-list>]]` ,

   where `<remainder>` is the final remainder and `<q-list>` is a list of quotients corresponding the divisors in `<p-list>`.

   Your procedure should implement the pseudo-code in Algorithm 8.1. You may use routines in your CAS to find leading terms and take apart polynomials.

2. **(Gröbner Basis Construction)** Write a procedure,

GrobnerBasis([<f-list>],<order>) ,

that constructs the *reduced Gröbner basis* for the list of polynomials (separated by commas) in <f-list>. As in Programming Project 1, the parameter <order> should be in the format your computer algebra system uses to specify monomial orders. The procedure should return a list of the basis polynomials, [g-list], again with the elements separated by commas.

You will need to write a utility routine

SPoly(<f-poly>,<g-poly>,<order>) ,

that produces the S-polynomial of <f-poly> and <g-poly> under the order specified. If you completed Programming Project 1, you can use your general division procedure to reduce polynomials. Otherwise, use the routine in your computer algebra system that performs these reductions.

Your code should be at the level of the pseudo-code in Algorithms 8.2 and 8.3, and you may use routines in your CAS to find leading terms and manipulate polynomials.

Chapter 9

Mathematical Correctness

An expert is someone who knows some of the worst mistakes that can be made in his subject, and how to avoid them.

Werner Heisenberg (1901–1976)
Physics and Beyond, 1971

The real danger is not that computers will begin to think like men, but that men will begin to think like computers.

Sydney J. Harris (1917–1986)

Most computer algebra systems report that $\int x^k \, dx = x^{k+1}/(k+1)$. This answer is not completely correct, as any first year calculus student has been drilled. The correct answer includes an arbitrary constant of integration. Moreover, the answer is in error for the special case of $k = -1$, when the solution is $\ln x$. This example illustrates how the implementers of CAS have opted to give what is probably the most useful result, rather than one that is strictly correct.

We have been schooled using calculators, which like to produce one result for each operation. For instance they invariably report $\sqrt{4} = 2$, ignoring the fact that -2 is also a square root of 4. For many years, the Maple kernel performed the analogous symbolic transformation, immediately simplifying $\sqrt{x^2}$ to x whenever a square root was encountered. While this facilitated the implementation of the system and increased its efficiency, users complained that the transformation often produced incomplete or incorrect results. The implementers eventually responded by fixing the problem.

Fractional powers, multivalued functions, and inverse functions pose problems for CAS. The transformation $\sqrt{ab} \to \sqrt{a}\sqrt{b}$ that we are taught in school is not valid when both a and b are negative. Most CAS reply that $\arctan(0) = 0$, taking the principal "branch cut" through this multivalued function, although a "more correct" answer is $\arctan(0) = n\pi$ for any integer n. The inverse functions sine and arcsine behave in such a way that $\sin(\arcsin(x)) = x$ for all x, but $\arcsin(\sin(x)) = x$ only if $-\pi/2 \le x \le \pi/2$. In recent years, there has been a trend in CAS to allow users to define the domains of variables to obviate such problems. This increases the complexity of the software and places an added burden on the user.

In this chapter, we concern ourselves with the applicability and correctness of various mathematical transformations. More generally, we examine potential sources of error in computer algebra systems as well as practical and theoretical limitations on their use.

9.1   Software Bugs

A software bug is an error or failure in a computer program that prevents it from working as intended or produces an incorrect result. Bugs can manifest themselves in a number of ways, with varying levels of inconvenience to the user of the program. Some bugs have only a subtle effect on a program's functionality and may lie undetected for a long time. More serious bugs may cause a program to crash or freeze.

Most computer scientists would agree that large, complicated programs inevitably contain bugs. Examples of large programs include operating systems, programming language compilers, the microcode on CPU chips and, of course, computer algebra systems. Despite their bugs, all of these programs are extremely useful and heavily used.

David Stoutemyer (1991), one of the principal authors of Derive writes,

> Be grateful when bugs are spectacular, such as producing obvious nonsense or crashing the program in a way that requires you to abandon the program and restart it or the computer. What we must guard against most is acceptance of an incorrect result because it superficially looks plausible—or we are too gullible or lazy to check it.

Bugs known to the developers of a computer algebra system serve a useful purpose. They can be corrected so the problem does not occur in future versions of the software. Commands exorcising old bugs are typically added to the test suites used prior to each new release.

Just as mathematical results obtained by hand or with the use of calculators should be checked, so too should the results produced by computer algebra systems. There are several ways to do this. One is to solve the problem on a second system and compare the results. It is extremely unlikely that two systems would have the same bug. They might, however, make the same assumptions about domains (e.g., real vs. complex), branch cuts, continuity and other such issues, and these assumptions might be inappropriate to the problem at hand. An added benefit of solving the problem on a second system is that one CAS may express the result in a more suitable or attractive way.

When an operation has a mathematical inverse, a result can be checked by applying the inverse operation. For example the result of a factorization can be checked by expansion, and that of an integration can be checked through differentiation. Sometimes it may happen that a CAS arrives at an expression that is not easily recognized as equivalent to the original input. In this case, the user should try to simplify the difference between the two forms to zero. CAS are generally much better at this than at transforming one form of an expression to another. Unfortunately, the question of determining whether an expression is equal to zero is, in general, unsolvable for most interesting classes of mathematical expressions.

A way around this difficulty is to substitute random values over the domain of interest for the variable(s). Expressions producing rational numbers should evaluate exactly to zero. Irrational expressions (those containing algebraic or transcendental quantities), however, may produce only approximations to zero due to the limits of finite precision and roundoff errors. In the latter case, the user is left with the problem of deciding whether the substitutions are sufficiently close to zero to be convinced of the equivalence of the expressions. Most of today's computer algebra systems provide a graphing capability. Obtaining a plot is a fast, convenient way to see the effect of a large number of substitutions at once.

9.2   Limitations of Computer Algebra

While computer algebra systems provide impressive computational capabilities, they are far from being a universal panacea for mathematical problem solving. There are theoretical limits on what they can do, as well as practical limitations of time and memory space that constrain the sizes of the problems they can solve.

One example of a theoretical limitation is the problem just mentioned of determining whether an arbitrary mathematical expression is equal to zero. Most CAS, for instance, are unable to perform the transformation

$$\ln\left(\tan\left(\frac{x}{2} + \frac{\pi}{4}\right)\right) - \text{arcsinh}(\tan(x)) \to 0,$$

which is valid for $-\pi/2 < x < \pi/2$. In fact, any CAS that makes this simplification probably has it programmed as a special case. The undecidability of the zero-equivalence problem and related issues are discussed in Sections 5.1 and 5.2.

Abel's Theorem, presented in Section 5.3, states that the solutions to quintic (degree-5) and higher degree polynomials cannot always be expressed in terms of radicals. The quintic $x^5 - x + 1$ is one such example. Abel's Theorem imposes another theoretical limitation on the capabilities of CAS. While closed form solutions can be found in principle for cubics and quartics, such results are of little value for most polynomials due to their size and complexity. Just ask your favorite CAS to solve a general cubic, $ax^3 + bx^2 + cx + d$, or quartic, $ax^4 + bx^3 + cx^2 + dx + e$.

Although it is possible, in principle, to factor any integer into a product of prime numbers raised to powers, actually doing so can be a computationally time intensive task. The security of the RSA public-key cryptosystem is based on the belief that it is not feasible to factor in a reasonable amount of time a number that is the product of two large (e.g., 100^+ decimal digits) primes.

Memory limitations pose another obstacle for CAS. The expanded determinant of an $n \times n$ matrix having n^2 distinct symbolic entries is a sum of $n!$ terms, each of which is a product of n symbols. So the number of symbols in the result is $n\,n!$, which exceeds the amount available of memory for even modest size matrices.

Algorithmic limitations present yet another barrier for CAS. The factorization of $x^{2458} + x^{1229} + 1$ brings most systems to a screeching halt. Yet polynomials of the form $x^{2p} + x^p + 1$ have exactly two factors for all primes $p \neq 3$. One is $x^2 + x + 1$, and its cofactor is a polynomial whose coefficients are all 1, 0, or -1. With this knowledge, it is trivial to coax a CAS to give the cofactor merely by calculating the quotient. Most CAS apply a general factorization algorithm, like those described in Chapter 6, to such polynomials. This can be time consuming for polynomials of moderate degree. Yet CAS often include heuristics for polynomials of the form $x^n - 1$, so $x^{2458} - 1$ factors quickly.

9.3   Sets of Measure Zero

Computer algebra systems are generally designed to give the answer that most users would like to see in most situations. In doing so, CAS typically report a result if it is true for all but a countable number of instances, thereby ignoring what are called *sets of measure zero*. For example, the kernel of a CAS typically applies the transformation $x^0 \to 1$ whenever x^0 is encountered. Yet this transformation is invalid when $x = 0$ (and only when $x = 0$) since 0^0 is considered by most mathematicians to be an undefined, or indeterminate, quantity. (More on this issue later.) The overwhelming majority of users would complain loudly if simplifications like $x^0 \to 1$ were not made automatically.

9.3.1   Cancellation of Polynomial GCDs

Consider the simplification

$$\frac{x^2 - 1}{x - 1} = \frac{(x-1)(x+1)}{x-1} = x + 1. \tag{9.1}$$

Are the expressions on the left and right equivalent? In introductory calculus and mathematical analysis, it is usually stressed that they are not. The ratio on the left is undefined when $x = 1$, while the expression on the right is defined everywhere. From an algebraic perspective, however, they are equal since their difference is $0/(x-1) = 0$. The expression on the right is certainly more meaningful and concise.

Some computer algebra systems, including Derive, automatically cancel greatest common divisors from rational functions, thereby gratuitously removing the singularity at $x = 1$ in (9.1). Other systems maintain the ratio on the left of (9.1) unless the user explicitly requests the simplification (e.g., with the `normal` command in Maple or the `Cancel` function in Mathematica).

9.3.2 Integrating $\int x^k \, dx$

As noted at the beginning of the chapter, computer algebra systems typically report that $\int x^k \, dx = x^{k+1}/(k+1)$, ignoring the only exponent for which the solution is not true: $k = -1$. While this special value constitutes a set of measure zero, it is certainly an important case that arises frequently in both pure and applied mathematics.[†]

There are several ways to deal with the special of case of $k = -1$ and return what may be a more satisfactory antiderivative. The most obvious is to return $\ln x$ if (and only if) k has been assigned -1 or this value can be inferred from domain information that has been supplied. Modern CAS generally perform this kind of checking automatically.

Another approach is to query the user concerning the value of the exponent. Most users would find it annoying if such questions were asked frequently. Moreover, if the integral arose during the course of another computation, such as the solution to a differential equation, the question might be posed in terms of dummy variables that the user has never seen.

Derive takes a third approach, returning the expression $(x^{k+1} - 1)/(k+1)$. This answer adds a specific non-zero constant, $-1/(k+1)$, to the solution returned by most CAS. Derive's antiderivative, however, has the advantage that its limit approaches $\ln x$ as $k \to -1$. Consequently, it gives correct numerical values for the integral everywhere as long as the limit function, rather than substitution, is used for evaluation.

A final solution is to return a conditional expression,

$$\text{if } k \neq -1 \text{ then } \frac{x^{k+1}}{k+1} \text{ else } \ln x \text{ end if.}$$

While this answer is correct, except for an arbitrary constant of integration, it is difficult for CAS to work with conditional results arising during the course of another computation. They may, for instance, cause a CAS to perform extra work exploring alternatives that are of no interest to the user. Furthermore, conditional expressions can grow quite complicated in a long sequence of calculations, challenging the ability of a CAS to make useful simplifications.

[†]Sets of measure zero like this one are frequently an important focus of mathematics.

9.4   Indeterminate Forms

The treatment of indeterminate forms presents a major difficulty for computer algebra systems. An *indeterminate form* is a mathematical expression that is not definitively or precisely determined. Certain forms of limits are indeterminate when knowing the limiting behavior of the individual parts of an expression is not sufficient to determine the overall limit. For example, the following limits of the form $0/0$ have two different values,

$$\lim_{x \to 0} \frac{x}{x} = 1, \qquad \lim_{x \to 0} \frac{x^2}{x} = 0.$$

Consequently, $0/0$ might be 1, or it might be 0 or, by appropriate examples, it might be made to be anything.

Common indeterminate forms include

$$\infty - \infty, \qquad \frac{0}{0}, \qquad \frac{\infty}{\infty}, \qquad 0 \cdot \infty, \qquad \infty^0, \qquad 1^\infty, \qquad 0^0.$$

Computer algebra systems generally allow users to substitute ∞ or $-\infty$ for a variable, or to evaluate expressions containing the symbol ∞. These computations are a frequent source of difficulty and possibly error.

The form 0^0 is a particularly vexing case. The problem here is that we get different limiting values depending on whether we regard this form as a power or an exponential,

$$\lim_{x \to 0} x^0 = 1, \qquad\qquad \lim_{x \to 0} 0^x = 0.$$

Several leading computational scientists including Kahan (1987) and Graham, Knuth and Patashnik (1994) argue, however, that 0^0 should ordinarily simplify to 1. Examining the function x^x as x approaches 0 justifies this view since

$$\lim_{x \to 0} x^x = 1.$$

To see this, we can express x^x as $e^{\ln x^x} = e^{x \ln x}$. L'Hôpital's rule gives

$$\lim_{x \to 0} x \ln x = \lim_{x \to 0} \frac{\ln x}{1/x} = \lim_{x \to 0} \frac{1/x}{-1/x^2} = \lim_{x \to 0} (-x) = 0,$$

and so

$$\lim_{x \to 0} x^x = \lim_{x \to 0} e^{x \ln x} = e^{\lim_{x \to 0} x \ln x} = e^0 = 1.$$

Figure 9.1 depicts the validity of this limit graphically. Nonetheless, the automatic simplification of 0^0 to 1 is a dangerous, error-prone transformation that must be used with great care, if at all, in CAS.

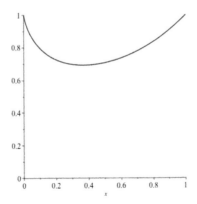

FIGURE 9.1: x^x as x approaches zero.

9.5 Spurious Solutions

When tackling mathematical problems, both humans and computer algebra systems sometimes use procedures that introduce *spurious solutions*. Spurious solutions are erroneous in that they do not satisfy the conditions of the original problem, but rather are artifacts of the computational procedures applied. One way that CAS can avoid reporting spurious solutions is by using additional calculations to verify that each potential solution is indeed a true one. In this section, we discuss three sources of spurious solutions: extraneous roots of equations, improper integrals, and antiderivatives having singularities that do not match those of the integrand.

9.5.1 Extraneous Roots

Equations involving square roots are typically solved by isolating the radical and squaring both sides of the resulting equation.[‡] The process of squaring may lead to a solution that is not actually one of the roots of the original equation. This extra answer is called an *extraneous root*. CAS reporting the solutions of the squared equation, without verifying that each indeed satisfies the original one, run the risk of returning bogus answers.

Example 9.1. Consider the equation

$$1 - x = \sqrt{x - 1}.$$

[‡]A similar procedure can be used for other roots.

Squaring both sides gives

$$(1-x)^2 = x - 1,$$

which has two roots: $x = 1$ and $x = 2$. Only $x = 1$, however, satisfies the original equation; the other root is extraneous. □

9.5.2 Improper Integrals

A standard way to compute the definite integral $\int_a^b f(x)\, dx$ is to determine the corresponding indefinite integral, $F(x) = \int f(x)\, dx$, and substitute the values of the endpoints,

$$\int_a^b f(x) = F(b) - F(a).$$

This method may fail for improper integrals. The definite integral $\int_a^b f(x)\, dx$ is *improper* if either: (1) the value of f is infinite at one or more points over the interval of integration, or (2) one or both of the limits of integration is infinite. To evaluate improper integrals, the interval must be broken into subintervals bounded by the singularities (undefined points in the domain) of the integrand. The definite integral is then evaluated over each subinterval using limits, instead of substitution, to determine the values at the endpoints. If each of these integrals converges, then the improper integral converges to the sum; otherwise, the integral diverges.

Example 9.2. Evaluate the improper integral $\displaystyle\int_0^4 \frac{1}{\sqrt{|x-1|}}\, dx$.

Solution. The integrand is infinite at $x = 1$, which lies between the limits of integration. The value of the corresponding indefinite integral is

$$\int \frac{1}{\sqrt{|x-1|}}\, dx = \begin{cases} -2\sqrt{1-x} & \text{if } x \le 1, \\ 2\sqrt{x-1} & \text{if } x > 1. \end{cases}$$

We evaluate the integral over the interval $[0, 1)$ as

$$\lim_{a\to 1^-} \int_0^a \frac{1}{\sqrt{|x-1|}}\, dx = \lim_{a\to 1^-} \left[-2\sqrt{1-a} + 2\sqrt{1-0} \right] = 2.$$

Similarly, the value of the integral over $(1, 4]$ is

$$\lim_{a\to 1^+} \int_a^4 \frac{1}{\sqrt{|x-1|}}\, dx = \lim_{a\to 1^+} \left[2\sqrt{4-1} - 2\sqrt{a-1} \right] = 2\sqrt{3}.$$

Since both pieces converge, the improper integral converges to $2 + 2\sqrt{3}$. □

Unfortunately, it is impossible in general for a computer algebra system to determine all singularities of an arbitrary function. Moreover, it takes a good deal of sophisticated code to decide a worthwhile percentage of cases. Consequently, some CAS make no attempt to verify the convergence of improper integrals, or fail to issue a warning when an attempt at verification is indecisive. Furthermore, a system may report that $\int_{-1}^{1} 1/x^2 \, dx$ diverges, while returning

$$\int_a^b \frac{1}{x^2} \, dx = -\frac{1}{b} + \frac{1}{a}$$

without warning the user that convergence depends on the values of a and b.

9.5.3 Discontinuities in Antiderivatives

The technique of integration by substitution is a standard topic in calculus textbooks and one that is employed by the integration routines in most computer algebra systems. An aspect of the technique that is rarely discussed is that, under certain conditions, substitutions can introduce spurious discontinuities in the integrals that are not present in the integrand.

Any substitution $u = s(x)$ can lead to antiderivatives with spurious discontinuities if $s(x)$ contains a singularity. For example, the substitution $u = 1/x$ has a singularity at the origin and so any integrals calculated with its aid should be checked for spurious discontinuities at $x = 0$. The following example illustrates this point.

Example 9.3. Evaluate $\displaystyle\int \frac{e^{1/x}}{x^2 (1 + e^{1/x})^2} \, dx$ with the substitution $x = 1/u$.

Solution. With $u = 1/x$, we have $du = -1/x^2 \, dx$ and

$$\int \frac{e^{1/x}}{x^2 (1 + e^{1/x})^2} \, dx = \int \frac{-e^u}{(1 + e^u)^2} \, du = \frac{1}{1 + e^u} = \frac{1}{1 + e^{1/x}}. \qquad (9.2)$$

While the value of the integrand is undefined at $x = 0$, the limit from both sides exists and is zero. Consequently, if we define the integrand to have value zero at $x = 0$, the resulting function

$$f(x) = \begin{cases} \dfrac{e^{1/x}}{x^2 (1 + e^{1/x})^2} & \text{if } x \neq 0, \\ 0 & \text{if } x = 0 \end{cases}$$

is continuous for all values of x.

The integral, however, has a jump discontinuity at $x = 0$. This situation runs counter to the Fundamental Theorem of Calculus.§ If a function f is continuous on some interval, then we also expect its integral F to be continuous.

---

§The Fundamental Theorem states that if the function f is continuous on a closed interval $[a, b]$, then its integral $F(x) = \int_a^x f(t) \, dt$ is continuous and differentiable on the open interval (a, b) and $F'(x) = f(x)$.

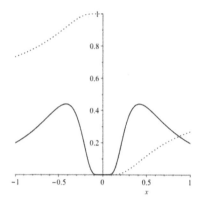

FIGURE 9.2: $\dfrac{e^{1/x}}{x^2\,(1+e^{1/x})^2}$ (—) and a discontinuous antiderivative ($\cdots$).

Our dilemma is illustrated in Figure 9.2, where the integrand and integral are plotted together near the origin. Note, however, that the expression on the right of (9.2) is a valid antiderivative in that sense that differentiation yields the original integrand.

The jump in the antiderivative is -1 at $x = 0$. This spurious discontinuity can be removed by "splitting the difference" and expressing the integral with the signum function as

$$\int \frac{e^{1/x}}{x^2\,(1+e^{1/x})^2}\,dx = \frac{1}{1+e^{1/x}} + \tfrac{1}{2}\operatorname{sgn}x, \qquad (9.3)$$

where

$$\operatorname{sgn}x = \begin{cases} 1 & \text{if } x > 0, \\ 0 & \text{if } x = 0, \\ -1 & \text{if } x < 0. \end{cases}$$

If we define the value of (9.3) to be $\frac{1}{2}$ at $x = 0$, which is the limit from both sides at this point, the resulting function is continuous for all x. This continuous integral is plotted together with the integrand in Figure 9.3. ☐

Computer algebra systems often employ the substitution $u = \tan(x/2)$ to solve trigonometric integrals. This powerful substitution converts any rational function of $\sin x$ and $\cos x$ into a rational function of u. The transformed integral can then be evaluated using the algorithms of Hermite, Horowitz, and Rothstein-Trager described in Chapter 7. We present an example to show how the substitution works.

Example 9.4. Evaluate $\displaystyle\int \frac{1}{2 + \cos x}\,dx$ using the substitution $u = \tan(x/2)$.

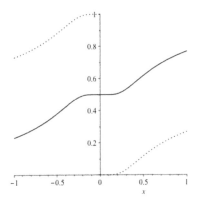

FIGURE 9.3: Continuous and discontinuous antiderivatives of $\dfrac{e^{1/x}}{x^2\left(1+e^{1/x}\right)^2}$.

Solution. The substitution may be rearranged as $x = 2\arctan u$ which, when differentiated, gives $dx = 2/(1+u^2)\,du$. The reference triangle in Figure 9.4 shows the effect of the substitution on the sine and cosine.

For our problem, the transformed integral is

$$\int \frac{1}{2+\cos x}\,dx = \int \frac{1}{2+\frac{1-u^2}{1+u^2}}\,\frac{2}{1+u^2}\,du$$

$$= 2\int \frac{1}{3+u^2}\,du.$$

Evaluating this integral and inverting the substitution, we find

$$2\int \frac{1}{3+u^2}\,du = \frac{2}{\sqrt{3}}\arctan\left(\frac{1}{\sqrt{3}}u\right)$$

$$= \frac{2\sqrt{3}}{3}\arctan\left(\frac{\sqrt{3}}{3}\tan\left(\frac{x}{2}\right)\right), \qquad (9.4)$$

where the arbitrary constant of integration is omitted.

The integrand is continuous for all x, while (9.4) contains spurious discontinuities whenever x is an odd multiple of π. These discontinuities are due to the singularities in $\tan(x/2)$. The integrand and (9.4) are plotted together in Figure 9.5. □

The substitution $u = \tan(x/2)$ is one of a collection of substitutions due to Karl Weierstrass which are useful in transforming an integral that is a rational function of sines and cosines into an integral that is a rational function of u. Other forms of the Weierstrass substitution are given in Table 9.1. The choice of which substitution to use when evaluating any particular integral is not

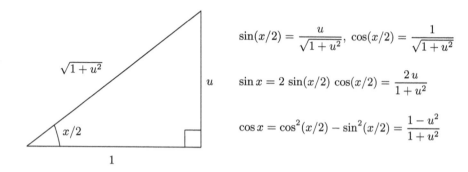

$$\sin(x/2) = \frac{u}{\sqrt{1+u^2}}, \ \cos(x/2) = \frac{1}{\sqrt{1+u^2}}$$

$$\sin x = 2\sin(x/2)\cos(x/2) = \frac{2u}{1+u^2}$$

$$\cos x = \cos^2(x/2) - \sin^2(x/2) = \frac{1-u^2}{1+u^2}$$

FIGURE 9.4: Reference triangle for the substitution $u = \tan(x/2)$.

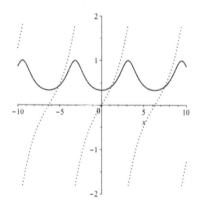

FIGURE 9.5: $\dfrac{1}{2 + \cos x}$ (—) and a discontinuous antiderivative $(\cdots)$.

Choice	u	$\sin x$	$\cos x$	dx	b	p
(a)	$\tan(x/2)$	$\dfrac{2\,u}{1+u^2}$	$\dfrac{1-u^2}{1+u^2}$	$\dfrac{2\,du}{1+u^2}$	π	2π
(b)	$\tan(x/2 + \pi/4)$	$\dfrac{u^2-1}{1+u^2}$	$\dfrac{2\,u}{1+u^2}$	$\dfrac{2\,du}{1+u^2}$	$\pi/2$	2π
(c)	$\cot(x/2)$	$\dfrac{2\,u}{1+u^2}$	$\dfrac{u^2-1}{1+u^2}$	$\dfrac{-2\,du}{1+u^2}$	0	2π
(d)	$\tan(x)$	$\dfrac{u}{\sqrt{1+u^2}}$	$\dfrac{1}{\sqrt{1+u^2}}$	$\dfrac{du}{1+u^2}$	$\pi/2$	π

u is the substitution, p its period, and b the principal discontinuity

TABLE 9.1: Forms of the Weierstrass substitution.

critical and affects only the form of the final result. Choice (a) works well for integrands not containing $\sin x$, while (b) is good for those not containing $\cos x$.

David Jeffrey and Albert Rich (1994) present an algorithm to remove spurious discontinuities from the integrals resulting from these substitutions. The steps in their procedure are as follows:

1. Transform the integral using one of the substitutions in Table 9.1. For example, with choice (a) we have

$$\int f(\sin x, \cos x) = \int f\left(\frac{2\,u}{1+u^2}, \frac{1-u^2}{1+u^2}\right) \frac{2\,du}{1+u^2}.$$

2. Integrate the rational function of u obtained in Step 1.

3. Invert the substitution performed at Step 1. Call the result $\hat{F}(x)$.

4. Calculate
$$K = \lim_{x \to b^-} \hat{F}(x) - \lim_{x \to b^+} \hat{F}(x),$$
where the point b is given in Table 9.1. The correct integral is

$$F(x) = \int f(\sin x, \cos x) = \hat{F}(x) + K \left\lfloor \frac{x-b}{p} \right\rfloor,$$

where p is the period taken from the table and $\lfloor \; \rfloor$ is the floor function.

5. Optionally, the floor function obtained in Step 4 can be expressed with an inverse tangent using the relation

$$2\pi \left\lfloor \frac{x+\pi}{2\pi} \right\rfloor = x - 2\arctan(\tan(x/2)).$$

Any tangents of half angles can be rewritten with the identity

$$\tan(x/2) = \frac{\sin x}{1 + \cos x}$$

to express the solution in terms of the sine and cosine of x, which were used in the original integrand.

Example 9.5. Remove the spurious discontinuities from the integral in Example 9.4.

Solution. The first three steps of the procedure were applied previously to give

$$\hat{F}(x) = \frac{2\sqrt{3}}{3} \arctan\left(\frac{\sqrt{3}}{3} \tan\left(\frac{x}{2}\right)\right).$$

At Step 4, we calculate

$$K = \lim_{x \to \pi-} \hat{F}(x) - \lim_{x \to \pi+} \hat{F}(x) = \frac{2\sqrt{3}}{3}\pi$$

and use K to form the corrected integral,

$$\int \frac{1}{2 + \cos x}\, dx = \frac{2\sqrt{3}}{3} \arctan\left(\frac{\sqrt{3}}{3} \tan\left(\frac{x}{2}\right)\right) + \frac{2\sqrt{3}}{3}\pi \left\lfloor \frac{x - \pi}{2\pi} \right\rfloor.$$

Eliminating the floor function, the integral can also be expressed as

$$\frac{2\sqrt{3}}{3} \arctan\left(\frac{\sqrt{3}}{3} \tan\left(\frac{x}{2}\right)\right) - \frac{2\sqrt{3}}{3} \arctan\left(\tan\left(\frac{x}{2}\right)\right) + \frac{\sqrt{3}}{3}x. \qquad (9.5)$$

Figure 9.6 graphs this formula together with the discontinuous antiderivative in (9.4). Their values are the same only on the interval $(-\pi, \pi)$. □

9.6   Branch Cuts

Fractional powers, logarithms, and inverse trigonometric functions return a single value on numeric calculators and numerical computer software. Based on this experience, most users expect and want the same behavior when a numerical expression or subexpression is evaluated by a computer algebra system. For example, users expect $8^{1/3}$ to simplify to 2 rather than the set $\{2, -1 + i\sqrt{3}, -1 - i\sqrt{3}\}$, which contains all three cube roots of 8. Similarly, users expect $\ln(1)$ and $\arctan(0)$ to simplify to 0 rather than the sets $\{2n\pi i \mid n \in \mathbb{Z}\}$ and $\{n\pi \mid n \in \mathbb{Z}\}$, respectively, where the notation $n \in \mathbb{Z}$ means

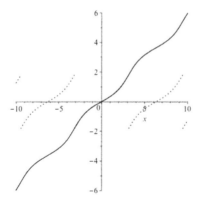

FIGURE 9.6: Continuous and discontinuous antiderivatives of $\dfrac{1}{2 + \cos x}$.

that n is some integer. In order to avoid inconsistencies and errors, CAS must exercise extreme care when manipulating such multivalued expressions.

Most mathematical texts use the notations $z^{1/n}$ and $\sqrt[n]{z}$ interchangeably. The choice is often based on convenience or æsthetics, with $\sqrt[n]{z}$ typically used when z is a short expression and $z^{1/n}$ favored when z is lengthy or contains fractions. Jeffrey and Norman (2004) identify four separate meanings that these notations might convey. To avoid ambiguity, they suggest a different notation for each definition.

1. *A set of values:* $\{z^{1/n}\}$

   With this notation, a fractional power is treated as a set of values. For any $z \in \mathbb{C}$ (i.e., for complex z),

   $$\{z^{1/n}\} = \{w \in \mathbb{C} \mid w^n = z\}.$$

   Each value in the set belongs to one branch of the function.

   Example. $\{(-8)^{1/3}\} = \{-2, 1 + i\sqrt{3}, 1 - i\sqrt{3}\}.$

2. *The principal value:* $\sqrt[n]{z}$

   A complex quantity $z = a + bi$ can be represented in polar form by the pair of values $(|z|, \arg z)$. The *absolute value*, $|z| = \sqrt{a^2 + b^2}$, is a real number giving the distance of z from the origin. The *argument*, $\arg z = \arctan(b/a)$, gives the polar angle with the convention that $-\pi < \arg z \le \pi.$[†] (See Section 5.4.4 for a more in-depth discussion of the representation of quantities in the complex plane.)

[†]The absolute value of a complex number is sometimes referred to as the *modulus*, the *norm*, the *magnitude*, or the *amplitude*. The argument is occasionally called the *phase*.

Intuitively, the *principal value* of $z^{1/n}$ is the value from the set $\{z^{1/n}\}$
with the smallest polar angle, measuring in the counter-clockwise direction. More formally, $\sqrt[n]{z}$ denotes the single valued, principal branch
function, which is given in complex exponential form by

$$\sqrt[n]{z} = |z|^{1/n} e^{i\,(\arg z)/n}.$$

Example. $\sqrt[3]{-8} = 1 + i\sqrt{3}$.

3. *The real branch:* $\text{surd}(z, n)$.

   The *surd* function is useful when it is desired that the odd n-th root of
   a negative real number denote a negative real value. For $x \in \mathbb{R}$ (i.e., for
   real x),

   $$\text{surd}(x, n) = \text{sgn}(x)\,\sqrt[n]{|x|},$$

   where the n-th root is a real quantity. In particular, if n is odd then
   $\text{surd}(x, n) = x^{1/n}$ when $x \geq 0$ and $\text{surd}(x, n) = -(-x)^{1/n}$ when $x < 0$.
   For complex z, $\text{surd}(z, n)$ is taken to be the complex n-th root whose
   argument is closest to z in the complex plane, with ties decided by
   moving in the counter-clockwise direction from z.

   Example. $\text{surd}(-8, 3) = -2$.

4. *Any one value from* $\{z^{1/n}\}$: $z^{1/n}$

   Like the principal value and the surd, this notation represents a single
   number. In this case, however, some indeterminacy remains because the
   branch is not specified. Given $z \in \mathbb{C}$,

   $$z^{1/n} = |z|^{1/n}\, e^{i\,(\arg z + 2\pi k)/n}$$

   for some unspecified $k \in \mathbb{Z}$. Again, $\arg z \in (-\pi, \pi]$ and $|z|^{1/n}$ is a unique
   real number.

   Example. $(-8)^{1/3} = 2\,e^{i\,(2k+1)\pi/3}$ for $k = 0, 1, 2$.

   With so many options, it's no wonder that confusion abounds—both on
the part of humans and computer algebra systems. So what's a CAS to do
when faced with an expression like $(-8)^{1/3}$? The safest alternative is to leave
the expression alone! Another safe approach is to simplify it conservatively to
$-2 \times 1^{1/3}$. The overwhelming majority of users, however, would be perplexed
or disgruntled that $1^{1/3}$ does not simplify to 1. To quote Stoutemyer (1991),
"We train students to seek *explicit* solutions as the holy grail, so such *implicit*
results are doomed to poor acceptance."

   Carrying all branches of a multivalued function is not only a correct approach, but it is a reasonable one for which there is already precedent. It
mirrors the way CAS handle implicit solutions, like the roots of $x^5 - x + 1$

that cannot be expressed with radicals (e.g., Maple's `RootOf` and Mathematica's `Root` notation).

A CAS should provide mechanisms that are explicitly invoked by the user to obtain the principal branch, the real branch, or some other some other branch of a fractional power. Unfortunately, there is not a standard way to do this. In Maple the `simplify` command transforms $(-8)^{1/3}$ to the principal branch, $1 + i\sqrt{3}$, while Mathematica's `Simplify` function returns the real branch, -2.

9.6.1 Fractional Powers

With all this said, we turn our attention to some useful simplifications involving fractional powers. First, consider the pair of transformations

$$(u\,v)^{1/2} \rightleftarrows u^{1/2}\,v^{1/2} \tag{9.6}$$

to distribute fractional powers over products and to collect exponents. These transformations are not valid whenever both u and v are negative. For example,

$$((-1)\,(-1))^{1/2} = 1^{1/2} = 1$$

but

$$(-1)^{1/2}\,(-1)^{1/2} = i^2 = -1.$$

As a result, systems that automatically apply (9.6) get into trouble.

One solution is for a CAS to supply two separate procedures, one to distribute and the other to collect exponents. The user explicitly invokes one of these procedures when a particular transformation is desired. When the transformation is required inside a program, however, it is inconvenient to suspend the computation and ask the user how to proceed. Another problem is that a user may unwittingly request a transformation that is not valid.

A second approach is for a system to provide a control variable named, say, `TransformFractionalPowers`. Its default setting could be `false`, with optional settings of `distribute` and `collect`. Whenever one of the transformations in (9.6) might be applied, the CAS consults the control variable and continues accordingly.

A third solution is to allow users to specify the properties of variables. Maple provides an `assume` facility for users to declare domain information, while the `Element` command plays a similar role in Mathematica. In Derive, such information is provided through dialog boxes which, of course, are translated into commands for its kernel to process. These systems consult the domain information to determine the validity of a particular transformation before applying it. Even with such facilities, it is an extremely difficult problem for a CAS to transform an expression like $(x\,y\,|z|^2)^{1/2}/(x^{1/2}\,|z|)$ only as far as $(x\,y)^{1/2}/x^{1/2}$.

The simplification

$$(u^2)^{1/2} \to u \tag{9.7}$$

is valid only when $\arg u \in (-\pi/2, \pi/2]$. For instance,

$$((-1)^2)^{1/2} = 1^{1/2} = 1 \neq -1.$$

(What is the argument in this simple example?) If u is real, then the left side of (9.7) can be validly transformed to $|u|$, and can be further simplified to u whenever $u > 0$. A similar difficulty arises with the transformation

$$\left(\frac{1}{u}\right)^{1/2} \to \frac{1}{u^{1/2}},$$

which is not valid when u is negative. For example, $(1/(-1))^{1/2} = (-1)^{1/2} = i$, but $1/(-1)^{1/2} = 1/i = -i$.

Although (9.7) is a special case of (9.6) with $v = u$, such transformations are generally implemented based on syntax and so they are likely to be programmed separately. Consequently, a CAS may implement one of the transformations correctly but not the other.

Finally, we consider the problem of distributing and collecting fractional powers over exponentials.

$$(e^z)^{1/2} \rightleftarrows e^{z/2}.$$

These transformations are valid for real z, but must be used with care when z is complex. In such cases, the simplifications are valid only when the imaginary part of z lies in the interval $(-\pi, \pi]$. For example with $z = 2\pi i$, $(e^{2\pi i})^{1/2} = 1^{1/2} = 1$ but $e^{\pi i} = -1$.

9.6.2   Logarithms

When working with logarithms in calculus, we are taught that the function $\ln x$ is defined only for real $x > 0$. In Section 5.4.4 we saw, however, that the natural logarithm can take negative, and even complex, arguments and that its value is complex in such cases. As a result, many identities involving logarithms that we take for granted in calculus must be reexamined to determine their applicability in computer algebra.

The basic transformations to expand and collect logarithms are

$$\ln(uv) \rightleftarrows \ln u + \ln v. \tag{9.8}$$

These transformations are typically valid in the setting of introductory calculus, where both $u, v > 0$, but do not hold when both u and v can be negative. For example,

$$\ln((-1)(-1)) = \ln 1 = 0,$$

whereas

$$\ln(-1) + \ln(-1) = \pi i + \pi i = 2\pi i.$$

The transformations

$$\ln u^2 \rightleftarrows 2 \ln u$$

are a special case and are not valid when u is negative. This is easily seen when $u = -1$: $\ln((-1)^2) = \ln 1 = 0$ but $2\ln(-1) = 2\pi i$. Similarly, the transformations

$$\ln(1/u) \rightleftarrows -\ln u$$

are also invalid for negative u. For instance, $\ln(1/(-1)) = \ln(-1) = \pi i$ while $-\ln(-1) = -\pi i$.

Now consider the sense in which the logarithm and exponential functions are inverses of one another. For real z, the simplifications

$$e^{\ln z} \to z, \tag{9.9}$$

$$\ln e^z \to z \tag{9.10}$$

are safe transformations for a CAS. The transformation (9.9) is also safe when z is complex, but (9.10) is of limited applicability for complex z. We now discuss the reasons.

The logarithm of a complex number is not a unique value in much the same way that the square root of a positive real number has two possible values—the positive (principal) value and its negative. Recall that a complex number $z = a + bi$ can be expressed in complex exponential (polar) form as

$$z = a + bi = |z|\,e^{i\theta}, \tag{9.11}$$

where the absolute value $|z| = \sqrt{a^2 + b^2}$ and $\theta = \arctan(b/a)$. The arctangent is a multivalued function. Its principal value is, by convention, the one for which θ lies in the interval $(-\pi, \pi]$ and this angle is the argument, $\arg z$. We can, however, add any integer multiple (positive or negative) of 2π to the argument and obtain the same vector representing z in the complex plane. Taking the logarithm of both sides of (9.11), we see that the logarithm of a complex number is a multivalued function,

$$\ln z = \ln(|z|) + \theta\,i = \ln(|z|) + (\arg z + 2\pi k)\,i,$$

where k is any integer. Consequently, (9.10) is valid only when the imaginary part of z lies in the interval $(-\pi, \pi]$. For example,

$$\ln e^{3\pi i} = \ln(-1) = \pi i \neq 3\pi i.$$

9.6.3 Trigonometric Functions

The inverse trigonometric functions are inherently multivalued. For example, if we restrict the domain of the tangent to the real numbers $\mathbb{R}^\ddagger$, then its range is also $\mathbb{R}$. This, in turn, is the domain of the arctangent. The periodicity of the tangent causes the arctangent to be multivalued. The principal

---

$\ddagger$Note that $\tan x$ is undefined at the points $x = \frac{1}{2}(2n + 1)\pi$ for any $n \in \mathbb{Z}$ (i.e., for $x = \ldots - \frac{5}{2}\pi, -\frac{3}{2}\pi, -\frac{1}{2}\pi, \frac{1}{2}\pi, \frac{3}{2}\pi, \frac{5}{2}\pi, \ldots$).

branch of the tangent is taken to be the interval $(-\pi/2, \pi/2)$, and the tangent assumes every real value on this interval. Consequently, the principal value of the arctangent also lies in this interval. Calculators and CAS invariably return the principal value, reporting $\arctan(0) = 0$, although it is true that $\arctan(0) = n\pi$ for any $n \in \mathbb{Z}$.

Now let's examine the sense in which the tangent and the arctangent are inverses of one another. The two transformations of interest are

$$\tan(\arctan u) \to u, \tag{9.12}$$

$$\arctan(\tan u) \to u. \tag{9.13}$$

Equation (9.12) is always valid. (Why?) Equation (9.13), however, is valid only when $u \in (-\pi/2, \pi/2)$, assuming the arctangent is implemented in the standard way and returns the principal value. For example, $\arctan(\tan \pi) = \arctan(0) = 0 \neq \pi$.

Branch cut issues also arise in the half angle formulas for the sine and cosine. When rearranging the double angle formula for the cosine,

$$\cos(2\,\theta) = 2\,\cos^2\theta - 1 = 1 - 2\,\sin^2\theta,$$

to isolate $\sin\theta$ or $\cos\theta$, the question arises as to what sign to place before the radical. Valid half angle transformations for all real x are

$$\sin\left(\frac{x}{2}\right) \rightleftarrows \operatorname{sgn}\left(\sin\left(\frac{x}{2}\right)\right)\sqrt{\frac{1-\cos x}{2}},$$

$$\cos\left(\frac{x}{2}\right) \rightleftarrows \operatorname{sgn}\left(\cos\left(\frac{x}{2}\right)\right)\sqrt{\frac{1+\cos x}{2}}.$$

In practice, few CAS provide these transformations either automatically or optionally due, perhaps, to the subtle branch cut issues involved.

9.7 Solution of Equations

An important capability of a computer algebra system is its ability to solve equations symbolically. This feature is valuable to users in its own right, but it also serves as a central building block for many of the internal algorithms used in the implementation of a CAS. Most of the issues discussed in this chapter arise in the construction of symbolic solvers. Consequently, equation solving provides an excellent case study for mathematical correctness.

Our focus is on problems involving a single variable, as it has been throughout this book. We begin with linear and polynomial equations, then move on to equations involving algebraic functions, exponentials and logarithms and, finally, trigonometric functions. Areas not covered—although important capabilities in most CAS—include numeric (approximate) solving, differential equations, and polynomial systems of equations (Gröbner basis computation).

Laurent Bernardin (1999) surveys and compares the equation solving capabilities of seven computer algebra systems: Maple, Mathematica, Derive, Axiom, Macsyma, MuPAD, REDUCE, and the TI-92 symbolic calculator. The systems were presented with a series of examples to evaluate their ability to solve equations correctly in various domains. While the problems are generally quite basic, many illustrate common pitfalls that a CAS might stumble upon when solving equations typical in a particular domain. Most of the equations were able to baffle one or more of the CAS, and a few of the problems stumped most or all of the systems. Each of the examples presented here is drawn from this test suite. Our focus is on the issues designers of CAS face in implementing equation solvers.

9.7.1 Linear and Polynomial Equations

Linear equations pose no difficulty whatsoever for CAS. Equations in a single variable are solved merely by rearrangement,

$$3x + 4 = 0 \longrightarrow x = -\tfrac{4}{3}.$$

Systems of equations are easily solved using standard linear algebra techniques such as Gaussian elimination.

Even as simple an equation as

$$ax + b = 0,$$

however, raises issues of mathematical correctness. When regarded as an equation in three unknowns, there are an infinite number of solutions. If a and b are thought of as fixed parameters, then the existence of a solution depends on whether a is zero. Despite this, CAS conveniently and invariably report

$$x = -\frac{b}{a},$$

even though this solution is not correct for $a = 0$, a set of measure zero. An exception is Mathematica. While its Solve function replies $x = -b/a$, the Reduce command correctly reports the parameterized solution.

In[1]:= Reduce[a x + b == 0, x]

Out[1]= (b == 0 && a == 0) || $\left(a \neq 0 \,\&\&\, x == -\frac{b}{a} \right)$

The solutions to polynomial equations of degrees 2, 3, and 4 can be expressed in terms of radicals. For example, the solutions to the general quadratic equation

$$ax^2 + bx + c = 0$$

are given by

$$x = \frac{-b + \sqrt{b^2 - 4\,a\,c}}{2\,a}, \quad x = \frac{-b - \sqrt{b^2 - 4\,a\,c}}{2\,a}.$$

So it is a simple matter for CAS to produce explicit solutions for quadratics. Closed form solutions involving radicals can also be given for general cubic and quartic equations,

$$a\,x^3 + b\,x^2 + c\,x + d = 0,$$
$$a\,x^4 + b\,x^3 + c\,x^2 + d\,x + e = 0,$$

although these formulas are long and messy. Thus, a closed form, as opposed to a numerical result, is not of much use for all but simple polynomials. Consequently, some CAS opt to give by default implicit solutions to cubics and quartics, but can produce explicit solutions when requested.

For polynomials of degree five or more, Abel's Theorem states that closed form solutions in terms of radicals may not exist. An example of a quintic solvable in terms of radicals is

$$x^5 + x + 1 = (x^3 - x^2 + 1)(x^2 + x + 1),$$

whose solutions are

$$x = -\frac{\alpha}{6} - \frac{2}{3\,\alpha} + \frac{1}{3},$$
$$x = -\frac{1}{2} \pm \frac{\sqrt{3}}{2}\,i,$$
$$x = \frac{\alpha}{12} + \frac{1}{3\,\alpha} + \frac{1}{3} \pm \frac{\sqrt{3}}{2}\left(\frac{2}{3\,\alpha} - \frac{\alpha}{6}\right)i,$$

where $\alpha = \sqrt[3]{100 + 12\sqrt{69}}$. Whereas the quintic

$$x^5 - x + 1,$$

though almost identical in appearance, is irreducible and its solutions cannot be expressed in closed form. David S. Dummit (1991) presents a procedure for deciding whether a quintic is solvable in terms of radicals and, if it is, to produce these solutions. While the algorithm is rather involved, it is suitable for implementation in a CAS. No similar procedures have been developed for polynomials of higher degree.

9.7.2 Equations with Algebraic Functions

We now turn our attention to equations involving algebraic functions. Equations containing square roots are typically solved by first isolating the radical, then squaring both sides of the result. The process of squaring often leads to an extraneous root that does not satisfy the original equation. This problem was considered briefly in Example 9.1.

Example 9.6. Solve $x + \sqrt{x} = 1$.

Solution. The "schoolbook" technique just described results in a quadratic, $x^2 - 3x + 1$, one of whose roots is spurious. The only true solution of the original equation is

$$x = \frac{3 - \sqrt{5}}{2}.$$ □

Equations with radicals other square roots can be solved in much the same way.

Example 9.7. Solve $x + \sqrt[3]{x+1} = 2$.

Solution. Here, isolating the radical and raising both sides to the third power yields a cubic, $x^3 - 6x^2 + 13x - 7$. Its two complex roots are spurious and

$$x = 2 - \frac{\alpha}{6} + \frac{2}{\alpha},$$

where $\alpha = \sqrt[3]{324 + 12\sqrt{741}}$, is the only solution to the original equation. □

The job of solving an equation containing two or more radicals of the same type is only slightly more difficult.

Example 9.8. Solve $\sqrt{x} - \sqrt{x-1} = 3$.

Solution. We begin by eliminating one of the radicals, say $\sqrt{x-1}$, by isolating it and squaring. This gives $6\sqrt{x} = 10$, whose only solution is $x = \frac{25}{9}$. The original equation, however, has *no solutions*. The left side is real only for $x \geq 1$ and its value is easily seen to be decreasing over the range $[1, 0)$. □

The solution process is more difficult when an equation involves different types of radicals.

Example 9.9. Solve $x + \sqrt{x} + \sqrt[3]{x} = 3$.

Solution. Isolating the cube root and cubing results in

$$x = 27 - 18x + 6x^2 - x^3 - \sqrt{x}\left(27 - 17x + 3x^2\right).$$

This, in turn, can be rearranged to isolate the product of the square root and its quadratic cofactor. Squaring this equation eliminates the remaining radical and yields a polynomial of degree six,

$$x^6 - 21x^5 + 176x^4 - 733x^3 + 1603x^2 - 1755x + 729,$$

having two real roots and four complex roots. $x = 1$ is the only (and obvious) solution of the original equation. □

9.7.3 Exponentials and Logarithms

We now move on to equations involving exponentials and logarithms. These are often solved by employing a substitution to transform them into a polynomial or algebraic equation. In addition to spurious solutions, issues of branch cuts arise frequently when dealing with exponentials and logarithms. It is not uncommon for computer algebra systems to report only the principal solutions, rather than the complete set of solutions, to such equations.

In our first two examples, both of which involve a single exponential, Euler's formula holds the key to obtaining a mathematically correct and complete set of solutions.

Example 9.10. Solve $e^{x^2+x+2} = 1$.

Solution. The naive way to tackle this problem is to take the natural logarithm of both sides,

$$x^2 + x + 2 = \ln 1 = 0,$$

and solve the resulting quadratic to get $x = \frac{1}{2}\left(-1 \pm \sqrt{7}\,i\right)$.

A better approach is to start with a version of Euler's formula, $e^{2\pi i} = 1$. Equating exponents, we see that the solutions to the original equation are given by the roots of

$$x^2 + x + 2 = 2\pi\,i.$$

This time the quadratic formula produces

$$x = \frac{-1 \pm \sqrt{-7 + 8\pi\,i}}{2}.$$

While we are closer to our goal, this is still not the complete solution.

To obtain the a complete and correct set of solutions, we observe that $e^{2n\pi i} = 1$ for any $n \in \mathbb{Z}$ and not just for $n = 1$. Consequently, the complete set of solutions to the original equation is

$$x = \frac{-1 \pm \sqrt{-7 + 8n\pi\,i}}{2},$$

where $n \in \mathbb{Z}$. $\qquad\square$

Example 9.11. Solve $e^{i\,x} = i$.

Solution. Euler's Formula expresses the complex exponential in terms of sines and cosines,

$$e^{i\,x} = \cos x + i\,\sin x.$$

Thus, the equation is satisifed whenever $\cos x = 0$ and $\sin x = 1$. (Why?) The principal solution is $x = \frac{1}{2}\pi$ and the set of all solutions is

$$x = \tfrac{1}{2}\pi + 2\pi\,n = \tfrac{1}{2}\pi\,(1 + 4\,n)$$

for any $n \in \mathbb{Z}$. $\qquad\square$

Next, we consider a pair of logarithmic equations.

Example 9.12. Solve $\ln x = a + i\,b$.

Solution. The obvious solution is

$$x = e^{a+i\,b}.$$

Substituting this solution into the original equation gives

$$\ln\left(e^{a+i\,b}\right) = a + i\,b.$$

However, the transformation $\ln e^z \to z$ for complex z given by Equation (9.10) is valid only when the imaginary part of z lies in the interval $(-\pi, \pi]$, as discussed in Section 9.6.2. Consequently, the equation has a solution only if

$$-1 < \frac{b}{\pi} \leq 1. \qquad \square$$

Example 9.13. Solve $\ln x = 17 + i\,\frac{75}{11}\pi$.

Solution. This instance of the previous example has *no solution*. Bernardin (1999) reports that several CAS unable to handle the general case correctly are able to deal successfully with this equation. $\qquad \square$

We are now ready for some more complicated equations that involve more than one exponential or logarithm.

Example 9.14. Solve $e^{2x} - 2\,e^x + 2 = 0$.

Solution. Setting $y = e^x$, the equation reduces to the quadratic $y^2 - 2y + 2$, whose solutions are $y = 1 \pm i$. Inverting the substitution, we have $x = \ln y = \ln(1 \pm i)$. These represent the principal solutions to the original equation. The complete family of solutions is given by adding $2\pi n\,i$ for any integer n to the principal solutions,

$$x = \ln(1 \pm i) + 2\pi n\,i$$

where $n \in \mathbb{Z}$. $\qquad \square$

Example 9.15. Solve $\ln(e^x + 1) = 2\,x$.

Solution. Exponentiating and substituting $y = e^x$ and leads to the quadratic equation $y^2 - y - 1 = 0$, whose solutions are $y = \frac{1}{2}\left(1 \pm \sqrt{5}\right)$. This gives solutions $x = \ln y = \ln\left(\frac{1}{2}\left(1 \pm \sqrt{5}\right)\right)$ for the original equation. The solution where the logarithm has a negative argument is extraneous since, when it is substituted into the original equation, the left side is real but the right is complex. Therefore,

$$x = \ln\left(\frac{1 + \sqrt{5}}{2}\right)$$

is the only solution. $\qquad \square$

Example 9.16. Solve $\ln(e^x + 1) = \ln(e^x - 5) + \ln(e^x - 2)$.

Solution. To make progress toward a solution we begin by combining logarithms using Equation (9.8),

$$\ln\left(\frac{(e^x - 5)(e^x - 2)}{e^x + 1}\right) = 0.$$

Consequently, we will need to check each solution found to be sure (9.8) is valid.

The logarithm can be eliminated by observing that our problem reduces to solving

$$\frac{(e^x - 5)(e^x - 2)}{e^x + 1} = 1.$$

since $\ln 1 = 0$. Letting $y = e^x$, the solutions to this equation are obtained from those of the quadratic $y^2 - 8y + 9$: $y = 4 \pm \sqrt{7}$. Reversing the substitution, $x = \ln y = \ln(4 \pm \sqrt{7})$.

The solution $x = \ln(4 - \sqrt{7})$, while real, is extraneous. Both $e^x - 5$ and $e^x - 2$ are negative and, therefore, the transformation of (9.8) is not valid.

The only real solution of the original equation is $x = \ln(4 + \sqrt{7})$ and (9.8) holds for this value. The complete family of complex solutions is

$$x = \ln\left(4 + \sqrt{7}\right) + 2\pi n\, i,$$

for any $n \in \mathbb{Z}$. $\square$

Example 9.17. Solve $(x^5 - 1)^x = 0$.

Solution. The solutions are given by the roots of $x^5 - 1$, provided the real part of any root is non-negative. (Why?) The factorization of $x^5 - 1$ is

$$(x - 1)(x^4 + x^3 + x^2 + x + 1).$$

Clearly, $x = 1$ is a solution to the original equation. The four roots of $x^4 + x^3 + x^2 + x + 1$ are each complex, and exactly two have non-negative real parts. Hence, the solutions to the original equation can be expressed in a notation reminiscent of Maple's RootOf and Mathematica's Root functions as

$$x = 1, \alpha \qquad \alpha \mid \alpha^4 + \alpha^3 + \alpha^2 + \alpha + 1 = 0 \text{ and } \Re(\alpha) > 0.$$

$\square$

9.7.4 Trigonometric Equations

Trigonometric equations occur frequently in college mathematics, as well as scientific and engineering applications. In this section, we study some of the issues that arise in their solution. While trigonometric equations are special cases of exponential equations, they deserve consideration in their own right because of their importance and the additional simplifications that apply.

Due to the periodicity of the trigonometric functions, the complete so-lutions of equations in which they occur generally involve issues relating to branch cuts.

Example 9.18. Solve $\tan\left(\frac{3}{2}x\right) = c$.

Solution. Applying the arctangent to both sides produces $x = \frac{2}{3}\arctan c$, but this is just the principal solution. The complete solution is given by

$$x = \frac{2}{3}\arctan c + \frac{2}{3}n\pi,$$

for any $n \in \mathbb{Z}$. □

While the sine and cosine are continuous, ratios of sines and cosines and the other trig functions (tangent, cotangent, secant, and cosecant) have sin-gularities that can create problems for symbolic solvers.

Example 9.19. Solve $\dfrac{\sin x + \cos x + 1}{\sin x + 1} = 0.$

Solution. The solutions to the numerator over the interval $(-\pi, \pi]$ are $x = -\pi/2$ and π. We must, however, remove the solution to the denominator, $x = -\pi/2$, since the ratio is undefined at this point. Therefore, the principal solution is $x = \pi$ and the full set of solutions is

$$x = (1 + 2n)\pi,$$

for $x \in \mathbb{Z}$. Bernardin (1999) reports that some of the CAS he tested give solutions from the family $x = \left(-\frac{1}{2}+2n\right)\pi$, where the denominator is zero. □

Example 9.20. Solve $\dfrac{1}{\tan x} = 0.$

Solution. A plot of $1/\tan x$ on the interval $(0, \pi)$ is shown in Figure 9.7. The curve would cross the x-axis at $x = \pi/2$, were it not for the fact that the function is undefined at this point—a set of measure zero. In terms of limits,

$$\lim_{x \to \pi/2} \frac{1}{\tan x} = 0.$$

This situation is termed an *asymptotic solution*. More generally, the set of asymptotic solutions over all branches of the function is given by

$$x = \left(\frac{1}{2} + n\right)\pi,$$

where $n \in \mathbb{Z}$. □

We turn our attention now to the solution of equations containing inverse trigonometric functions. These can usually be transformed into algebraic equa-tions that may introduce spurious solutions unless sufficient care is given.

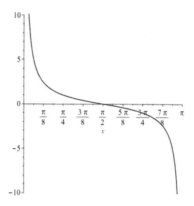

FIGURE 9.7: $\dfrac{1}{\tan x}$ on the interval $(0, \pi)$.

Example 9.21. Solve $2 \arcsin x = \arccos(3x)$.

Solution. Taking the cosine of both sides and applying the double angle formula to $\cos(2 \arcsin x) = 1 - 2 \sin^2(\arcsin x)$, we obtain a quadratic, $2x^2 + 3x - 1 = 0$, to solve. One solution, $x = -\frac{1}{4}\left(3 + \sqrt{17}\right) \approx -1.78$ lies outside the real domains of the inverse trig functions and is spurious. Consequently,

$$x = \frac{-3 + \sqrt{17}}{4}$$

is the only solution to the original equation. □

Example 9.22. Solve $\arccos x = \arctan x$.

Solution. Taking the cosine of both sides and referring to the reference triangle in Figure 9.8, we have

$$x = \cos(\arctan x) = \frac{1}{\sqrt{1 + x^2}}.$$

Squaring both sides and rearranging produces a quartic, $x^4 + x^2 - 1 = 0$, to solve. Letting $y = x^2$, we have $y = \frac{1}{2}\left(-1 \pm \sqrt{5}\right)$ from the quadratic formula. Taking the square root gives four values for x, $\pm\frac{1}{2}\sqrt{\pm 2\sqrt{5} - 2}$. The two complex solutions and the negative real solution are extraneous, so

$$x = \frac{1}{2}\sqrt{2\sqrt{5} - 2}$$

is the only root of the original equation. □

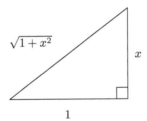

FIGURE 9.8: Reference triangle for $\cos(\arctan x) = 1/\sqrt{1 + x^2}$.

9.8 Concluding Remarks

Computer algebra systems are extraordinarily powerful mathematical assistants. Nonetheless, they should never be regarded as a panacea for mathematical computation. First, there are theoretical limitations with regard to the types of problems CAS can solve. Perhaps more importantly, practical considerations of time and storage limit the sizes of the computations they can feasibly perform. And while faster and more powerful computers continually appear, they still cannot overcome the inherent computational complexity of some of the critical algorithms underlying CAS.

Like other large software systems, CAS inevitably contain programming bugs. While the implementers do their best to avoid errors, it is the responsibility of users to be aware of the underlying assumptions and limitations. Just as hand calculations should be checked, users of CAS bear the responsibility to verify that the results obtained make sense. You will probably be alarmed if you try some of the experiments proposed in the exercises at the end of the chapter.

All this said, the designers and implementers of CAS continually strive to improve their systems. Meanwhile, researchers are at work developing new and improved algorithms. So enjoy what is already available while awaiting the advances that inevitably lie ahead.

9.9   Exercises

1. A typical computer algebra system will return the result shown for the following summation.

$$\sum_{k=0}^{n} a^k \to \frac{a^{k+1} - a}{a - 1}$$

Comment on the correctness of this result. Is it true for all values of a? How might a computer algebra system handle this situation correctly?

2. Solve each of the following equations by hand, being mindful of extraneous (spurious) roots.

(a) $x + 2 + \dfrac{x}{x - 2} = \dfrac{2}{x - 2}$

(b) $\dfrac{1}{x - 2} = \dfrac{3}{x + 2} - \dfrac{6x}{(x - 2)(x + 2)}$

(c) $\sqrt{x} = 1 - x$

(d) $\sqrt{2x - 5} - \sqrt{x - 1} = 1$

(e) $\sqrt{x + 7} + \sqrt{x + 2} = 1$

(f) $x = \sqrt[4]{5x^2 - 4}$

(g) $\ln(x + 1) + \ln(x - 1) = 0$

(h) $\log_2(x - 5) + \log_2(x - 3) = 3$

(i) $|2x - 1| = 3x + 6$

(j) $|x - 3|^2 - 4|x - 3| - 12 = 0$

3. (a) Explain how the solution to an equation involving rational expressions can lead to an extraneous root.

   (b) Repeat part (a) for an equation containing radicals.

   (c) Repeat for an equation with logarithms.

   (d) Repeat for an equation involving absolute values.

4. Evaluate the following definite integrals. Examine the integrand over the interval of interest, breaking the interval into subintervals as required. Plotting may help. Use limits to evaluate the corresponding antiderivatives at singularities and infinity. Note that some of the integrals may diverge.

(a) $\displaystyle\int_{1}^{\infty} \frac{1}{x^2}\, dx$

(b) $\displaystyle\int_{0}^{4} \frac{1}{(x - 1)^2}\, dx$

(c) $\displaystyle\int_{0}^{1} \frac{1}{\sqrt{x}}\, dx$

(d) $\displaystyle\int_{0}^{\infty} \frac{1}{\sqrt{x}\,(x + 1)}\, dx$

(e) $\displaystyle\int_{-1}^{1} \frac{1}{x^{2/3}}\, dx$

(f) $\displaystyle\int_{-1}^{1} \frac{1}{x}\, dx$

(g) $\displaystyle\int_{0}^{1} \ln x\, dx$

(h) $\displaystyle\int_{0}^{4\pi} x \sin x\, dx$

(i) $\displaystyle\int_{0}^{\infty} e^{-x} \sin x\, dx$

5. The following antiderivatives were found using the substitution $u = \tan(x/2)$. In each case, the integrand is a continuous function but the antiderivative shown is not. Find a continuous antiderivative to express each of the integrals.

(a) $\displaystyle\int \frac{1}{2 + \sin x}\, dx = \frac{2\sqrt{3}}{3} \arctan\left(\frac{\sqrt{3}}{3}\left(2 \tan\left(\frac{x}{2}\right) + 1\right)\right)$

(b) $\displaystyle\int \frac{1}{5 - 3\cos x}\, dx = \frac{1}{2} \arctan\left(2 \tan\left(\frac{x}{2}\right)\right)$

(c) $\displaystyle\int \frac{\cos x}{2 + \cos x}\,dx = x - \frac{4\sqrt{3}}{3}\arctan\left(\frac{\sqrt{3}}{3}\tan\left(\frac{x}{2}\right)\right)$

6. The antiderivatives shown below were found with a Weierstrass substitution. While both integrands are continuous, the antiderivatives given are not. Find continuous antiderivatives to express the integrals.

(a) $\displaystyle\int \sqrt{1 + \cos x}\,dx = \frac{2 \sin x}{\sqrt{1 + \cos x}}$

(b) $\displaystyle\int \sqrt{1 - \cos x}\,dx = -2\sqrt{1 - \cos x}\,\cot\left(\frac{x}{2}\right)$

7. (a) Solve the equations

$$z^n = 1 \qquad \text{and} \qquad z^n = -1.$$

for $n = 2, 3, 4, 5$. These are called the *n-th roots of unity* and the *n-th roots of minus one*, respectively. In each case, give:

  i. the set $\{z^{1/n}\}$ of all solutions in both polar (i.e., complex exponential) and Cartesian (i.e., $a + bi$) form
  ii. the principal solution, $\sqrt[n]{z}$
  iii. the real solution, $\operatorname{surd}(z, n)$, where applicable
  iv. any one arbitrary solution, $z^{1/n}$

(b) Repeat part (a) for the n-th roots of the imaginary numbers i and $-i$,

$$z^n = i \qquad \text{and} \qquad z^n = -i.$$

8. Solve the following equations giving the information requested in Exercise 7. Relate the solutions to the roots of 1, -1, i, $-i$ in the previous exercise.

(a) $z^2 = 25$ (b) $z^2 = -25$ (c) $z^2 = 81/64$ (d) $z^2 = -17/32$

(e) $z^3 = 27$ (f) $z^3 = -27$ (g) $z^3 = 100$ (h) $z^3 = 81/64$

(i) $z^4 = 100$ (j) $z^4 = -100$ (k) $z^5 = 32$ (l) $z^5 = -32$

(m) $z^2 = 36\,i$ (n) $z^3 = 8\,i$ (o) $z^3 = -8\,i$ (p) $z^4 = 16\,i$

9. Find *all* solutions to the following equations.

(a) $|x - 1| = 2$ (b) $(x + 1)^{x+a} = (x + 1)^3$ (c) $e^{x^2 - 1} = 1$

(d) $e^{x^2 - 1} = i$ (e) $e^{3x} + 1 = 0$ (f) $e^{2x} - 4e^x + 5 = 0$

(g) $e^{2ix} = -1$ (h) $4^{x^2} \cdot 2^x = 8$ (i) $5 + \dfrac{i}{2^x} = 0$

10. Find *all* solutions to these trigonometric equations. Indicate any "asymptotic" solutions.

(a) $\cos x = \sin(2x)$ (b) $\dfrac{\sin(2x)}{\sin x + 1} = 0$ (c) $\dfrac{\sin x + \cos(2x)}{\cos(3x)} = 0$

(d) $\dfrac{1}{\sec x} = 0$ (e) $\dfrac{1}{\cot x} = 0$ (f) $\arcsin x = \arctan x$

(g) $\arcsin x = \arccos x$ (h) $\arccos x = \arcsin(2x)$ (i) $\sinh x = \tanh x$

11. Assuming x is real, state the values of x for which the following "inverse relations" hold.

(a) $\dfrac{1}{1/x} = x$

(b) $e^{\ln x} = x$, $\ln e^x = x$

(c) $\sqrt{x^2} = x$, $\left(\sqrt{x}\right)^2 = x$

(d) $\left(x^3\right)^{1/3} = x$, $\left(x^{1/3}\right)^3 = x$

(e) $\sin(\arcsin x) = x$, $\arcsin(\sin x)$

(f) $\tan(\arctan x) = x$, $\arctan(\tan x) = x$

(g) $\sinh(\operatorname{arcsinh} x) = x$, $\operatorname{arcsinh}(\sinh x) = x$

(h) $\tanh(\operatorname{arctanh} x) = x$, $\operatorname{arctanh}(\tanh x) = x$

Exercises 12 to 16 give "derivations" of mathematical fallacies. Find the problem with each of these derivations. Most of the difficulties are related to issues discussed in this chapter.

12. Find the error in the following derivation which shows that $1 = 0$.

$x = 1$	Let $x = 1$
$x^2 = x$	Multiply both sides by x
$x^2 - 1 = x - 1$	Subtract 1 from both sides
$\dfrac{x^2 - 1}{x - 1} = \dfrac{x - 1}{x - 1}$	Divide both sides by $x - 1$
$\dfrac{(x + 1)(x - 1)}{x - 1} = \dfrac{x - 1}{x - 1}$	Factor
$\dfrac{(x + 1)\cancel{(x - 1)}}{\cancel{(x - 1)}} = \dfrac{\cancel{x - 1}}{\cancel{x - 1}}$	Cancel like terms
$x + 1 = 1$	Subtract 1
$x = 0$	

13. Where is the problem in the following erroneous derivation that $x + 1 = x$? Start with $(x + 1)^2$, expand and rearrange the equation.

$$(x + 1)^2 = x^2 + 2x + 1$$

$$(x + 1)^2 - (2x + 1) = x^2$$

Subtract $x(2x + 1)$ from both sides, then remove the common factor from the last two terms on the left.

$$(x + 1)^2 - (2x + 1) - x(2x + 1) = x^2 - x(2x + 1)$$

$$(x + 1)^2 - (2x + 1)(x + 1) = x^2 - x(2x + 1)$$

Add $\frac{1}{4}(2x + 1)^2$ to both sides and factor.

$$(x + 1)^2 - (2x + 1)(x + 1) + \tfrac{1}{4}(2x + 1)^2 = x^2 - x(2x + 1) + \tfrac{1}{4}(2x + 1)^2$$

$$\left[(x + 1) - \tfrac{1}{2}(2x + 1)\right]^2 = \left[x - \tfrac{1}{2}(2x + 1)\right]^2$$

Finally take the square root of both sides, then add $\frac{1}{2}(2x+1)$.

$$(x+1) - \frac{1}{2}(2x+1) = x - \frac{1}{2}(2x+1)$$

$$x + 1 = x$$

(And if we subtract x from both sides, we once again have that $1 = 0$.)

14. Here's a paradox resulting from the (incorrect) manipulation of infinite series. Can you explain the problem?

Start with the Taylor series for $\frac{1}{1-x}$ expanded about $x = 0$.

$$1 + x + x^2 + \cdots + x^n + \cdots = \frac{1}{1-x} \tag{1}$$

Multiply both sides of (1) by x

$$x + x^2 + x^3 + \cdots + x^{n+1} + \cdots = \frac{x}{1-x} \tag{2}$$

Substitute $1/x$ for x in (1) and simplify

$$1 + \frac{1}{x} + \frac{1}{x^2} + \cdots + \frac{1}{x^n} + \cdots = \frac{1}{1 - \frac{1}{x}} = \frac{x}{x-1} \tag{3}$$

Now add the series in (2) and (3), and simplify.

$$\cdots + \frac{1}{x^2} + \frac{1}{x} + 1 + x + x^2 + x^3 + \cdots = \frac{x}{1-x} + \frac{x}{x-1}$$
$$= \frac{x}{1-x} - \frac{x}{1-x} = 0$$

If $x > 0$, all of the series terms on the left are positive, yet their sum is 0. How can this be?

15. (a) Here's a simplification performed by many early releases of Maple. Can you spot the fallacy?

Simplify: $\dfrac{\tan^2 x - \sec^2 x + 1}{\sin^2 x + \cos^2 x - 1}$

First convert all trigonometric functions to sine and cosines.

$$\frac{\tan^2 x - \sec^2 x + 1}{\sin^2 x + \cos^2 x - 1} = \frac{\dfrac{\sin^2 x}{\cos^2 x} - \dfrac{1}{\cos^2 x} + 1}{\sin^2 x + \cos^2 x - 1}$$

Combine terms in the numerator, then rearrange the expression

$$\frac{\dfrac{\sin^2 x - 1 + \cos^2 x}{\cos^2 x}}{\sin^2 x + \cos^2 x - 1} = \frac{\sin^2 x - 1 + \cos^2 x}{\sin^2 x + \cos^2 x - 1} \cdot \frac{1}{\cos^2 x}$$

and cancel like terms using the Pythagorean identity for sines and cosines.

$$\frac{\cancel{\sin^2 x - 1 + \cos^2 x}}{\cancel{\sin^2 x + \cos^2 x - 1}} \cdot \frac{1}{\cos^2 x} = \frac{1}{\cos^2 x}$$

(b) At one time, Mathematica simplified the original expression to zero. Can you explain how this likely came about? What should a computer algebra system return for this simplification?

16. Find the flaw in the following "proof" by induction.

   To prove: $a^n = 1$ for all $n \geq 0$, where $a \neq 0$.

   Basis step. For $n = 0$, $a^0 = 1$.

   Induction step. Assume $a^0 = a^1 = a^2 = \cdots = a^n = 1$ and consider a^{n+1}.

$$a^{n+1} = \frac{a^n \cdot a^n}{a^{n-1}}$$

$$= \frac{1 \cdot 1}{1} \qquad \text{by the induction hypothesis}$$

$$= 1$$

Bibliography

Books are the compasses and telescopes and sextants and charts
which other men have prepared to help us navigate the dangerous seas
of human life.

Jesse Lee Bennett (1885–1931)
What Books Can Do for You, 1923

General References

Akritas, Alkiviadis G. 1989. *Elements of Computer Algebra with Applications.*
New York: Wiley-Interscience.

Cohen, Joel S. 2002. *Computer Algebra and Symbolic Computation: Elementary Algorithms.* Natick, Mass.: A K Peters.

— . 2003. *Computer Algebra and Symbolic Computation: Mathematical Methods.* Natick, Mass.: A K Peters.

Davenport, James H., Yvon Siret, and Evelyne Tournier. 1993. *Computer Algebra: Systems and Algorithms for Symbolic Computation.* Second edition. London: Academic Press.

Gathen, Joachim von zur, and Jürgen Gerhard. 2013. *Modern Computer Algebra.* Third edition. Cambridge, UK: Cambridge University Press.

Geddes, Keith O., Stephen R. Czapor, and George Labahn. 1992. *Algorithms for Computer Algebra.* Norwell, Mass.: Kluwer.

Grabmeier, Johannes, Erich Kaltofen, and Volker Weispfenning, eds. 2003. *Computer Algebra Handbook: Foundations, Applications, Systems.* Berlin: Springer.

Knuth, Donald E. 1997. *The Art of Computer Programming, Volume 2: Seminumerical Algorithms.* Third edition. Boston: Addison-Wesley.

Liska, Richard, et al. 1999. "Computer Algebra: Algorithms, Systems and Applications". http://kfe.fjfi.cvut.cz/~liska/ca/.

Wester, Michael J., ed. 1999. *Computer Algebra Systems: A Practical Guide.* Chichester, UK: John Wiley & Sons.

Chapter 1, Introduction

Barden, William Jr., and Gregory Whitten. 1980. *muMATH/muSIMP*. Honolulu: The Soft Warehouse.

Barton, David, Stephen R. Bourne, and John P. Fitch. 1970. "An Algebra System (CAMAL)". *The Computer Journal* 13 (1): 32–39.

Bourne, Stephen R., and J. R. Horton. 1971. "The Design of the Cambridge Algebra System (CAMAL)". In *SYMSAC '71: Proceedings of the Second ACM Symposium on Symbolic and Algebraic Manipulation*, 134–143.

Brown, William S. 1966. "A Language and System for Symbolic Algebra on a Digital Computer (ALTRAN/ALPAK)". In *SYMSAC '66: Proceedings of the First ACM Symposium on Symbolic and Algebraic Manipulation*, 501–540.

Char, Bruce W., et al. 1991. *Maple V Language Reference Manual*. New York: Springer-Verlag.

Chou, Shang-Ching. 1988. *Mechanical Geometry Theorem Proving*. Dordrecht, Holland: Springer Netherlands.

Collins, George E. 1966. "PM, A System for Polynomial Manipulation". *Communications of the ACM* 9 (8): 578–589.

— . 1971. "The SAC-I System: An Introduction and Survey". In *SYMSAC '71: Proceedings of the Second ACM Symposium on Symbolic and Algebraic Manipulation*, 144–152.

Delaunay, Claude-Eugène. 1860, 1867. *Théorie du Mouvement de la Lune*. Two volumes. Paris: Mallet-Bachelier.

Deprit, André, Jacques Henrard, and Arnold Rom. 1970. "Lunar Ephemersis: Delaunay's Theory Revisited". *Science* 168 (3939): 1569–1570.

Engelman, Carl. 1965. "MATHLAB: A Program for On-line Machine Assistance in Symbolic Computation". In *AFIPS '65: Fall Joint Computer Conference, Part I*, 117–126.

— . 1971. "The Legacy of MATHLAB 68". In *SYMSAC '71: Proceedings of the Second ACM Symposium on Symbolic and Algebraic Manipulation*, 29–41.

Fuchssteiner, Benno et al. 1996. *MuPAD User's Manual*. Wiesbaden, Germany: Springer Fachmedien.

Gonnet, Gaston H., Dominik W. Gruntz, and Laurent Bernardin. 2000. "Computer Algebra: Systems". In *Encyclopedia of Computer Science*, fourth edition, ed. by Edwin D. Reilly, Anthony Ralston, and David Hemmendinger, 287–301. London: Nature Publishing Group.

Griesmer, James H., and Richard D. Jenks. 1971. "Scratchpad/1: An Interactive Facility for Symbolic Mathematics". In *SYMSAC '71: Proceedings of the Second ACM Symposium on Symbolic and Algebraic Manipulation*, 42–58.

Hall, Andrew D. Jr. 1971. "The Altran System for Rational Function Manipulation: A Survey". *Communications of the ACM* 14 (8): 517–521.

Hearn, Anthony C. 1982. "REDUCE: A Case Study in Algebra System Development". In *EUROCAM '82: Proceedings of the European Conference on Computer Algebra*. Berlin: Springer-Verlag.

Hearn, Anthony C., and Rainer Schöpf. 2017. *REDUCE User's Manual.* reduce-algebra.com. http://www.reduce-algebra.com/manual/manual.pdf.

Jenks, Richard D., and Robert Sutor. 2005. *Axiom: The Scientific Computation System.* axiom-developer.org. http://axiom-developer.org/axiom-website/bookvol0.pdf.

Lamagna, Edmund A., and Michael B. Hayden. 1998. "NEWTON: An Interactive Environment for Exploring Mathematics". *Journal of Symbolic Computation* 25 (2): 195–212.

Lamagna, Edmund A., Catherine W. Johnson, and Michael B. Hayden. 1992. "The Design of a User Interface to a Computer Algebra System for Introductory Calculus". In *ISSAC '92: Proceedings of the International Symposium on Symbolic and Algebraic Computation*, 358–368. ACM Press.

Martin, William A. 1967. "Symbolic Mathematical Laboratory". Project MAC Report MAC-TR-36, Massachusetts Institute of Technology.

Martin, William A., and Richard J. Fateman. 1971. "The MACSYMA System". In *SYMSAC '71: Proceedings of the Second ACM Symposium on Symbolic and Algebraic Manipulation*, 59–75.

McLaughlin, Don. 1982. "Symbolic Computation in Chemical Education". In *Applications of Computer Algebra*, ed. by Richard Pavelle, 119–146. Springer.

Moses, Joel. 1967. "Symbolic Integration". PhD thesis, Massachusetts Institute of Technology.

Norman, Arthur C. 1983. "Algebraic Manipulation". In *Encyclopedia of Computer Science and Engineering*, second edition, ed. by Anthony Ralston and Edwin D. Reilly, 41–50. New York: Van Nostrand Reinhold.

Pavelle, Richard, Michael Rothstein, and John Fitch. 1981. "Computer Algebra". *Scientific American* 245 (6): 136–152.

Rich, Albert, et al. 1988. *Derive 1.0: A Mathematical Assistant for Your Personal Computer.* Honolulu: The Soft Warehouse.

Sammet, Jean E., and Elaine R. Bond. 1964. "Introduction to FORMAC". *IEEE Transactions on Electronic Computers* EC-13 (4): 386–394.

Slagle, James R. 1961. "A Heuristic Program that Solves Symbolic Integration Problems in Freshman Calculus: Symbolic Automatic Integrator (SAINT)". PhD thesis, Massachusetts Insititue of Technology.

Winkler, Franz. 2000. "Computer Algebra: Principles". In *Encyclopedia of Computer Science*, fourth edition, ed. by Edwin D. Reilly, Anthony Ralston, and David Hemmendinger, 282–287. London: Nature Publishing Group.

Wolfram, Stephen. 1988. *Mathematica: A System for Doing Mathematics by Computer*. Reading, Mass.: Addison-Wesley.

Zimmermann, Paul et al. *Computational Mathematics with SageMath*. Seattle: SageMath, Inc. `http://sagebook.gforge.inria.fr/english.html`.

Chapter 2, Computer Algebra Systems

Bernardin, Laurent, et al. 2012. *Maple Programming Guide*. Waterloo, Ontario: Maplesoft.

Heck, André. 2003. *Introduction to Maple*. Third edition. New York: Springer.

Maeder, Roman E. 1997. *Programming in Mathematica*. Third edition. Reading, Mass.: Addison-Wesley.

Wolfram, Stephen. 2003. *The Mathematica Book*. Fifth edition. Champaign, Ill.: Wolfram Media.

— . 2017. *An Elementary Introduction to the Wolfram Language*. Second. Champaign, Ill.: Wolfram Media.

Chapter 3, Big Number Arithmetic

Akritas, Alkiviadis G. 1989a. *Elements of Computer Algebra with Applications*. Chap. 1.2, Classical Integer Arithmetic: Algorithms and Their Complexity. John Wiley & Sons.

Dirichlet, P. G. Lejeune. 1849. "*Über die Bestimmung der mittleren Werte in der Zahlentheorie*". Abh. Akademie der Wissenschaften, Berlin 2:69–83.

Geddes, Keith O., Stephen R. Czapor, and George Labahn. 1992a. *Algorithms for Computer Algebra*. Chap. 3.6, Data Structures for Multiprecision Integers and Rational Numbers; 4.3, Fast Arithmetic Algorithms: Karatsuba's Algorithm. Norwell, Mass.: Kluwer.

Karatsuba, Anatolii, and Yu. Ofman. 1962. "Multiplication of Many-Digital Numbers by Automatic Computers". *Proceedings of the USSR Academy of Sciences* 145 (2): 293–294.

Knuth, Donald E. 1997a. *The Art of Computer Programming, Volume 2: Seminumerical Algorithms.* Third edition. Chap. 4.3.1, Multiple-Precision Arithmetic: The Classical Algorithms; 4.5.1, Rational Arithmetic: Fractions; 4.5.2, The Greatest Common Divisor; 4.6.3, Evaluation of Powers. Boston: Addison-Wesley.

Schönhage, Arnold, and Volker Strassen. 1971. "Fast Multiplication of Large Numbers". *Computing* 7:281–292.

Stein, Josef. 1967. "Computational Problems Associated with Racah Algebra". *Journal of Computational Physics* 1 (3): 397–405.

Chapter 4, Polynomial Manipulation

Aho, Alfred V., John E. Hopcroft, and Jeffrey D. Ullman. 1974. *The Design and Analysis of Computer Algorithms.* Chap. 8, Integer and Polynomial Arithmetic. Reading, Mass.: Addison-Wesley.

Brown, William S. 1971a. "On Euclid's Algorithm and the Computation of Polynomial Greatest Common Divisors". *Journal of the ACM* 18 (4): 478–504.

Brown, William S., and Joseph F. Traub. 1971b. "On Euclid's Algorithm and the Theory of Subresultants". *Journal of the ACM* 18 (4): 505–514.

Collins, George E. 1967. "Subresultants and Reduced Polynomial Remainder Sequences". *Journal of the ACM* 14 (1): 128–142.

Geddes, Keith O., Stephen R. Czapor, and George Labahn. 1992b. *Algorithms for Computer Algebra.* Chap. 2.7, The Primitive Euclidean Algorithm; 2.8, Quotient Fields and Rational Functions; 3.7, Data Structures for Polynomials, Rational Functions, and Power Series; 7.1, Polynomial GCD Computation; 7.2, Polynomial Remainder Sequences. Norwell, Mass.: Kluwer.

Karatsuba, Anatolii, and Yu. Ofman. 1962. "Multiplication of Many-Digital Numbers by Automatic Computers". *Proceedings of the USSR Academy of Sciences* 145 (2): 293–294.

Knuth, Donald E. 1997b. *The Art of Computer Programming, Volume 2: Seminumerical Algorithms.* Third edition. Chap. 4.6, Polynomial Arithmetic; 4.6.1, Division of Polynomials. Boston: Addison-Wesley.

Chapter 5, Algebraic Simplification

Berele, Allan, and Stefan Catoiu. 2015. "Rationalizing Denominators". *Mathematics Magazine* 88 (2): 121–136.

Bhalotra, Y., and Sarvadaman Chowla. 1942. "Some Theorems Concerning Quintics Insoluble by Radicals". *Mathematics Student* 10:110–112.

Borodin, Allan, et al. 1985. "Decreasing the Nesting Depth of Expressions Involving Square Roots". *Journal of Symbolic Computation* 1 (2): 169–188.

Brown, William S. 1969. "Rational Exponential Expressions and a Conjecture Concerning π and e". *American Mathematical Monthly* 76 (1): 28–34.

Cohen, Joel S. 2002a. *Computer Algebra and Symbolic Computation: Elementary Algorithms.* Chap. 7, Exponential and Trigonometric Transformations. Natick, Mass.: A K Peters.

Collins, George E. 1969. "Algorithmic Approaches to Symbolic Integration and Simplification". *ACM SIGSAM Bulletin*, no. 12: 5–16.

Jeffrey, David J., and Albert D. Rich. 1999. "Simplifying Square Roots of Square Roots by Denesting". Chap. 4 in *Computer Algebra Systems: A Practical Guide*, ed. by Michael J. Wester, 61–73. Chichester, UK: Wiley.

Landau, Susan. 1994. "How to Tangle with a Nested Radical". *The Mathematical Intelligencer* 16 (2): 49–55.

— . 1998. "$\sqrt{2} + \sqrt{3}$: Four Different Views". *The Mathematical Intelligencer* 20 (4): 55–60.

Maxfield, John E., and Margaret W. Maxfield. 1971. *Abstract Algebra and Solution by Radicals.* Philadelphia: W. B. Saunders.

Moses, Joel. 1971a. "Algebraic Simplification: A Guide for the Perplexed". *Communications of the ACM* 14 (8): 527–537.

Mulholland, Jamie, and Michael Monagan. 2001. "Algorithms for Trigonometric Polynomials". In *ISSAC '02: Proceedings of the 2001 International Symposium on Symbolic and Algebraic Computation*, 245–252.

Richardson, Daniel. 1968. "Some Unsolved Problems Involving Elementary Functions of a Real Variable". *Journal of Symbolic Logic* 33 (4): 514–520.

Risch, Robert. 1979. "Algebraic Properties of the Elementary Functions of Analysis". *American Journal of Mathematics* 101 (4): 743–759.

Rosenlicht, Maxwell. 1976. "On Liouville's Theory of Elementary Functions". *Pacific Journal of Mathematics* 65 (2): 485–492.

Chapter 6, Factorization

Akritas, Alkiviadis G. 1989b. "Elements of Computer Algebra with Applications". Chap. 6, Factorization of Polynomials over the Integers. New York: Wiley-Interscience.

Berlekamp, Elwyn R. 1967. "Factoring Polynomials over Finite Fields". *Bell System Technical Journal* 46 (8): 1853–1859.

Cantor, David G., and Hans Zassenhaus. 1981. "A New Algorithm for Factoring Polynomials over Finite Fields". *Mathematics of Computation* 36 (154): 587–592.

Davenport, James H., Yvon Siret, and Evelyne Tournier. 1993a. *Computer Algebra: Systems and Algorithms for Symbolic Computation.* Second edition. Chap. 4.2.1, Factorization of Polynomials in One Variable; 4.2.2, Hensel's Lemma. London: Academic Press.

Geddes, Keith O., Stephen R. Czapor, and George Labahn. 1992c. *Algorithms for Computer Algebra.* Chap. 6.5, The Univariate Hensel Lifting Algorithm; 8.2, Square-Free Factorization; 8.6, Distinct Degree Factorization; 8.7, Factoring Polynomials over the Rationals. Norwell, Mass.: Kluwer.

Knuth, Donald E. 1997c. *The Art of Computer Programming, Volume 2: Seminumerical Algorithms.* Third edition. Chap. 4.6.2, Factorization of Polynomials. Boston: Addison-Wesley.

Lenstra, Arjen K., Hendrik W. Lenstra, and László Lovász. 1982. "Factoring Polynomials with Rational Coefficients". *Mathematische Annalen* 261 (4): 515–534.

Mignotte, Maurice. 1974. "An Inequality about Factors of a Polynomial". *Mathematics of Computation* 28 (128): 1153–1157.

Murty, M. Ram. 2002. "Prime Numbers and Irreducible Polynomials". *American Mathematical Monthly* 109 (5): 452–458.

Yun, David Y. Y. 1976. "On Square-Free Decomposition Algorithms". In *SYMSAC '76: Proceedings of the Third ACM Symposium on Symbolic and Algebraic Manipulation,* 26–35.

Zassenhaus, Hans. 1969. "On Hensel Factorization". *Journal of Number Theory* 1 (3): 291–311.

Chapter 7, Symbolic Integration

Bronstein, Manuel. 1998. *Symbolic Integration Tutorial.* Tech. rep. INRIA Sophia Antipolis.

— . 2006. *Symbolic Integration I: Transcendental Functions.* Second edition. Berlin: Springer.

Bronstein, Manuel, James H. Davenport, and Barry M. Trager. 1998. *Symbolic Integration is Algorithmic!* Short course, Computers & Mathematics Conference, Massachusetts Institute of Technology.

Collins, George E. 1969. "Algorithmic Approaches to Symbolic Integration and Simplification". *ACM SIGSAM Bulletin*, no. 12: 5–16.

Geddes, Keith O., Stephen R. Czapor, and George Labahn. 1992d. *Algorithms for Computer Algebra.* Chap. 11.1, Rational Function Integration: Introduction. Norwell, Mass.: Kluwer.

Hermite, Charles. 1872. "Sur l'Intégration des Fractions Rationelles". *Annales Scientifiques de l'École Normale Supérieure (Série 2)* 11:145–148.

Horowitz, Ellis. 1971. "Algorithms for Partial Fraction Decomposition and Rational Function Integration". In *SYMSAC '71: Proceedings of the Second ACM Symposium on Symbolic and Algebraic Manipulation*, 441–457.

Marchisotto, Elena A., and Gholam-Ali Zakeri. 1994. "An Invitation to Integration in Finite Terms". *College Mathematics Journal* 25 (4): 295–308.

Mead, D. G. 1961. "Integration". *American Mathematical Monthly* 68 (2): 152–156.

Moses, Joel. 1967. "Symbolic Integration". PhD thesis, Massachusetts Institute of Technology.

— . 1971b. "Symbolic Integration: The Stormy Decade". *Communications of the ACM* 14 (8): 548–560.

Risch, Robert H. 1969. "The Problem of Integration in Finite Terms". *Transactions of the AMS* 139:167–189.

Ritt, Joseph F. 1948. *Integration in Finite Terms.* New York: Columbia University Press.

Rosenlicht, Maxwell. 1968. "Liouville's Theorem on Functions with Elementary Integrals". *Pacific Journal of Mathematics* 24 (1): 153–161.

Rothstein, Michael. 1976. "Aspects of Symbolic Integration and Simplification of Exponential and Primitive Functions". PhD thesis, University of Wisconsin, Madison.

Slagle, James R. 1963. "A Heuristic Program that Solves Symbolic Integration Problems in Freshman Calculus". *Journal of the ACM* 10 (4): 507–520.

Subramaniam, T. N., and Donald E. G. Malm. 1992. "How to Integrate Rational Functions". *American Mathematical Monthly* 99 (8): 762–772.

Trager, Barry M. 1976. "Algebraic Factoring and Rational Function Integration". In *SYMSAC '76: Proceedings of the Third ACM Symposium on Symbolic and Algebraic Manipulation*, 219–226.

Chapter 8, Gröbner Bases

Buchberger, Bruno. 2006. "An Algorithm for Finding the Basis Elements of the Residue Class Ring of a Zero Dimensional Polynomial Ideal". *Journal of Symbolic Computation* 41 (3-4): 475–511.

Buchberger, Bruno, and Manuel Kauers. 2010. "Groebner Basis". *Scholarpedia* 5 (10): 7763. http://www.scholarpedia.org/article/Groebner_basis.

— . 2011. "Buchberger's Algorithm". *Scholarpedia* 6 (10): 7764. http://www.scholarpedia.org/article/Buchberger's_algorithm.

Cox, David A., John Little, and Donal O'Shea. 2015. *Ideals, Varieties, and Algorithms: An Introduction to Computational Algebraic Geometry and Commutative Algebra*. Fourth edition. Cham, Switzerland: Springer.

Mencinger, Matej. 2013. "On Groebner Bases and Their Use in Solving Some Practical Problems". *Universal Journal of Computational Mathematics* 1 (1): 5–14.

Monagan, Michael B. 2014. "Groebner Bases: What Are They and What Are They Useful For?" *Maple Application Center.* http://www.maplesoft.com/applications/view.aspx?SID=153693.

Paprocki, Mateusz. 2010. "Design and Implementation Issues of a Computer Algebra System in an Interpreted, Dynamically Typed Programming Language". MA thesis, University of Technology of Wrocław, Poland. http://mattpap.github.io/masters-thesis/html/index.html.

Chapter 9, Mathematical Correctness

Bernardin, Laurent. 1999. "A Review of Symbolic Solvers". Chap. 7 in *Computer Algebra Systems: A Practical Guide*, ed. by Michael J. Wester, 101–120. Chichester, UK: Wiley.

Bunch, Bryan. 1982. *Mathematical Fallacies and Paradoxes*. New York: Van Nostrand Reinhold.

Dummit, David S. 1991. "Solving Solvable Quintics". *Mathematics of Computation* 57 (195): 387–401.

Graham, Ronald L., Donald E. Knuth, and Oren Patashnik. 1994. *Concrete Mathematics: A Foundation for Computer Science*. Second edition. Upper Saddle River, N.J.: Addison-Wesley.

Jeffrey, David J. 1994. "The Importance of Being Continuous". *Mathematics Magazine* 67 (4): 294–300.

Jeffrey, David J., and Arthur C. Norman. 2004. "Not Seeing the Roots for the Branches: Multivalued Functions in Computer Algebra". *ACM SIGSAM Bulletin* 38 (3): 57–66.

Jeffrey, David J., and Albert D. Rich. 1994. "The Evaluation of Trigonometric Integrals Avoiding Spurious Discontinuities". *ACM Transactions on Mathematical Software* 20 (1): 124–135.

Kahan, William. 1987. "Branch Cuts for Complex Elementary Functions, or Much Ado About Nothing's Sign Bit". In *The State of the Art in Numerical Analysis*, ed. by Arieh Iserles and Michael J. D. Powell, 165–212. Oxford, UK: Clarendon Press.

Stoutemyer, David R. 1991. "Crimes and Misdemeanors in the Computer Algebra Trade". *Notices of the AMS* 38 (7): 778–785.

Index

EDMUND ARTHUR LAMAGNA is a professor of computer science at the University of Rhode Island. He earned a doctorate from Brown University, where he completed both his undergraduate and graduate studies while concentrating in computer science, applied mathematics, and engineering.

Ed's professional interests lie at the intersection of computer science and mathematics. In particular, he has contributed to the fields of computer algebra and to the design and analysis of algorithms. In recent years, he has become interested in technical and societal aspects of cybersecurity, personal privacy, and election validation and security.

Throughout his career, Ed has also been involved in the development of innovative approaches for the teaching and learning of mathematics and computer science. This work includes the use of technology to facilitate learning in college calculus, and the use of puzzles and games to develop analytical thinking skills.